Agriculture and the State in Soviet and Post-Soviet Russia

Pitt Series in Russian and East European Studies

Jonathan Harris, *Editor*

Agriculture and the State in Soviet and Post-Soviet Russia

Stephen K. Wegren

University of Pittsburgh Press

Published by the University of Pittsburgh Press, Pittsburgh, Pa. 15261
Copyright © 1998, University of Pittsburgh Press
All rights reserved
Manufactured in the United States of America
Printed on acid-free paper
10 9 8 7 6 5 4 3 2 1

Wegren, Stephen K. 1956–
    Agriculture and the state in Soviet and post-Soviet Russia / Stephen K. Wegren
      p. cm—(Pitt series in Russian and East European studies)
    Includes bibliographical references and index.
    ISBN 0-8229-4062-0 (cloth : acid-free paper)
    1. Land reform—Russia (Federation)   2. Agriculture—Economic aspects—Russia (Federation)   3. Agriculture and state—Russia (Federation)   I. Title.   II. Series in Russian and East European studies.
    HD1333.R9W44   1997
    338.1'847—dc21   97-45318
                                                          CIP

A CIP catalog record for this book is available from the British Library.

In memory of my father

Contents

Figures and Tables ix
Acknowledgments xi
List of Abbreviations xv

1 Introduction: Agriculture and the State 1

2 State Interventions in Rural Social Policy During the Soviet Period 18

3 Reform of the Collective Agricultural Sector 59

4 Financial Levers and the End of the Social Contract 108

5 Land Reform and the Development of Private Farming 148

6 Financial Levers and the Impact on Private Farming 182

7 The State and Agrarian Reform 227

Notes 243
Index 287

Figures and Tables

Figures

1 Urban-Rural Wage Relationship, 1985–1992   53
2 Urban-Rural Wage Relationship, 1985–1995   125
3 Rate of Private Farm Creation, 1991–1996   209

Tables

1   Wage Rate Increases by Occupation, 1965–1969   21
2   Income Distribution by Raion in Kostroma Oblast, 1965–1982   22
3   Income Distribution by Raion in Kostroma Oblast, 1980–1990   24
4   Average Monthly Collective Farm Wages as a Percentage of Average Monthly State Farm Wages, 1985–1991   25
5   Growth of Urban Centers, USSR, 1959–1970   39
6   Livestock Holdings on Personal Plots, USSR, 1980–1989   45
7   The Development of Peasant Farms in the RSFSR, 1990–1991   47
8   Average Monthly Collective Farm Wage Related to Wage of State Industrial Worker, 1965–1984   51
9   Distribution of Rural Migrants to City of Kostroma in 1982, by Age and Sex   56
10 State and Collective Farm Reorganization in Russia, 1992–1994   82
11 The Reorganization of State and Collective Farms in European Russia as of January 1, 1994   83
12 Construction of Rural Infrastructure, 1990–1994   123
13 A Comparison of Urban-Rural Housing Amenities, Russia, 1993   124

x **Figures and Tables**

14 Selected Production Indices of the Russian Agricultural Sector, 1990–1995   128
15 Selected Food Consumption Indices in Russia, 1990–1995   130
16 Economic Indicators of Novosel'skoye Sovkhoz, Kaluga Oblast, 1993   132
17 Grain Harvests and State Purchases in Russia, 1986–1996   136
18 Structure of Food Trade, 1991–1994   139
19 Policy Positions on Agrarian Reform   142
20 Percentage of Food Production and Food Sales from Private Farmers in Russia, 1992–1996   171
21 Creation Rate of Private Peasant Farms, Russian Federation, January 1991–July 1994   172
22 Development of Private Farms by Region in European Russia, 1990–1994   173
23 Average Size of Private Peasant Farms in European Russia, 1991–1997   177
24 Private Farm Revenues, Expenditures, and Credits by Region, 1992   190
25 Summary of Private Farm Indicators: Rank Order of Regions, 1992   195
26 Summary of Private Farm Indicators: Rank Order of Regions, 1993   196
27 Comparison of Rank Order of Regions, 1992 and 1993   197
28 Development of Private Farming and Goals of Federal Program for Private Farming, 1996–2000   202
29 Financing the Development of Private Farming, 1996–2000   203
30 Land Privatization Patterns in Kostroma and Moscow Oblasts as of July 1, 1994   206
31 Percentage of Agricultural Land Registered for Private Farms by Oblast, Central Region, as of October 1, 1994   207
32 Private Farm Creation Rate by Region, European Russia, 1993–1996   210
33 Number of Private Farms per Thousand Persons by Region, European Russia, 1992–1996   211
34 Number of Private Farms per Thousand Persons, Central Region, on January 1, 1996   213
35 Private Farm Failure by Region, European Russia, 1993–1995   218

Acknowledgments

As is true for virtually any book, a great number of individuals and institutions contributed, both directly and indirectly, to the completion of this project. It would be impossible to thank each individually, but a few warrant special mention. This book has been more than six years in the making—a learning process that has been rewarding and at times exhausting. The intellectual origins of this book date to two wonderful years I spent at Duke University where I held first a MacArthur predoctoral and then a MacArthur postdoctoral fellowship at the invitation of Jerry Hough and his Center on East-West Trade, which he founded and directed. During those two years at Duke I was fortunate to work with Jerry and with Robert Bates. Jerry's high standards for scholarship, tenacious work habits, and keen intellect contributed to my growth as a scholar specializing in the Soviet Union and Russia. His constant admonition to keep in mind the big questions and the larger issues was—and remains—invaluable advice. It was also during my time at Duke that Jerry established an exchange with the agricultural institute in Kostroma, and so, in March 1990, I ventured to that previously closed city.

The initial trip to Kostroma was important because a primary focus of this book is on the Russian non–black earth, specifically the Central region. During the next six years I conducted fieldwork in Kaluga, Kostroma, Moscow, and Rostov Oblasts. In particular, Kostroma Oblast plays a dominant role in my analysis, as from March 1990 to July 1996 I returned nine times for research.

There are scholarly reasons for an emphasis on the Central region. First, the Central region is in the heart of Russia, which translates into political sensitivity. The Soviet leadership since the mid-1970s has expressed consistent concern for the fate of the Russian non–black earth zone, and the postcommunist government under President Yel'tsin has continued this emphasis. Second, Soviet agricultural policies benefited economically weak and high-cost farms and raions. Therefore, a perfect method for testing the reversal of past rural egalitarianism is to examine areas where "the weak" benefited most during Soviet times.

My study of Russian agrarian reform would not be intellectually whole without a framework. Robert Bates introduced me to the literature on comparative peasant studies and urban-rural issues. The personal tutorial I had with him was one of the most useful experiences I ever had and was exactly what I needed as I branched out from a regional specialization to a broader literature. The influence of these two men on my understanding of Russian politics and my approach to agrarian reform in Russia is much greater than is reflected in citations; to them I hope this book is a small down payment on the intellectual debt I owe.

Despite the opening of Russia, my research still required contacts and cooperative individuals. Literally dozens of Russians have aided me in one way or another, and several have been so instrumental that without them this book would have been impossible to complete. Very special acknowledgments go to Pyotr Stepanovich Tutun, Galina Leonidovna Soboleva, Nikolay Nikolaiovich Nikitov, and Nikolay Ivanovich Solov'yev in Kostroma Oblast, and my friends Vasiliy Olonichev and Andrei Shepetov. To each of them I extend my heartfelt thanks, and I hope that economic reforms are successful so that the quality of their lives will improve.

Institutional support was provided by a variety of sources. The most important financial support came from the Social Science Research Council, which provided me with a three-year postdoctoral fellowship. This support gave me a semester free of teaching and made it possible for me to make frequent trips to Russia. It was during this three-year fellowship that I collected the bulk of the material for this study. Southern Methodist University provided funds for research during two summers and, likewise, granted me a semester away from teaching to complete the manuscript. The John G. Tower Center for Political Studies at Southern Methodist University provided financial support for summer research. I thank all of these institutions for their support. Needless to say, none of these institutions is resonsible for the views expressed herein.

I would like to acknowledge and thank the Slavic librarians at Duke University, the University of California, Berkeley, the University of Illinois, the Lenin Library (now renamed the Russian State Library), and the wonderful staff on the sixth floor of the Kostroma State Archive. Without the assistance and cooperation of these individuals the quality of this research effort surely would have suffered. Certain chapters in the book draw upon documents from the Kostroma State Archive where I conducted research during the summers of 1994 and 1995. The footnoting system used to identify those documents is the normal system used in historical research. For those unfamiliar with this method, I first give the name of the document, the name of the archive, the fond number (f.), opis number (o.), dela number (d.), and page number (l., or ll. if plural). Unless otherwise noted, all translations from Russian sources and documents are mine. I also conducted my own interivews in Russian. The name of the person or organization, the location, and the date of the interview are indicated in the notes.

Finally, thanks are due to a number of individuals who commented on chapters in various stages of completion. I extend my gratitude for his collegiality and assistance to W. David Patterson in the Political Science Department at SMU, even though he is not a Russian specialist. David Patterson and Thomas P. Bernstein read the theoretical chapter and made useful comments. Frank Durgin and Peter Rutland read the entire manuscript and offered many valuable comments. Frank Durgin in particular went far beyond the call of duty and requirements of friendship by reading several of the chapters more than once. I add the usual disclaimer that full responsibility is mine alone for shortcomings in the book. Last but not least, I extend my deep appreciation to my wife, Zhanna, who patiently tolerated a husband who thought and talked about his book for entirely too long.

List of Abbreviations

AKKOR	Association of Peasant Farms and Cooperatives of Russia
APK	agroindustrial complex
APR	Agrarian Party of Russia
CDSP	*Current Digest of the Soviet Press*
CIS	Commonwealth of Independent States
ERS	Economic Research Service, USDA
FYP	five-year plan
IFC	International Financial Corporation
khozraschet	self-accounting
kolkhoz(y)	collective farm(s)
kolkhoznik(i)	collective farmworker(s)
kray	territorial/administrative subdivision within a republic, larger than an oblast
MTS	machine tractor station
nadbavki	price supplements
NEP	New Economic Policy
OECD	Organization for Economic Cooperation and Development
ogorody	personal vegetable gardens
okrug	administrative district based on nationality
RDI	Rural Development Institute, Seattle
RFE/RL	Radio Free Europe/Radio Liberty
Rossel'khozbank	Russian Agricultural Bank
RSFSR	Russian Soviet Federated Socialist Republic
sovkhoz(y)	state farm(s)
sovkhoznik(i)	state farmworker(s)
trudoden'	workday
USDA	United States Department of Agriculture

Agriculture and the State in

Soviet and Post-Soviet Russia

1

Introduction
Agriculture and the State

Agrarian reform has been central to every one of Russia's attempts at political and economic transformation in the twentieth century. The Stolypin reforms were linked to Russia's attempt to industrialize and to introduce limited democracy in the early twentieth century. War communism and forced grain requisitions led to urban depopulation and widespread urban hunger, which almost undermined the survival of the nascent Bolshevik regime. The New Economic Policy (NEP), which forged a class alliance between the Party and the peasantry, allowed Lenin to consolidate the Communists' hold on the country and facilitated a rebound in economic production. Collectivization was central to the imposition of Stalinism in the countryside and to rapid industrialization, which helped prepare the nation for war. Khrushchev's agrarian policies did not deconstruct Stalinist rural institutions, but they were important in creating a rural economy to feed an urbanized industrial work force.

Today in Russia agriculture remains an important sector, much more so than in other developed nations. Some 38 percent of the Russian population live in rural areas. More than 15 percent of the work force is employed in agricultural occupations. In contrast, the United States has less than 3 percent of its work force employed in agriculture; France has just over 5 percent; Japan has under 6 percent; even India has approximately 5.6 percent of its population employed in agriculture. As late as 1990 the contribution of agriculture to Russia's gross domestic product was about 15 percent of GDP. Subsequently, it declined to about 6.5 percent in 1994, but even so this was a somewhat higher percentage than in the United States, where agriculture accounted for $221 billion of a GDP that totaled $4.7 trillion (4.4 percent of GDP in 1992, in constant 1987 dollars).[1]

Just as agriculture remains an important part of the Russian economy, agrarian reform is central to the transformation of post-Soviet Russia. Economically, agrarian reform is important in order to increase farm productivity and production, to improve economic competitiveness, and to help the nation become more food self-reliant. Politically, agrarian reform is important in order to create rural institutions that complement and deepen societal democratization. Barrington Moore, in *Social Origins of Dictatorship and Democracy*, reminds us that the nature of a nation's rural sector and the manner in which it develops is linked to the nature of the political regime.

This book is an analysis of Russian agrarian reform. But to focus on microeconomic aspects such as food production would not contribute much to an understanding of reform processes or the nature of post-Soviet society. Instead, this is a study of the relationship between state actions and reform outcomes. The intent is not only to analyze agrarian reform but to contribute to our understanding of "the state" in transitional post-Soviet Russia.

The key to understanding the rural economy is to focus on rural social policy. Rural social policy is defined as the *incentives* confronted by rural actors; the *economic environment* in which food producers operate; and the *rural social conditions* in which rural dwellers exist. The manner in which state interventions affect these variables will influence the performance of the agricultural sector, as well as political-economic relationships among societal actors.

This study is organized around a set of analytical questions concerning the role of the state in the rural economy:

1. How has the state intervened in the past to influence rural social policy?
2. How has the post-Soviet state intervened to change rural social policy? To what degree have these interventions been effective?
3. To what degree have patterns of Soviet rural social policy been reversed?

This study sees a Russian state that is able to influence key aspects of rural social policy—incentives, the economic environment of food producers, and rural social conditions—in its rural economy.

The Nature of State Interventions

All states intervene in their rural economies, but the duration, intensity, and objectives will differ. There is nothing deterministic about the nature of state interventions, although common general objectives may be identified. These objectives include retaining the power to govern and the pursuit of political stability. State interventions are often grounded in political and ideological reasons, meaning that states will pursue economic rural policies that correspond with political beliefs and values.

In Latin American countries, interventions in the form of agrarian reform were undertaken in response to peasant pressure for land distribution or to address problems of rural inequality. Because land was concentrated in the hands of a few wealthy families, politically motivated land reforms often originated from the peasantry.[2] A landless restless peasantry is profoundly revolutionary, and several revolutions in the twentieth century may be understood as peasant attempts to redress issues of inequity. Faced with rural unrest, governments had the option of supporting the peasant movement in the hope of maintaining (or gaining) rural support, as in Mexico. Governments who oppose land reform run the risk of rural alienation, civil war, and defeat, as in the case of Nicaragua during the 1980s.[3]

One can see the use of interventions to pursue larger political objectives in Africa as well. There, states have intervened in order to protect urban interests. The protection of urban interests and the discrimination against rural interests resulted in a theory of "urban bias." The urban bias theory argues that financial, productive, and human resources are deliberately pumped out of the countryside in order to benefit urban dwellers. Economically, urban bias is often characteristic of developing states, those Third World nations making the transition from agriculture to industry.[4]

Why does urban bias arise? According to urban bias theory, the countryside is economically poor because it is politically weak against state interventions.[5] Urban bias holds not that the state is a neutral institution but rather that political institutions are dominated by urban elites who use their political power to discriminate against the countryside. The urban political elite use the state and the instruments of power as vehicles to discriminate against rural dwellers by intervening in markets and discriminating against food producers in order to ensure urban support. A politically strong countryside would act to prevent or block urban bias.

Urban interests are politically important to the state because the threat of political instability is greater among urban dwellers. Urban dwellers have the ability to organize more easily than rural dwellers and have a common interest that unites them: cheap food. Proponents of urban bias, therefore, argue that the political core of agricultural policy revolves around the way the government intervenes in (or controls) markets. One author summarizes governments' strategy in the following way:

Governments face a dilemma: urban unrest, which they cannot successfully eradicate through co-optation or repression, poses a serious challenge to their interests as employers and sponsors of industry. Their response has been to try to appease urban interests not by offering higher money wages but by advocating policies aimed at reducing the cost of living, and in particular the cost of food. Agricultural policy thus becomes a byproduct of the political relations between governments and their urban constituents.[6]

At the same time, because the rural population is usually more numerous in Third World nations, the state has to devise political strategies in order to develop rural alliances. Lipton and others argue that a common strategy is to buy off big farmers by providing subsidized inputs and tax advantages. In return large farmers support the regime even though they are discriminated against on price.[7] Thus, urban bias discriminates against food producers, but not all food producers. The eloquence of the model does not make it immune to attack, but the model has been attractive to a number of scholars who apply its basic ideas to different regions and individual countries.[8] Urban bias would appear to apply to the former Soviet Union as well. At various times in Soviet history, but particularly under Stalin, the Soviet Union essentially fulfilled the main characteristics of the urban bias model.

States intervene, therefore, for different reasons. In contemporary Russia, we shall see not only that the state has intervened in the rural economy, but also that the nature of the state's interventions has had an enormous impact on agrarian reform. State interventions in contemporary Russian agriculture are often politically and ideologically motivated, but this is unique neither to Russia nor to other agrarian sectors in other countries.

Beyond the motivation for state action, interventions affect rural institutions, agricultural policies, and rural social relationships. There is common agreement that state intervention is necessary in order to provide public goods and to correct market failures. There is less agreement over the nature and impact of state interventions, which gives rise to controversy in evaluating the costs and benefits of state interventions. Less intensive state interventions may affect only one or at most two of these variables. More intensive state interventions affect all three of these variables.

The State as a Unit of Analysis

Because states frequently intervene in their rural economies, a common thread running through the literature on peasant-state relations is the state as the unit of analysis. What do we mean by "the state"? One author has argued that "states may be viewed as organizations through which official collectives may pursue distinctive goals, realizing them more or less effectively given the available resources in relation to social settings."[9] During the Soviet period the state was all-encompassing, including government and party organizations and governing bodies. The Soviet state was characterized by the lack of a civil society and the domination of a single party. During most of the Soviet period, the state is relatively simple and straightforward, with minimal emphasis on political conflict and maximum emphasis on the impact of state-defined rural social policies.[10]

During the post-Soviet period, politics assume greater significance as policy conflicts and group bargaining become more important. Within agricultural policy, for example, the post-Soviet period is characterized by the rise of rural opposition, the creation of "loyal" rural groups, and the defection of bureaucratic agricultural interests from state policy. In the post-Soviet period the state is defined as central decision-making institutions, including "the government and the bureaucratic agencies that derive their authority from it."[11] The state as the unit of analysis does not assume that the state is always a unified actor. In using the state as a unit of analysis we assume that the "state" is an identifiable actor in society, an actor that is distinct from society, and one that is able to pursue independent policies ("independent" does not mean that state and societal interests are unrelated, simply that state policies are identifiable and are not solely a function of societal or interest group pressure).

For most of the Soviet period, scholars concentrated on the state over society because that is where the locus of power was found. During the Stalinist period, the state was all-encompassing; in this ideal type there were few if any methods for civil society to influence state actions, and for this reason the Soviet Union was considered an example of a strong state/weak society.[12] While this ideal-type model does not fully account for all Soviet reality, it does accurately capture the enormous control over society exercised by the Communist Party, and the Party's penetration of all societal institutions (during the Soviet period the state was essentially synonymous with the Communist Party). The Party enjoyed an absolute monopoly on power over society, the military, other institutionalized groups, and all policy initiatives.

This totalitarian image for many years dominated Western analyses of Soviet society. Yet some scholars were aware of sources of societal change. For example, even in the 1950s, one author suggested that the dilemma between a terrorized society and the need for modernization would create pressures for a diminution of terror and control. Over time, other scholars such as Jerry Hough looked at the changing educational background of the top Soviet elite and social-demographic changes in society to predict a new orientation for Soviet leaders. By the early Brezhnev period, scholars were describing "interest groups" in the Soviet Union, a term probably misapplied, but correctly capturing the emergence of differing views among state (and nonstate) organizations. Over time the "state" became weaker while society became relatively stronger. As Soviet society evolved, scholars began to deemphasize the control aspect of the party-state and instead began to examine how the party acted as the arbiter among organizational interests in society.[13] Despite these developments, even in its post-totalitarian period before Gorbachev it would be fair to say that Soviet society remained "mobilized" and reform impulses came from above.

Under Gorbachev there were several notable trends that began to change the role of the party-state: a curtailment of state (read "party") control over society—a conscious decision emanating from the Nineteenth Party Conference during May–June 1988;[14] a Communist Party that was increasingly pluralistic;[15] the rise of independent, nonstate organizations and movements;[16] a growing regional autonomy from the center;[17] and the emergence of a civil society.[18] The combination of these changes left the Soviet state too weak to implement reform and the Party too fragmented to withstand internal conservative opposition to reform. For a variety of reasons (the loss of legitimacy, the loss of will to hold the nation together, nationalist rivalries, and the political struggle between the president of the USSR and the president of Russia), the USSR as a formal political, economic, military union imploded in late 1991.

With the demise of the Soviet state, it is fair to ask whether a continued emphasis on the state is warranted. Should we be looking elsewhere to understand the Russian state and its effect on agrarian reform? There is no question that the contemporary Russian state is more open and pluralist than its Soviet predecessor. For example, some forty-three parties offered candidates during the December 1995 parliamentary elections, even though only four parties cleared the 5 percent threshold and were assigned seats from party lists in the State Duma. Despite its pluralist characteristics, a statist approach is an appropriate framework for our understanding of Russian reform in general and agrarian reform in particular, for several reasons.

First, a statist approach is appropriate because Russian agrarian reform has originated from above, not from below, which is to say that the state was not responding to pressures for reform but rather acted independently. The historical legacy—both before and during Soviet communism—was reform from above, and post-Soviet Russia is no exception. The impact of state-sponsored reform initiatives is, therefore, a logical focus of analysis.

Second, the state as the unit of analysis is appropriate because the state has most of all influenced rural institutions, policies, and social relationships in the rural economy.[19] By influencing these aspects of the rural economy the state is able to influence the structure, organization, and operation of the agricultural sector. No other actor has been able to influence all aspects of reform in the manner the Russian state has. This is hardly surprising given the historical legacy of the strong Soviet state and weak civil society. One could focus on separate rural groups, but that would yield only a partial understanding of reform. The broadest and deepest understanding of reform must begin with the state and its interventions. Thus, the logical question is: If not the state, then who should be focus of analysis? The post-Soviet state may (or may not) be in the transitional phase to a "weak state," but it certainly is *not* weak in the agricultural sphere.

Third, the state as unit of analysis fits Russian reality. During the many interviews I conducted with land reform officials, farm chairmen, academics, and even private farmers, I was impressed—and somewhat surprised—by continual statements about the need to "wait for the center" to act. The periphery continued to expect the center to lead. Admonitions from a Westerner to form grassroots movements or to take local initiative encountered the typical Russian passivity. Few if any people felt that local initiative could be effective, at least in agrarian reform.

At the same time that a statist approach is best suited for understanding Russian agrarian reform, this study is cognizant that by using a state-centric approach certain choices have to be made. On the one hand, a statist approach emphasizes the way state institutions influence the rural sector and considers state actions as being of central importance to the outcome of agrarian reform. On the other hand, nuances and aspects of reform are lost with this emphasis. For example, the state approach acknowledges but does not emphasize various policy debates and downplays interest group conflict during the policy formation process. One could perhaps write a book about policy conflicts over contemporary agrarian reform among individuals and interest groups, but that is not a primary focus of this book. Again, due to the scope of state interventions and the importance of those interventions a statist approach is justified. Because no single approach can be all-encompassing or can include all available information and data, choices have to be made.

The Importance of Rural Social Policy

In the Soviet agrarian system, much attention was given to rural institutions and policies. On the institutional level, we are familiar with the organization and functioning of state and collective farms, with the organization and impact of the machine tractor stations (MTS), and with the food-procurement and trade system. On the policy level, Stalin's collectivization and antirural orientation are well known. We are familiar with Khrushchev's collective farm amalgamation, the Virgin land program, and elimination of the MTS. Brezhnev's massive rural investment program has been well documented, as were his policies to reclaim land and to mechanize production processes.

Whereas most scholarly attention has been devoted to changes in rural institutions and policies, little emphasis has been placed on rural social policy or the resultant social relationships in the rural sector. This relative neglect is unfortunate, for social relationships in the Soviet period were as important and as integral to the agrarian system as were rural institutions and policies. At present, the mistakes of the past would seem to be repeating, despite new research opportunities. Most analyses of contemporary agrarian reform consider only the institutional and policy components, even though social relationships are equally—if not more—important to the success or failure of Russian agrarian reform. A focus on institutions and policies limits our understanding of contemporary reform in two ways: first, for agrarian reform to be successful, social relationships must change from their Soviet patterns. Second, a focus on only institutions and policies channels the analyst to certain frameworks while ignoring others. This study will analyze rural institutions and policies but will also take the analysis a step further.

Ignoring social relationships within the agrarian system means that contemporary agrarian reform simply cannot be fully understood. It is necessary to understand the ways state interventions affect incentives, the rural economic environment, and rural social relationships if we are to understand both the processes of and the requisites for successful reform. Why? New social relationships give rise to new social groups and new economic classes. New social groups and classes in turn facilitate the decentralization of power from the center, provide political diversity in ways that complement democratic institutions, and often serve as the "engine" for economic development. Furthermore, an understanding of state interventions leads to broader insights into the nature of the political system and state-societal relations.

Why is rural social policy important? Rural social policy is important for a number of reasons. First, the nature of state interventions will influence the incentives faced by rural actors. Incentives act as basic motivations for behavior. The definition of incentive structures influences which behaviors will be engaged in, and the quality of those behaviors. Thus incentives drive actions, which are important to results.

Second, rural social policy is important because it defines the economic environment of rural dwellers. Economists who study agricultural policy often overlook the social-political ramifications of rural social policy, instead emphasizing production and factors in the production and trade cycle. But production cannot be divorced from the economic environment. States can create conditions that are advantageous, hostile, or essentially neutral for food producers.

Third, rural social policy influences sociopolitical conditions in which rural dwellers exist. Social conditions in turn affect the relationships among different groups of the population and affect the degree of political support for state policies. State interventions can therefore create a range of different sociopolitical environments. Social groups may be conflictual or cooperative. Political support may bolster or weaken state support. Depending on the state's strategy and intent, different environments will arise. One would expect states to desire a cooperative and supportive rural sector. But the state holds the levers to influence actual conditions.

Last, rural social policy is important as a measure of state priorities. The Soviet leadership was more interested in using rural policies as a tool of social engineering than in creating rational economic incentives. The Soviet state intervened in ways that were uneconomical and in the long term politically destabilizing. The regulation of the rural economy led to disincentives to be efficient and created an environment that was hostile to initiative and individual enterprise. A central argument of this book is that Russian rural social policy must change from its Soviet past. Rural incentives must change in order to address the efficiency and cost problems inherited from the Soviet era. Without fundamental changes in rural incentives agricultural production will continue to be high-cost and highly subsidized.

Contemporary Russian agrarian reform, therefore, is not merely about changing economic policies, although those are undoubtedly important. Agrarian reform is inherently political because it attempts to redefine political and economic relationships and to change how different actors fare within the rural sector. Successful agrarian reform requires new answers to the questions, Who gets what? and Who can do what? Given the lack of market structures to guide these processes, the Russian state intervened to provide ostensibly new answers. The criterion for successful reform should be the extent to which rural social policies have changed from their Soviet past.

Weak and Strong States

A statist approach gives rise to the questions concerning state strength. States are often categorized as strong or weak. How are strong and weak states usually distinguished? Traditionally, two main factors have been identified in a "weak" state. Stephen Krasner argues that "the weakest kind of state is one that is completely permeated by pressure groups." This view suggests that the state is hostage to group and individual interests and is unable to pursue independent policies. A second factor relates to policy outcomes. Joel Migdal points out that proponents of the "weak state" often study social policy implementation, "especially the difficulties state leaders have had in ensuring intended widespread changes in people's social behavior and planned transformation in social relations." If a state cannot implement what it has enacted, then it must be weak.[20]

The traditional approach by proponents of the strong state concentrated on "the state's penetration of society and extraction of resources." Krasner defined a strong state as one that can "change economic institutions, values, and patterns of interaction among private groups."[21] Here too the emphasis is on policy outcomes.

The traditional approach to the weak or strong state suffers from measurement and circular logic problems. According to the literature, a strong state is one that is able to implement reforms, but the ability to implement reforms is offered as a measurement of state strength. We have independent and dependent variables that are, in essence, one and the same. Thus, policy implementation is not a good measure of state strength. An analytical approach that focuses primarily on policy implementation distorts our understanding of reality and leads us to incorrect conclusions regarding state strength.

Using the Soviet Union as an example, one could pick up any number of specialized case studies and one of the conclusions invariably would be that the center was unsuccessful in fully implementing a given policy or set of policies. For example, it is common knowledge that the state was unsuccessful in rooting out undesired behaviors on the part of the population such as religion or black market activities. Social historians who study the Stalinist period have shown that the social environment and individual behaviors were not as controlled from the center as was originally thought.[22] Some have argued that even during the monstrous purges of the 1930s the excesses stemmed from out-of-control local officials and Party organizations rather than from the center.[23] Do these examples mean the Stalinist state was weak? Of course not. Would anyone seriously argue that the state under Brezhnev was weak, even in his latter years? Probably not. Even as Brezhnev the leader weakened after a series of strokes, the capacities and instruments of the state remained strong. In the rural sphere, the state continued to define the economic environment for actors in the food system, input and output prices, trade policies, and the flow of resources to the countryside. In short, despite a weak leadership there was very little of civil society that escaped the influence of the state.

Attempting to distance ourselves from the traditional approach to the strong state, it is not solely policy implementation but *autonomy, capacities,* and *instruments* of the state that determine whether a state is weak or strong. The Russian state is autonomous because it undertook agrarian reform independently, not as the result of pressure by rural groups wanting greater equality. Richard Sakwa has written that "parties emerged but not a party system; . . . it was the government, standing as it were above politics, that formulated its own policies independent of any party."[24] As long as a state has the capacity to influence—if not determine—incentive structures that affect individual and enterprise decisions and behaviors, the state cannot be considered weak. Instruments of the state give the capacity to implement reforms; the lack of implementation does not mean the state is weak, however.[25] State capacities influence policy and reform outcomes, but policy implementation is only one capacity of the state, not the full extent of its levers.

Is Russia a Weak State in Agriculture?

In the wake of the collapse of the USSR, contemporary conventional wisdom holds that the Russian state is "weak." Indeed, there is little question that it is weaker than its Soviet predecessor in terms of its penetration of society, the existence of numerous nonstate organizations and associations, the reduced scope of state authority, the degree of decentralization, the growth of regional and local autonomy, and the ability of the center to implement policy. On a theoretical level we should not be surprised that the Russian state is weaker, for as Samuel Huntington argued, social and economic changes undermine traditional sources of authority and political institutions.[26] But whether "weaker" equals "weak" is another question.[27]

Historically, the state has had an identifiable impact on the agrarian sector in Russia. When the Soviet Union existed and the Communist Party dominated society, the influence of the state on the rural economy was axiomatic.[28] Debate arises when assessing the influence of the Russian state today. In the contemporary era, the Russian state, though weaker than in its Soviet past, still is relatively stronger than other actors in the rural sphere. Why is a strong state evident in the rural economy? The contemporary Russian state enjoys an advantaged position in the rural economy because rural political parties and groups in general remain weak. The state's strength allows it to intervene in ways that other power contenders cannot, and using its advantages the state is able to influence reform outcomes.

How would state strength and rural group weakness be measured? In examining this question we could first look at the degree to which the state had been penetrated by rural group interests, a common strategy among analysts of the state. However, this measure is not effective and lends very little insight. During 1992–1993, for example, rural liberals did not penetrate state governing bodies to any significant degree, but yet inside accounts reveal that they enjoyed considerable access and participated in the writing of legislation. Furthermore, during 1994–1995, rural conservatives did indeed penetrate governing bodies to a significant extent and still were ineffective. Rural conservatives were unable to increase financing or to reverse the decline of resources to the rural sector. The best they could do was to delay state-initiated legislation. Thus, an important qualifier to the penetration of governing bodies is ideological compatibility, which in turn will influence access.

But this still leaves the question of how we measure state strength and rural group weakness. Two useful measures would be, first, a comparison of rural policy preferences with actual governmental policy to see who won in the case of incompatibility. Second, one could look at trends in the flow of resources and finances. (Both of these measures will be examined in more detail in subsequent chapters.)

Some authors have argued that agrarian reform was not successful in Russia because the state was too weak, a view that emphasizes implementation of policy.[29] This view implied dual powers by an "agrarian lobby." On one hand, it was argued, the agrarian lobby opposed reform and hindered its implementation. On the other hand, it was argued, the agrarian lobby had permeated the state and its governing institutions to such an extent that it was able to control state agricultural policies. The discrepancy between the two presumed powers of the lobby was never reconciled: why would an all-powerful lobby allow reform policies to be adopted that later had to be blocked? There is no doubt that an agrarian lobby existed and that elements within this movement opposed aspects of agrarian reform. Yet the weak state–strong agrarian lobby explanation obscures much more than it explains about agrarian reform.

There are several sound reasons why the weak state is not an appropriate approach to understanding Russian agrarian reform. First, advocates of the weak state approach are fundamentally mistaken in their equating the strength of the leader with the strength of the state. Advocates of the weak state approach have failed to keep state capabilities and leadership weakness analytically separate. According to the weak state view, if Yel'tsin (or whoever) is weak, if his decrees go unimplemented, then he is weak and ergo the state is weak. The view that the leader in essence equals the state flows from the totalitarian model and our understanding of the Soviet system. However, the applicability of this view has long since been outdated. The implicit connection between the leader and the state ignores the breadth and scope of "the state"— and here I use the term not in the narrow sense as it evolved in Soviet studies but in the broadest sense of the term as used by comparativists in political science.[30]

Second, the weak state approach overlooks the ability of the state to define incentives to which individuals and collectives react. The Russian state controls several institutional and policy levers that allow it to define and influence the incentives confronting rural actors. Third, the rural sector in Russia is part of a weak society and a weak civil structure that traditionally faced an exceptionally strong state. Although this is less true today than in the past, the state-societal relationship remains unequal. Fourth, dating from collectivization, rural reform originated from above, not from below, and contemporary agrarian reform is no exception.[31] The rural sector has historically been the object of reform: not an equal partner either in the formulation of policy or in the way peasants were treated. This condition remains true today.

Furthermore, those who argue that the Russian state is weak have confused basic policy *implementation* with the *performance* of reform institutions. It would be difficult to deny that a significant number of wide-ranging reforms have not been implemented in the agricultural sector. Reviewing early reform goals, one is struck by how much institutional change was implemented during the first five years of agrarian reform. In October 1991, Russia's President Boris Yel'tsin indicated that he would work for the privatization of unprofitable state and collective farms, work for land reform and the development of a private farming sector, and introduce the freedom of sale and purchase of land.[32] In 1992 the former minister of agriculture V. Khlystun further sketched the four main directions of agrarian reform: (1) land reform and the creation of private farms; (2) the reorganization of state and collective farms; (3) the privatization of food-processing plants, material-technical supply enterprises, and construction enterprises; and (4) the creation of a market infrastructure.[33]

During the next few years these ideas were transformed into concrete policies and laws. By 1997 one could point to a number of changes in the Russian countryside. For example, by mid-1997 there were about 280,000 private peasant farms that had been allotted more than 12 million hectares of land (about 6 percent of agricultural land). Former rural workers, discharged military personnel, and even urban dwellers had been allocated land plots for the purpose of beginning private farming operations. In addition, tens of millions of small-scale land plots—such as private plots or plots held within collective settings—had been privatized. Land reform legislation introduced the right to buy and sell land, to barter, exchange, and bequeath it, and a rudimentary land market had emerged. State and collective farms reorganized, the vast majority of which converted into joint stock farms, with only about one-third retaining their previous status. To facilitate and deepen farm transformation, a program of farm privatization was introduced in Nizhniy Novgorod and selected other regions. Wholesale trade organizations charged with purchasing foodstuffs from farms were privatized, as were food-processing plants, food retail outlets, and rural service enterprises. Last, a market infrastructure was introduced through the creation of commodity exchanges, which, combined with changes in state procurement policies, meant that the state's direct control over food production was significantly decreased. It is clear, then, that early agrarian reform ideas were transformed into policy and implemented, although not always perfectly.

When commentators complain about the ineffectiveness of agrarian reform, they often lament that reform was not implemented, but what they are really talking about is the poor performance of reform institutions. It took a strong state to establish or implement reform because in many cases the nature of reform was bitterly contested by agrarian interests. Those who argue that the Russian state is weak ignore how state interventions affected the operation and performance of reform policies.

Last, those who argue that the Russian state is weak vastly overstate the political power of the "agrarian lobby." Those who argue that reform was not successful because of the agrarian lobby cannot account for several factual realities. For example, on every single policy issue on which the conservative agrarian lobby and the government disagreed during 1991–1995, the agrarian lobby lost and the government won.[34] A view that sees the agrarian lobby as omnipotent fails to explain the financial destruction of the countryside, fails to explain why resource flows have decreased to the rural sector during every year since reform was begun (in constant rubles), and fails to explain how the state has been able to enforce unequal terms of trade on food producers of all types. The ignoring of state instruments and of their impact on reform has meant that the importance of state-sponsored incentives has been misunderstood and conservative opposition overstated.

Agriculture and the State in Post-Soviet Russia

The Soviet state was able to influence incentives, the economic environment of food producers, and rural social conditions. The legacy of the Soviet state is that it intervened to regulate the private sector, defined economic relationships between food-producing enterprises and among rural individuals, and influenced the rural social environment. Although the post-Soviet state is weaker than its predecessor, the weak state approach distorts reality and blinds us to the dominant role the state has played in Russian agrarian reform. This study sees a Russian state that is able to influence key aspects of rural social policy in the rural economy. State instruments and capacities are manifest in the following aspects of the rural economy.

The Impetus for Reform

Ever since collectivization in the Soviet Union, rural reform has come from above: the Stolypin reforms, NEP, and collectivization serve as examples. In present-day Russia the impetus for rural change has once again come from the state. The state independently and autonomously defined and implemented reform policies. Thus agrarian reform was not the result of rural demands. A strong state was necessary to establish reform institutions. Indeed, it is inconceivable that agrarian reform institutions could be established without a strong state.[35]

The political ramifications of top-down reform have been enormous in Russia. It is hard to overstate the importance of agrarian reform from above. First, the state's autonomy has meant that, rather than acting in concert with rural desires, its program of agrarian reform often ran contrary to rural interests, as evidenced by rural popular opinion about the course of agrarian reform. Second, alliances between urban interests and the government took advantage of a divided and politically weak rural sector in order to adopt policies that damaged rural interests and weakened the financial condition of food producers. Because of these occurrences, we should not be surprised that election results in 1993, 1995, and 1996 indicated that support for "liberal" proreform political candidates was lower in rural areas than in urban areas.

The origin of reform also influenced the nature of agrarian policy. If the impetus for reform comes from below, and if a government joins this peasant movement, then its agrarian policies cannot be antirural (terms of trade, taxes, investments, industrial-agricultural price relationships, and so on). However, if agrarian reform originates from above, the state is less dependent on rural support and the state can pursue agrarian policies that may be harmful to the rural sector. In such an instance, a proreform rural group is put in an uncomfortable position of enduring this discriminatory policy or joining a nonstate antireform organization, the latter option running counter to the ideals and values of the proreform group. We shall see that this is exactly what has happened in Russia.

The State and the Rural Sector

State interventions have a significant impact on how the rural sector fares vis-à-vis other sectors. Rural interests are said to benefit from the development of competitive party systems in which "political competition for votes leads to a shift in policy in favor of rural interests."[36] Subsequent chapters, however, will demonstrate an intensification of budgetary, resource, and monetary discriminations against the rural sector. It is one thing to have an antirural bias in traditional societies where agricultural interests are not well organized. But the strength of the Russian state is apparent in that agricultural interests are highly organized, and still the rural sector has fared the worst of all branches in the economy during reform.

Why has this occurred? The post-Soviet Russian state is pluralist, with a multiparty system and competitive elections. In practice, however, the Russian state remains essentially corporatist.[37] Rural interests have been weak in the post-Soviet state, although the degree of weakness varies with the specific rural group. Rural liberals have had access to policy makers but poor representation. Rural conservatives have had less access but better representation in the legislative branch. In general, however, rural interests have not enjoyed the corporatist relationship that existed during the Soviet era, have not had significant access to the most powerful policy makers, and have not been well represented where political influence matters most, with consequences that have been felt by the entire rural sector.

The State and Privatization

A state defines the economic conditions in which private enterprises will operate. At first glance it might appear that privatization—whether it be of land or agricultural enterprises—would diminish the ability of the state to implement reform from above. In fact, to a certain degree this has happened. But there is also a flip side to privatization that, in my opinion, better fits the Russian case. This alternative views privatization as a change in the way the state exercises power in the countryside; this is different from an overall reduction in state power.[38] In fact, privatization has the potential to increase political power in at least three ways:

1. Privatization offers the state new means of control: budgetary, tax, and credit instruments.

2. Rural interests and farmers who benefit from new budgetary, tax, and credit policies develop a stake in privatization and become wedded to the state. Thus privatization is not merely a process that weakens state power but also a process that builds state support.

3. When, as in the case of Russia, many millions of people acquire privatized land plots, their ability to mount a cohesive counterforce to state power decreases. In other words the cost of collective action rises. With such a large group, getting the group to act by offering "selective incentives" becomes nearly impossible and the free rider problem becomes acute.[39] Rather than a large unified rural group able to act independently of state wishes, instead, rural interests are so fractured that the state is rather easily able to dominate rural groups.

Thus, assuming the state uses its instruments and capacities effectively—and manages other aspects of the political relationship wisely—privatization need not lead to a weakening of state strength.

Conclusion

Post-Soviet rural policies and institutions have been influenced by state interventions. In influencing policies and institutions, state interventions also influenced behavioral incentives, the rural economic environment, and social relationships in the countryside. These three aspects are critically important because they shape the behaviors and decisions of rural dwellers, how new rural actors and enterprises will fare, and the political-economic relationships that affect agricultural performance. As we move our analysis to Russian agrarian reform in subsequent chapters, the main arguments will revolve around the following themes:

1. Agrarian reform and privatization offered the state new ways to exercise power in the countryside. Even as state political power waned in the countryside, levers such as budgetary controls, taxes, and credits influenced the development of the private sector and reform in general.

2. The nature of reform legislation and economic levers created incentives that affected the effectiveness of reform initiatives.

3. The state was central because it greatly influenced the economic environment in which farms and farmers operated.

4. The role of the state retained principal importance in defining social relationships. The standard of living in the countryside declined significantly during Russian agrarian reform, giving rise to rural opposition which further complicated efforts at rural reform.

5. Social legacies of the Soviet period, in particular rural collectivism and egalitarianism, were not reversed sufficiently to allow for successful reform.

Although too much institutional and policy change has occurred in the Russian countryside to conclude that reform has completely failed, it is fair to conclude that important aspects of the Soviet legacy remain. The Soviet legacy has not been fully overcome, and the purpose of this study is to explore why.

2

State Interventions in Rural Social Policy During the Soviet Period

According to their level of economic development and the nature of their political system, states pursue different strategies of intervention in their agricultural sector. That states intervene in their rural economy is neither surprising nor strange. What is interesting, however, is how these interventions define the nature of the agrarian sector in general. Ever since collectivization, the Soviet state—seeking societal control—has intervened in the rural economy and defined the structure, organization, and operation of the agricultural system. State interventions are important because they influence three critical variables that affect agricultural performance. State interventions (1) influence the incentives that farms and farmworkers confront, (2) define the economic environment in which rural food producers operate, and (3) affect social conditions that have implications for rural support for agrarian policies.

Any assessment of contemporary Russian agrarian reform requires an understanding of the historical context from which Russian reforms were born. The purpose here, then, is to review the nature of state interventions in Soviet rural social policy. Aside from control over input and output prices and the food trade system, three areas of state interventions are particularly important in defining the nature of the Soviet rural sector, as they made Soviet-type agricultural sectors unique. The specific focus will be on state interventions that influenced rural incentives through the pursuit of egalitarianism within the collective sector. The economic environment was affected by limitations on nonstate agricultural activities. Finally, state financial policies affected rural social conditions and rural support. The nature of these interventions constituted the legacy in Soviet rural social policy.

Incentives: The State and Rural Egalitarianism

During the Soviet period, state interventions created and enforced conditions of egalitarianism within the rural sector. A basic thrust of post-Stalin-leadership rural social policy until Gorbachev was to reduce differences across rural occupations, regions, and within farms. Toward this end the state used various policy, financial, and economic levers. Wage policy, investment policy, and pricing policy in particular were used in the pursuit of rural egalitarianism.

Wage Policy

The basic form of labor organization in state and collective farms in the USSR was the production brigade, a collective form of organization.[1] Production brigades varied by geographical location and by the type of crop that was grown. For example, integrated tractor-fieldwork brigades would tend, on average, about 1,100 hectares of arable land and would have 120 workers. Simple fieldwork brigades would work on 400 hectares of arable land and have 100–110 workers. Vegetable workers would work on 50–60 hectares of arable land and have 40–60 workers. Mechanized brigades would work on average 400–600 hectares and use 4–6 tractors, although the actual "optimal size" varied by economic region.[2] Each brigade had a leader who was personally responsible for the work of his brigade; the brigade leader had the highest wage rate. Within brigades, labor organization was further subdivided into "links" *(zvena)*. In general, links were organized into manual and mechanized labor, depending on the nature of the work performed; thus, brigades were remunerated at different wage rates. In general, more skilled tasks were paid slightly more than manual tasks.

The brigade system suffered from two main problems. First, wage rate differentials among work categories were small and penalized (that is, underpaid) skilled workers. Sociologist Tatyana Zaslavskaya noted that during 1954–1955 the average daily oblast wage differential for a collective farmer was six times greater in the highest paid region as compared to the lowest paid region.[3]

Second, small wage differentials were not the worst aspect of the system. Archival data from Kostroma Oblast for 1959 show that the valuation of workdays *(trudni)* had little to do with farm income received from production.[4] For example, Palkinskiy district (raion) had the lowest level of income per 100 hectares of arable agricultural land, whereas Kostromskoy raion had the highest, more than five times the level of income per 100 hectares. Yet, a workday *(trudoden')* was valued less than twice as high in Kostromskoy raion (from 7.43 to 4.67 rubles per workday). Moreover, the raions that paid more per workday than Kostromskoy raion in fact had income levels 20 percent lower per 100 hectares of arable land than Kostromskoy raion.[5]

From the mid-1950s onward, wage differences decreased as part of a deliberate policy. In particular, Khrushchev and Brezhnev attempted to narrow income differences between state and collective farmworkers, and within farms. Khrushchev also dealt with state-collective farm wage rate differentials. Khrushchev attempted to raise the level of development of collective farms. State farms had more skilled and mechanized workers and more tractors, combines, and other mechanized machinery than the average collective farm. As a result, a higher percentage of state farmworkers qualified for higher wage rates, which in turn increased differences in average wage rates.[6]

Khrushchev attempted to address state–collective farm wage differentials beginning in 1953 when he promised to increase the number of skilled personnel, engineers, technicians, and various agricultural specialists assigned to machine tractor stations and to increase their wages.[7] Years later, a chairman of the collective farm Ukraina remembered the effect of the September 1953 plenum:

> it is impossible to overstate the influence of the decisions [made at the plenum]. These were cardinal decisions not only for economic questions. After the plenum specialists and organizers were sent to the farm. We received a livestock expert . . . and by 1954 the output of milk per cow increased by 800 kilograms and exceeded the output of neighboring farms.[8]

Wage rate data in general are difficult to obtain on any systematic basis during the Khrushchev era, but from available examples it appears that rates for state farmworkers were about double those of collective farmworkers (even accounting for the fact that collective farmworkers were paid by the trudoden' earned, which was less than a full day, whereas state farmworkers often were paid by norms achieved during an eight-hour workday). In 1957, for example, the average collective farm payment per workday (cash and payment in-kind) was 7.5 rubles, whereas earnings per worker in state farms averaged about 20 rubles per day, or 14 rubles for the equivalent of a trudoden'.[9] Although state farm wage rates remained higher overall, wage differences between state and collective farmworkers narrowed during the Khrushchev period (not including income from private plots). As a by-product of increasing purchase prices, wage rates, and assigning more skilled personnel to service collective farms, average collective farm wages closed the gap with average state farm wages. Soviet sources indicated that, between 1960 and 1965, collective farm wages increased from 53 to 69 percent of the average state farm wage. Overall, between

Table 1. Wage Rate Increases by Occupation, 1965–1969 (in rubles per *trudoden'*)

Category of Worker	1965	1969	Percentage Increase from 1965
Tractorist-machinist	4.25	5.10	20
Milkmaid	2.98	3.78	27
Manual field worker	2.35	3.13	33
Mechanizers	2.69	3.56	33
Kolkhoz chairman	8.54	9.70	14
Chief specialists	6.77	7.15	6
Brigadiers	3.82	4.73	24
Specialists and service personnel	2.50	3.17	27

SOURCE: A. Zinochkin and E. Moykin, "Vazhnyy faktor razvitiya kolkhoznogo proizvodstva," *Ekonomika sel'skogo khozyaystva*, no. 5 (May 1971): 75.

1960 and 1975, rural wages increased faster than urban wages, but within the rural sector the wages of collective farmworkers *(kolkhozniki)* increased by 3.2 times and those of workers and employees on state farms by 2.4 times.[10] Thus, the wage relationship remained unequal, but significantly less so by the end of the Khrushchev era, and one Western author reported that in 1966 several union republics collective farms paid higher average wages than did state farms.[11]

Brezhnev continued and expanded the trends begun under Khrushchev. During the Brezhnev era there was an emphasis on increasing wages more for those who previously were at the bottom of the wage scale. In addition to benefiting from the introduction of a minimum wage on collective farms,[12] access to state pensions, a higher percentage of income paid in cash, wage advances, as well as other benefits,[13] low-skilled and low-wage collective farmworkers saw their wages increase at a faster rate than high-wage earners, most notably farm chairmen.[14] These trends are indicated in table 1 for the years 1965–1969.

The table shows that farm chairmen and specialists continued to earn higher wages than other farmworkers. However, as a result of the trends indicated, wage differentials decreased. For example, wage differentials between farm chairmen and manual labor in the fields decreased from 3.4 to 3.1 from 1965 to 1969.

Not only were differences between farm chairmen and manual laborers leveled under Brezhnev, there was wage leveling among all farmworkers. Using unpublished wage data, table 2 shows the standard deviation in wages from 1965 to 1982 for collective farms by raion in Kostroma Oblast.

Table 2. Income Distribution by Raion in Kostroma Oblast, 1965–1982 (collective farms only)

	Mean Monthly Wage, in Rubles (23 raions)	Standard Deviation	Ratio of Standard Deviation to Mean
1965	35	15	43
1966	49	8	16
1967	60	9	15
1968	61	8	13
1969	63	8	13
1970	69	9	13
1975	103	11	11
1980	135	11	8
1981	143	11	8
1982	151	11	7

SOURCE: Author's calculations from unpublished Goskomstat data, Kostroma Oblast.
NOTE: The oblast had twenty-four raions, but one raion had no collective farms and thus was not included in the data analysis. (Numbers have been rounded up.)

The table shows an interesting aspect to Brezhnev's collective farm wage policy. According to these data, the ratio of the standard deviation to the mean decreased over time, meaning that income variance decreased. In other words, income differences among farmworkers grew progressively narrower across raions. Whereas the variance in farmworker income was relatively high in 1965, by 1982 it was six times less. (Unfortunately, I was not able to obtain data for the 1960–1964 period to show if 1965 was an aberration or part of a trend toward greater equality in income distribution.) In any event, the data clearly show that the distribution of income on collective farms in Kostroma Oblast became more equal over time. Between 1965 and 1970 drastic progress was made in lessening differences in income distribution, and starting in 1975 the analysis suggests that a norm might have been introduced stipulating that differences in farm income distributions were not to vary by much more than 10 percent. Throughout the 1970s and into the 1980s, farmworker incomes did not vary by more than 8 percent across the twenty-three raions—a rather extraordinary finding given that about three hundred collective farms were in existence.[15]

Finally, wage differences between state and collective farmworkers were also narrowed. Under Brezhnev collective farm wages increased as a result of massive agricultural investments, which increased collective farm mechanization, and as a deliberate policy to lessen differences between state and collective farms. Beginning in 1965, the rate of growth in wages for kolkhozniki outstripped that of state farmworkers. Nationwide, in 1965 the average kolkhoznik wage was 69 percent that of the average state farm wage; in 1975 it was 72 percent, and by 1980 the average kolkhoznik wage was 84 percent that of state farmworkers.[16]

Gorbachev set out to link remuneration to final product and to instill wage differentiation in a way that would reward the productive and penalize the lazy. Toward this end, basic wage rates were revised to correspond more closely to differences in the skill, experience, and responsibility required by various jobs. In a resolution adopted by the Central Committee and USSR Council of Ministers in September 1986, basic wage rate increases were differentiated for industry: manual workers received an increase of 25–30 percent, white-collar employees received 30–35 percent, and high-ranking specialists received even larger increases, beginning in 1987.[17] Each enterprise was required to finance these wage raises from its own resources, and in general remuneration was to be tied to the gross income of an enterprise. The intent was to force enterprises to improve production processes and the organization of labor, or to release redundant labor.

In agriculture, wages for managers, specialists, and employees were to be increased by an average of 30–35 percent in order to widen the gap with average worker salaries. Salaries for directors of food-processing plants—poultry, meat, and milk—were to increase to an average of 330 rubles a month, an increase of over 53 percent. Different wage rates were also introduced for workers within the agroindustrial complex. For example, at food-processing plants, grade 1 workers had their wage rates increased 16 percent, whereas grade 6 workers (a higher classification signifying more skill and responsibility) had their wages increased 33 percent.[18]

In addition, supplements to basic wage tariffs were retained. These supplements were calculated into and paid out of the enterprise wage fund and in fact were considered "necessary expenses of production."[19] Bonuses up to 50 percent of the basic salary were offered for "achievement or for the fulfillment of important and responsible work." Bonuses for management personnel were established not to exceed nine months' salary, an increase over the previous maximum of 2.6 months' salary.[20] With the conversion to self-financing, state enterprises—including collective farms—were henceforth given the right to set their own wage rates.[21]

Did the distribution of incomes within farms become less equal? Using unpublished wage data for Kostroma Oblast during the Gorbachev period, we are able to ascertain whether income distribution under Gorbachev differed from existing patterns under Brezhnev.

According to the wage data depicted in table 3, the ratio of the standard deviation to the mean increased from the Brezhnev period to the Gorbachev period. In other words, the distribution of income differences among farmworkers grew moderately. By the end of the Brezhnev period, variance in farmworker income across Kostroma's twenty-three raions (encompassing about three hundred collective farms) was only 7 percent. From that point, under Gorbachev, differences in wage distributions increased to 10 percent in 1985 and to 12 percent in 1988 before declining again to 10 percent in 1989 and 1990. One way to interpret these findings is to say that Gorbachev's reforms did indeed increase differences in income distribution. On the other hand, the data clearly show that the distribution of income on collective farms in Kostroma Oblast remained relatively equal over time: even after the "success" of Gorbachev's reforms, differences in income distribution were equal to the levels existing in 1975. Thus, it would not be inaccurate to conclude that income distribution in Kostroma under Gorbachev was slightly more differentiated than during the late Brezhnev period, but less differentiated than during the first ten years of Brezhnev's rule.

The second issue of rural egalitarianism concerned income differences between collective and state farms. Despite Gorbachev's wage reform, basic wage trends between state and collective farmworkers continued their pre-Gorbachev trends, which is to say that, between 1985 and 1990, average monthly

Table 3. Income Distribution by Raion in Kostroma Oblast, 1980–1990 (collective farms only)

	Mean Monthly Wage, in Rubles (23 raions)	Standard Deviation	Ratio of Standard Deviation to Mean
1980	135	11	8
1981	143	11	8
1982	151	11	7
1985	192	19	10
1986	211	23	11
1987	219	25	11
1988	225	26	12
1989	241	24	10
1990	265	27	10

SOURCE: Author's calculations from unpublished Goskomstat data, Kostroma Oblast.
NOTE: The oblast had twenty-four raions, but one raion had no collective farms and thus was not included in the data analysis. (Numbers have been rounded up.)

Table 4. Average Monthly Collective Farm Wages as a Percentage of Average Monthly State Farm Wages, 1985–1991

	RSFSR only	Kostroma Oblast
1985	82	106
1986	84	97
1987	84	98
1988	85	109
1989	84	106
1990	85	89
1991	87	n.a.[a]

SOURCE: *Narkhoz Rossii 1992*, pp. 405, 408; unpublished wage data for Kostroma Oblast; and author's calculations.

NOTE: Gross income differences, after factoring transfer payments and income from private plots, were smaller. See *Narkhoz Rossii 1992*, p. 144.

a. Beginning in 1991, the Kostroma branch of Goskomstat stopped reporting state and collective farm wages separately and instead aggregated farm incomes.

wages on collective farms increased faster than monthly wages on state farms. In the RSFSR from 1985 to 1991, for example, monthly collective farm wages rose from an average of 166 to 265 rubles (an increase of 59 percent), while monthly state farm wages increased from 202 to 313 rubles (an increase of 55 percent).[22] The relationship between average monthly state and collective farm wages is illustrated in table 4.

The fact that interfarm wage trends continued post-1953 patterns is another indicator that Gorbachev's attempt to reform wage relationships was not fully successful.

Investment and Financial Policies

During the Soviet period, another state lever used to pursue rural egalitarianism took the form of investment and financial policies. Of particular note were policies toward weak farms. With finite resources, the question was, Where would state investments be most effective? A unique feature of Soviet-style agriculture was the protection of economically weak farms.

During Brezhnev's rule, in particular, priority was given to aiding economically weak farms and poor agricultural regions. Various advantages were introduced to improve economic conditions for weak farms during Brezhnev's era. Although directed assistance to poor regions and farms may be dated to the

1960s, the state program "On Measures for the Increasing of Production of Agricultural Products and Strengthening Weak Collective Farms during 1971–1975" defined many of the measures that characterized this program. Because Kostroma Oblast lies in the Russian non–black earth zone and was a poor agricultural area, examples from that oblast illustrate the various advantages that Brezhnev's rural egalitarianism conveyed upon weak farms and regions. This program was introduced in Kostroma Oblast in August 1971.

Archival documents make it clear that weak farms in general received specialized attention from Party, oblast, and farm officials. Specialized statistical summaries, meetings, and "verification" reports were generated in order to monitor economic progress (or lack thereof). Weak farms and regions were shown priority assistance, even to the detriment of more productive farms. Weak farms were granted loans and credits—even if past debts had not been paid and there was little prospect of future repayment of loans.[23] The oblast agricultural administration *(upravleniye sel'skogo khozyaystva)* provided help to weak farms by distributing machinery, mineral fertilizers, feed for social livestock, and directing cadre specialists to organize economic work and wages.[24] One verification *(proverka)* campaign in Kostroma Oblast was conducted to verify special-purpose uses of the oblast budget devoted to low-profit and loss-making collective farms. The report from that verification criticized specific raions for spending too little of the allocated monies for economically weak farms.[25]

Another resource flow to weak farms was personnel. During the Soviet period, farms and agricultural organizations submitted personnel needs to their oblast agricultural administration, who in turn submitted oblast-level requests to an "upravleniye kadrov" within the Ministry of Agriculture. The national administration of cadres would then transmit general plans for personnel assignments to oblast-level upravleniye, which in turn would make specific personnel assignments for individual raions and farms. The best students from agricultural institutes were given first choice of which farm they would like to work. The worst students, with fewer choices, ended up going to the poorest farms in the most remote regions, a fact that affected both quality of performance and length of employment.

In addition to assignments, there was also a contract system. If a student entering an agricultural institute or college signed a contract with a farm, then he was admitted to the agricultural institute without an entrance examination. Otherwise, students had to take competitive exams in order to be admitted. In return, upon graduation the student had to work for two years at his sponsoring farm, after which he was free to transfer to other farms. The contract system was popular with students from remote raions who otherwise would not fare well on competitive exams. Because students were obligated to work for only two years, farms with remote locations, which often were among the poorest, had difficulty retaining skilled personnel and therefore had vacancies that went unfilled. Those factors, plus a general rural out-migration that affected all farms but especially weak farms, led farm managers of weak farms to appeal to superiors for additional skilled labor. In the Kostroma archives there were numerous letters from farm chairmen requesting additional labor for their farms, usually of those possessing skills.[26] In addition, it was (and remains) common for nonagricultural labor to be mobilized to help with the spring sowing and the fall harvest, and weak farms seemed to benefit the most from this practice.[27]

Another benefit for weak farms in Kostroma Oblast was a special monetary fund established for bonus payments to farms with low profits and those who worked at a loss. Raion agricultural officials in poor agricultural regions could apply for special increased bonus rates for especially weak farms, that is, preferential treatment.[28] Other special funds for weak agricultural regions distributed benefits in-kind. For example, in 1971 the RSFSR Council of Ministers created a special insurance fund for cereals and fodder, "from which state and collective farms which were experiencing shortages in feed could withdraw cereals." By September 1984 it was reported that state and collective farms in the RSFSR "owed" this fund 208,000 tons of feed, of which only 23,000 tons had been repaid. Several oblasts, all located in non–black earth regions, including Kostroma, had not repaid a single ton.[29]

Flowing from his new conception of "social justice," Gorbachev tried to address Brezhnev's egalitarianism by introducing wage differentiation.[30] In a similar vein, the treatment of weak farms represented one of the biggest tests for Gorbachev's new rural social policies. By the early Gorbachev period, weak state and collective farms—defined as farms who were unprofitable or had a profit of less than 10 percent—numbered about 23,500 in 1987, or 48 percent of all farms in the USSR. Weak farms had about one-half of all agricultural land and animal stocks, yet contributed only 25–30 percent of gross production. Output per 100 hectares on weak farms was one-fifth to one-sixth the level of farms with profitability rates of 25 percent or more.[31]

There is evidence that the leadership continued to favor support for weak farms, despite the economic cost. An interview with V. S. Murakhovskiy, head of Gosagroprom, in 1987 reflected a continuation of past approaches in which resources were to be directed to weak farms to increase their productive capacity. In that interview he stated:

Bringing laggards into line is probably the most difficult area of restructuring in agriculture; however, the time has come to tackle it in earnest and with every means available to us. This is not only an economic but also a political task.... If we do not halt the process of economic stratification we will not only reduce our opportunities for increased food production but also undermine the faith in the positive changes on the part of a very considerable section of rural workers.... I am convinced *that it is necessary to introduce special moral and material incentives* for leaders and specialists who achieve a radical improvement in the economy of laggard farms.... Laggard farms must be given real priority in construction work, land reclamation, and the allocation of machinery and other resources.[32]

Furthermore, in order to aid weak farms—and agriculture in general—at the end of 1989 a decision was adopted to write off a large portion of the debt incurred by enterprises and organizations within the agroindustrial complex.

Why would Gorbachev want to continue support for weak farms? It is likely that a number of factors contributed to this decision. First, many of the weakest farms were located in oblasts and raions in Central and Northern European Russia, that is, in proximity to Moscow, which made closing them politically risky. Second, the sheer number of farms and people involved rendered any drastic step economically problematic. Closing nearly one-half of all state and collective farms would have created a new rural exodus and exerted unbearable demand on already insufficient housing supplies in Russian towns and cities.

High-level political pressure made it difficult to break from the old conception of social justice. In the Politburo it is well known that Yegor Ligachev was a conservative rival of Gorbachev. Ligachev defended state and collective farms, stating that if they had tools and equipment equal to Europe and the United States the USSR could be self-sufficient in food. At the second Congress of People's Deputies in December 1989, he argued that farms often were unprofitable because they did not receive enough resources, claiming that they operated on a technical level found in the United States some forty years ago.[33] Ligachev also argued that it would be almost four times cheaper to equip unprofitable farms than to convert them to private farms.[34] After being named the head of the Central Committee Commission on Agrarian Policy in the fall of 1988, he continually advocated higher investments into agriculture. On Moscow television in April 1990 he commented:

To obtain sufficient quantities of food, of high quality and variety, society must ensure that our peasants are well equipped technically and well provided with their social requirements. After all, peasants who do not have the equipment they need, whose life in the village is difficult, who are obliged to live in uncomfortable apartments without any facilities, without the all-around development of social infrastructure, cannot feed the country.[35]

Months before being defeated at the Twenty-Eighth Party Congress and removed from the Politburo, Ligachev continued to hammer away at the need to devote more resources to agriculture and criticized plans to establish the rate of growth in capital investments lower than expected growth rates in food production.[36]

The existence of political pressures from above and below undermined attempts to reform economically weak farms. For example, even with reform of the purchase price system, weak farms were given certain advantages. With the conversion of agricultural enterprises to self-accounting *(khozraschet)*, the purchase price system was changed to include differentiated price supplements *(nadbavki)*. Previously, a uniform 50 percent of the base purchase price was paid to poor farms, which increased to 75 percent in 1983. The new system paid differentiated price supplements to farms that had profitability rates below established norms. The lower the level of profitability (or higher the level of nonprofitability), the higher the differentiated price supplement; and thus high cost continued to be rewarded. In 1989 the fund to pay for differentiated price supplements exceeded 30 billion rubles from the state budget.[37] This system of special bonus payments to financially weak farms was discontinued in 1990, supplanted by a special government investment fund created in January 1991 to assist weak state and collective farms.

State support targeted for economically weak farms was reduced from 3.3 billion rubles in 1987 to a planned 900,000 rubles in 1990. However, in January 1991 a new investment fund to help financially weak state and collective farms was formed, to be financed out of the state budget. This fund, with some 17 billion rubles (one-third of USSR farm investment during that year), was intended to offset increased prices of productive inputs, services, and credit costs.[38] Also in January 1991 the USSR Council of Ministers adopted a resolution to deliver motor fuels, tractors, automobiles, and other machinery from central supplies to farms, and to create reserves of automobile and diesel fuels during the harvest. These allocations—whose use was controlled by the central government—were made without regard to farm profitability.[39] In short, little progress was made to decrease the resource and input use of high-cost farms.

Pricing Policies

The third state intervention used to level differences across regions and farms was pricing policy. As is well known, the state controlled the entire financial process, including the granting of loans and credits to farms, the allocation and price of farm inputs, and the purchase prices for farm outputs. However, the state also defined price zones and offered bonus payments to farms that benefited economically weak farms.

In 1953, Khrushchev introduced purchase prices that were uniform within zones but that differed across zones. The problem was that Khrushchev's price zones were very large, sometimes encompassing the equivalent of whole nations in Western Europe. Under Khrushchev inequalities occurred due to the large size of zones. Farms with vastly different climatic and soil conditions, as well as productive capacities, received similar purchase prices under the Khrushchevian system. Thus, as time went on, a growing divergence between farms within zones resulted, with some farms becoming more "intensive" and others remaining backward. Brezhnev's answer to this problem was to introduce subzones (or, as they were sometimes called, microzones). Subzones would include raions with equal soil and climatic conditions.

Under Brezhnev a three-tiered zonal price system was introduced. Zonal pricing was designed to aid state and collective farms in less suitable agricultural areas, and it was in these areas that economically weak farms were most commonly found. First, at the national level a single purchase price existed for the entire country for some products, such as wool, flax, corn, and most beans. Second, within republics purchase prices were differentiated by food product, although for small republics such as Belorussia and Moldavia single prices for all agricultural products were used. Third, within oblasts, *krays,* and autonomous republics subzone purchase prices were used to differentiate farms with differing soil and climatic conditions and other "objective factors."[40]

This pricing system thus led to an increase in the number of price zones. For example, prior to 1965 there were 8 price zones for grain. By the early 1970s in the RSFSR alone there were 35 zones and subzones for grains. By 1978 there were 199 zones for wheat alone. The number of zones for livestock and animal husbandry products increased as well: for cattle the number of price zones throughout the USSR increased from 41 to 57.[41] By 1982 there were 106 milk zones. Because of the number of zones, special bonuses, and other quirks of the Brezhnev procurement system, different producers were paid different prices for the same products (an implicit land rent differential), a system that came under attack and was criticized for stifling production.[42]

The increased number of pricing zones was intended to link production costs to purchase prices, the idea being that if production cost more, then purchase prices should be higher. Obviously this system benefited oblasts in the higher-cost regions such as the Northern, Northwestern, Central, and Volga-Vyatka economic regions, but it did nothing to stimulate the efficient use of resources nor did it restrain production costs. Instead, farms with higher production costs were rewarded with higher purchase prices.

Purchase prices were a lever used by Brezhnev to pursue rural egalitarianism. Under Brezhnev it was felt that, in order to lessen differences between regions and farms, "farms located with poorer natural conditions should have purchase prices for agricultural products that are higher" than for regions with a more developed economic base and better growing conditions.[43] As one author argued, "the absence of differentiated prices for separate products, in our opinion, is unjustified. Differences in expenditures for one centner of this or that agricultural product by oblast, kray, and autonomous republic are significant and determined most of all by objective reasons."[44]

Thus, the idea was to narrow differences among farms by paying higher purchase prices to weak farms in poor agricultural regions.[45] Through a system of nadbavki to the basic purchase price, weak farms in poor regions obtained an effectively higher purchase price than strong farms. Although all farms were eligible to receive nadbavki, over time even these bonus payments were differentiated in order to pay weak farms more.[46]

At first, weak farms or farms in poor agricultural regions were at a disadvantage because they were less able to earn large bonus payments, which in turn kept them poor. But with higher procurement prices for farms in unfavorable zones and with the large number of subzones, poor farms actually received more for their production than the more efficient farms. One Soviet commentator summarized Brezhnev's zonal pricing scheme:

With the introduction of price supplements to purchase prices for low-profit and loss-making farms a majority of them practically had their own individual prices. Besides that, purchase prices for agricultural products were extraordinarily differentiated by zones, not only separate oblasts and raions, but even separate farms.[47]

The Brezhnev zonal and pricing system led to a *widening* of price differentials in favor of weak farms, and these differentials were significant. Under Khrushchev, oblasts located in unfavorable zones in the RSFSR were paid 35 percent more for soft wheat than Northern Caucasus oblasts; under Brezhnev this difference widened to 80 percent.[48] (These differentials refer to base purchase prices only; nadbavki would add 50 percent of the base price per unit of above-plan production sold to the state.)

Finally, other financial levers were used to benefit weak farms. The July 1978 plenum changed the rules governing income tax of collective farms so that farms with a profitability of less than 25 percent were exempted from income tax.

Gorbachev tried to move away from the wasteful and ineffective financial policies of his predecessors. Collective farms were converted to self-financing and self-accounting *(khozraschet)* during 1988–1989, a move that had first been considered by the Politburo in December 1986. The conversion to a new system of farm financing would make farms responsible for meeting their own costs; no longer could they assume that central subsidies would bail them out. According to Viktor Nikonov, then CPSU secretary for agriculture, self-financing and khozraschet would improve labor productivity and farm efficiency. But before self-financing could be implemented, one of the most important conditions was farm profitability. Nikonov maintained that profitability levels of 35–40 percent were required for self-financing to be successful. However, Nikonov revealed that in 1986 some 6,400 state and collective farms worked at a loss (about 13 percent of the total), and another 53 percent of all farms had profitability levels of less than 25 percent. Only about 20 percent of state and collective farms were in a position to convert to self-financing.[49] In order to make farms "profitable," that is, to make them eligible for conversion to self-financing, at the beginning of 1988 nadbavki were increased dramatically in order to improve farm finances, so much so that nadbavki often exceeded the purchase price by several times.[50]

The effect of increased price supplements (which, it will be remembered, benefited poor farms the most) can be seen in Kostroma Oblast. The increase in nadbavki dramatically decreased the number of loss-making farms. In the oblast in 1986, there were forty state farms (31 percent) and eighty-eight collective farms (44 percent) that were unprofitable. In 1987, these numbers increased to fifty-one state farms and ninety-three collective farms, so the financial condition of farms was worsening. In 1988, however, after nadbavki were increased, the number of unprofitable state farms dropped to six (4.5 percent) and the number of unprofitable collective farms declined to one (0.5 percent).[51] This remarkable turnaround was accomplished by a significant increase in the level of price supplements to farms. During 1983–1987 about 20 percent of total oblast expenditures to agriculture were used for nadbavki in Kostroma Oblast. During 1988–1989, however, the level of nadbavki increased to 45.5 percent of total agricultural expenditures, leading to an increase in the number of profitable farms, an increase in farm profitability levels, and almost a fourfold increase in net farm income even though nearly all categories of production declined.[52] Although some authors claimed that self-financing and khozraschet were responsible for the decline in the number of unprofitable farms, in point of fact the decrease was due to new infusions of state monies to prop up poor farms.[53]

Second, the number of price zones was decreased—a system that had benefited weak farms. The number of price zones was reduced from forty-nine zones to five for most meats, and from ninety-eight zones to four or five for milk. Uniform prices were adopted for most grains, although republics could establish their own intrarepublican price zones but had to fund price differences from their own budgets. In addition, subsidies for farm inputs and machinery were reduced.

Economic Environment: The Nonstate Agricultural Sector

A second important state intervention, one that helped define the overall economic environment, was regulation of the rural private sector. This regulation was important because it minimized domestic competition to the collective sector, and because it complemented efforts to keep the countryside egalitarian. Although many rural dwellers invested enormous amounts of time in their plots, state regulations limited production potential, trading opportunities, and income for much of the Soviet period.

The rural "private" sector during the Soviet period commonly referred to "private plots," dacha plots, or some similar small-scale nonstate agricultural activity. The term "private plot" *(lichnoye podsobnoye khozyaystvo)* refers to the personal land plot outside the social sector of agriculture.[54] The Soviet "private" sector was not really private at all, at least in the sense that the land was owned and could be bought or sold. Some analysts argue, however, that it was appropriate to call them private plots since they were in essence and in practice more private than social and the land plot was outside the collective sector of agriculture. In order to avoid confusion with genuinely private land in the post-Soviet period we shall refer to such plots as "personal plots."

Personal plots denoted individual or family agricultural activities whose output was used to supplement the family's diet. Personal plots were quite small. In post-Stalin years the overall size of a plot could not exceed one-half of a hectare, including the land on which a dwelling was built, or one-fifth of a hectare on irrigated land, although regional variations did exist.[55] Both urban and rural dwellers were eligible to operate a personal plot. In the rural sphere personal plots were much more prevalent among collective farmers than state farmworkers.[56] Personal plots were commonly located around an individual's house. However, those without houses could also obtain land for such activities. Apartment dwellers were able to obtain a plot of land either from the local soviet or through their place of work. Pensioners could also keep personal plots.

There were also other forms of private farming operations, all of which were small-scale. Urban dwellers could rent small land plots *(ogorody)*, often from a nearby state or collective farm, in order to grow a few vegetables to supplement family supplies. In addition, there were the plots associated with enterprises and factories, which grew food to sell to their workers or to use in the dining hall, called subsidiary farming *(podsobnoye khozyaystvo)*.[57] Enterprises would obtain land from nearby state or collective farms or would be assigned land by the city soviet. The land would then be used to grow food, with the output sold to enterprise workers in the enterprises' own food stores at below-market prices and used in the enterprise cafeteria.[58] Subsidiary agricultural operations were conducted also by the army, in order to grow food for the soldiers.[59]

A third type of private agriculture was collective gardens and orchards. These forms of small-scale agriculture were especially popular with urban residents who had migrated from rural areas but who wanted to remain in touch with the land. These people could obtain land within a collective setting to grow food to supplement family supplies, just as long as the person remained employed by the enterprise that organized the collective garden or orchard. Although collective in name, the labor was in fact individual. In urban areas the land for these endeavors was allocated by the city soviet from enterprises, and in rural areas the land often came from forestry funds or the state land fund; in other cases land could be assigned from state or collective farms.[60] Collective gardening faced numerous restrictions regarding size and the keeping of livestock, but under Gorbachev these restrictions were greatly relaxed.[61] Thus, in the Soviet Union "private" farming took on a number of different forms, and the popularity of these endeavors is evident in that more than one-half of the nation's population engaged in some form of private agriculture in the early to mid-1980s.[62] Because personal plots have historically been the most common and productive form of "private" agriculture, our focus is on *lichnoye podsobnoye khozyaystvo* by individuals rather than collective gardening or subsidiary agriculture of enterprises, organizations, and factories.[63]

The Personal Plot from Stalin to Gorbachev

Dating from the onset of collectivization, Stalin's policies toward the nonstate sector may be divided into three phases. The first phase was one of begrudging tolerance, indicated by the Model Collective Farm Charter of 1930, which allowed the retention of personal plots of limited size within the village. In 1932 chaotic channels of private food trade were replaced with a legalized system of collective farm markets where peasants could sell their personally grown produce.[64] In 1935 the Model Collective Farm Charter guaranteed the retention of personal plots because of their importance to the national food supply. For example, during the first two five-year plans over 60 percent of the cattle, 70 percent of the cows, 55 percent of the pigs, 80 percent of the goats, and 53 percent of the horses were held by personal plot operators.[65] Nonetheless, in the years just prior to the war Stalin argued that peasants were devoting more time to their plots than to communal work, and in a decree of May 1939 he attacked personal plots by limiting the size of a plot and the number of livestock that could be raised, and by instituting a minimum number of days that must be worked in the socialist sector.[66]

The second phase during the Stalin period was during World War II, and this period was characterized by leniency toward personal plots. During the war collectivization was interrupted as farm members were allowed to leave state and collective farms, and the ability of the state to control and manage the food sector declined. During the war, the role of nonstate agricultural activities expanded. First, and probably most important, personal plots were crucial in feeding the peasantry itself. This local self-reliance was important because it lowered overall demand on supplies coming from the state sector. Food from personal plots was used to provide for the massive number of people who were evacuated to the east, especially during the first two years of the war. Finally, personally grown food was sold to urban residents on the collective farm market.[67]

The third phase of Stalin's plot policy occurred in the postwar period and witnessed the return of Stalinist bias against personal plots, and in fact against all agriculture. In the collective sector, state investments into agriculture as a percentage of all investments were less during 1946–1950 than they had been during the first five-year plan. Stalin returned to his emphasis on superindustrialization and reconstruction. As a result, agriculture received one-fifth the level of government investments in heavy industry, and most of the investment in agriculture was used for growing food, not reconstruction or expansion of rural infrastructure.[68]

For rural dwellers in the postwar period, the personal plot had often meant the difference between survival and starvation. For example, as a result of policies during the war, produce from personal plots provided over 65 percent of a kolkhoznik family's income in 1946, up from 53 percent in 1940.[69] Stalin, however, initiated a crackdown on "liberal" policies that had been tolerated during the war.[70] The personal plot, which had been an important source of food during World War II, came under particular attack. In 1948 a tax in-kind was introduced on individual households and on production from personal plots.[71]

Toward the end of his rule Stalin renewed his attack on personal plots with the intent to eliminate this sector, despite their importance to the food supply.[72] Stalin's last direct attack on kolkhozy and collective farm markets came in his "will" on economic policy, published in 1952 as *Economic Problems of Socialism in the USSR*. In this short book Stalin called for the elimination of the kolkhoz (collective farm) market and the establishment of an exchange system between industry and agriculture.[73] The effects of Stalin's barter system would have been devastating for the peasants. In the process of converting collective farm property to public property, the personal plot (which peasants needed to survive) would have been further curtailed, if not eliminated altogether. But even if plots had survived, peasants would have had no legal outlet to sell their surplus produce.

After Stalin's death in 1953 Khrushchev attempted to moderate Stalin's exploitative policy toward agriculture.[74] Despite Khrushchev's distaste with the Stalinist approach to agriculture, Khrushchev was a deeply ideological leader whose agricultural policies were driven by his outlook on the development of communist society. At the same time, Khrushchev faced some very real problems. Agricultural production had not recovered to the same extent as had industry, even though Khrushchev was faced with an increased demand for food. In 1952 industrial production was 67 percent higher than 1940 levels and 2.5 times higher than prewar levels, whereas even in the good harvest year of 1952 agricultural production was barely above prewar levels. The total number of all livestock in 1953 was lower than in 1928, and the number of cattle remained below the 1916 level.[75]

In his very first speech on agriculture in September 1953, Khrushchev expressed concern that demand for food was growing faster than production and that "the most important task" was to improve the dietary structure and increase the production of meat and vegetables.[76] In fact, the population of the USSR had rapidly become less rural and more urban (thus increasing urban demand for food): the urban population in the USSR increased from 60.4 to 99.9 million between 1939 and 1959, whereas the rural population decreased from 130.2 to 108.8 million during the same period.[77] During the Khrushchev years alone, 1953–1964, the urban population of the USSR increased by more than 37 million whereas the rural population decreased by more than 1.5 million.[78] The theme of linking improvements in agricultural production to urbanization was one that Khrushchev returned to time and again in his most important speeches.[79]

In order to improve the food supply and to meet growing urban demand, Khrushchev pursued a number of different strategies. Khrushchev's first strategy was to induce higher agricultural production by increasing material incentives to farms. To achieve the production increases he desired, Khrushchev announced at the September 1953 plenum that procurement prices would be increased for livestock and poultry by a factor of 550 percent, prices for milk and butter increased by 220 percent, potato prices more than doubled, and prices for vegetables increased by 25–40 percent.[80] In addition, higher above-quota purchase prices were introduced for meat and milk. The strategy to give farms production incentives through higher state purchase prices was pursued until the end of the Khrushchev period. In subsequent years additional purchase price increases were announced for a series of animal husbandry products. In 1962 purchase prices for meat from cattle, pigs, and poultry were increased an average of 35 percent throughout the country.[81] In 1963 purchase prices for raw cotton, potatoes, and sugar beets were increased, and in 1964 purchase prices for milk were raised 18 percent.

Khrushchev's second strategy to improve food supply was to expand the area under cultivation, in a program known as the Virgin Land Program. He believed that food production could be rapidly increased by expanding the amount of land under cultivation. During the second half of the 1950s more than 30 million hectares of virgin and idle land in northern Kazakhstan and parts of the Russian Republic were assimilated by farms, and by 1960 the amount of arable land under cultivation increased to 220 million hectares, up from 188 million in 1953.[82] Furthermore, in the hope of reaping economies of scale he set in motion a program of merging small kolkhozy to form larger ones. Over the ten-year period 1953–1962, the number of kolkhozy was reduced from 97,000 to 39,500; the average number of households per farm rose from 208 to 440; the sown area per farm rose from 1,693 to 2,896; and the capital stock per farm increased from 66,900 to 779,000 rubles.[83]

His third strategy was to promote personal plot production. Khrushchev's approach to personal plots clearly falls into two time periods. The period 1953–1958 was a time of relaxation toward the personal sector. Khrushchev's plot policy, however, was merely tactical. His approach to personal plots was indicated by his statement that they were necessary "only as long as the social economy of the collective farm is insufficiently developed."[84] Similar to Stalin's, Khrushchev's antipathy toward the nonstate sector was tempered by the necessity for the food produced by that category of farming. His basic attitude favored the development of state farms and the communal sector.

Early in his tenure, 1953–1958, Khrushchev sought to increase personal plot production by easing a number of restrictions. For example, he ended obligatory deliveries to the state in the form of taxes from plot production. The amount of land used by personal plot operators of all types—kolkhozniki and workers and employees—increased from 6.88 to 7.32 million hectares between 1953 and 1958, thus reversing a long-term historical trend. Personal livestock herds were expanded by making it easier to acquire young animals. As a result, the number of cattle held in personal herds increased from 23.3 to 29.2 million head between 1954 and 1959,[85] and cows increased in number from 14.9 to 18.5 million.[86]

Even during the early "liberal" phase toward personal plots, state regulation of personal plots did not cease. Personal plots were limited in size, usually less than half an acre. The number of livestock or other animals was also restricted.

In 1953 a firm tax rate per *sotka* was established (a sotka is one hundred square meters), meaning that a tax was assessed according to the amount of land used, and this tax was due irrespective of actual output. In 1956 collective farms were given the right to regulate the size of personal plots, which introduced considerable conflict of interest—farm managers were resistant to give up "good" land and further wanted farmworkers to be working on collective endeavors, not their personal plots. In 1958, although obligatory sales to the state were ended, prices for personal produce sold to the state were pegged to state purchase prices.[87]

The second phase of Khrushchev's policy, after 1958, strengthened state regulation of personal plots and might even be considered hostile toward personal plots. In December 1958, before a plenary session of the USSR Central Committee, Khrushchev argued that "the existence of large personal plots and livestock holdings has become a serious obstacle to the further production of state farm *(sovkhoz)* production. . . . Sovkhozy must gradually . . . purchase the livestock from their workers and employees. . . . Then there will be no longer any need for sovkhoz workers to occupy themselves with a personal plot."[88] Sensing what was expected of them, farm managers and local officials set out to obtain personal livestock with particular alacrity, often violating the principle of "voluntary sales." By the spring of 1959, Khrushchev had to call for restraint, and an editorial in *Pravda,* March 5, 1959, argued against excesses. I. A. Benediktov, the minister of agriculture from 1938 to 1958, characterized Khrushchev's attitude toward plots after 1958:

Khrushchev . . . considered personal plots, and also the activities of cooperatives in the country as "remnants of the past and antiquated" which supposedly "diverted" the peasant from collective labor and somehow prevented the manifestation of the enormous potential of the "superiority of socialism" in the countryside.[89]

After 1958 state interventions restricted personal plots through a multi-pronged strategy. The first was a direct attack on personally owned livestock. Khrushchev made it clear that he wanted to lessen state dependence on personally raised meat, and in subsequent speeches he envisioned most of the growth in livestock herds as occurring in the socialized sphere.[90] Plot operators were prohibited from purchasing grain products to use as feed, and the sale of feed grain and concentrates to personal plot holders was discontinued. Payments in-kind to collective farmers, which had consisted mainly of grain for fodder, were also reduced. Fodder and grazing rights were restricted; and in conjunction with the reduction of payments in-kind of feed and fodder it became very difficult for the farmer to feed his livestock.

The second part of Khrushchev's strategy witnessed the sale of personal live-

Table 5. Growth of Urban Centers, USSR, 1959–1970

Size of City	Number of Residents at Beginning of Year, 1959 (in millions)	Percentage of Total Population	Number of Residents at Beginning of Year, 1970 (in millions)	Percentage of Total Population
Up to 50,000	40.4	40	47.4	35
50,000–100,000	11.0	11	13.0	10
100,000–500,000	24.4	24	38.3	28
500,000–2 million	14.8	15	26.3	19
More than 3 million	9.4	9	11.0	8
All cities	100.01	—	36.0	—

SOURCE: V. Perevedentsev, "Migratsiya naseleniya i ispol'zovaniye trudovykh resursov," *Voprosy ekonomiki*, no. 9 (September 1970): 37; and author's calculations.

stock to the collective or state farm. Although these sales were ostensibly "voluntary," they were often compulsory. The consequences of this policy led to a mass slaughter of livestock, although not on the same scale as during Stalin's collectivization. A former Central Committee member noted, "millions of head of cattle were destroyed. Neither kolkhozy nor sovkhozy were able to accept such a mass of cattle because of the poor state of the fodder base. It was advantageous for the peasants to kill the cattle and use the meat for themselves or for sale on the market."[91] As a consequence, between 1960 and 1964 the number of livestock held personally declined.

When Brezhnev came to power he shared Khrushchev's idea that increased food production was necessary. He was confronted with the same large increases in urban population, a result of continuing rural-urban migration. Whereas under Khrushchev the urban population had grown by 19.3 million between 1956 and 1966, under Brezhnev the urban population increased by 23 million between 1966 and 1981.[92] Between 1959 and 1969, for example, the net urban population of the USSR grew by 36 million, and the rural population declined by 3.1 million.[93] The bulk of the increase to urban centers came from rural areas: 16 million people migrated from rural localities, 5 million were converted to urban dwellers as a result of redefining villages as urban settlements, and 14.6 million resulted from the natural increase of the urban population. In particular, cities of all sizes increased between 1959 and 1970, but cities with between a hundred thousand and 2 million residents grew the fastest. Population increases in urban centers is indicated in table 5.

As a result of urban in-migration, the 1979 census revealed that the rural population comprised 38 percent of the population, down from 52 percent just twenty years earlier.[94] Although all industrializing nations experience rural-urban migration, the Soviet Union was unique in the speed with which a rural nation turned into an urban one.

The importance of these demographic trends was threefold. First, by the time Brezhnev took power the urban population was growing faster than increases in food production. Second, the urban population had experienced a significant increase in food per capita consumption during the Khrushchev years and expected those trends to continue; thus Brezhnev was faced with modest pressure from rising expectations. Third, because most rural migrants came from collective farms, their departure for the cities meant significantly fewer personal plot operators. Fewer personal operators in turn translated into less food from personal plots and more demand on food supplies from the socialized sector.[95]

Brezhnev adopted a much more tolerant policy toward personal plots than Khrushchev. The official historical record written during the Brezhnev period asserts that "unjustified limitations" had been introduced on personal plots while Khrushchev was in power. Brezhnev almost immediately reversed Khrushchev's restrictions and liberalized conditions for plot operations. Two weeks after Leonid Brezhnev assumed the post of First Secretary of the Communist Party of the Soviet Union, a plenum of the Central Committee instructed republican central committees and the Council of Ministers to review and "resolve the question" about limitations on personal plot production, in particular the norms that governed head of livestock.[96] At that plenum of October 27, 1964, the Central Committee of the CPSU passed a resolution entitled "On the Elimination of Unfounded Limitations on Personal Subsidiary Farming of Collective Farmers, Workers, and Employees."[97] Although only excerpts of this resolution were published, in part it resolved: "To instruct the CC of Communist parties and Council of Ministers of union republics to consider and resolve the question of removing the restrictions which were placed on personal subsidiary farming by collective farmers, workers, and employees in past years."[98] On November 4, 1964, the tax on livestock owned by urban dwellers (which had been in existence since 1956) was repealed. On November 6, 1964, in a speech commemorating the forty-seventh anniversary of the Bolshevik revolution, Brezhnev argued that previously "it was incorrect to disregard the possibilities of personal subsidiary farming of collective farmers, workers, and employees. . . . In past years in this sphere groundless restrictions were permitted, although economic conditions still have not developed for such a step. These restrictions have now been abolished."[99]

Thereafter, other measures were adopted that reflected the relaxation of

Brezhnev's plot policy. For example, in mid-November 1964 plot sizes that were in effect prior to 1956 were restored in the RSFSR. The model charter adopted at the Third All-Union Congress of Collective Farmers in late November 1969 devoted Section 10 to personal plot farming. According to the Model Statutes, the plot could extend to one-half a hectare, including land occupied by buildings, or up to one-fifth a hectare on irrigated land.[100] The Principles of Land Legislation—adopted the following month in December 1969—formalized the right to operate a personal plot. Articles 24, 25, 26, and 27 discussed the terms and conditions for collective farmers, state agricultural enterprises, and workers, employees, and other citizens working in the rural countryside.[101]

Brezhnev's plot policies paid dividends. Between 1958 and 1963 the value of gross production from personal plots (valued in constant 1965 prices) had declined by 2 billion rubles but during 1965–1967 increased by 3 billion rubles.[102] Put another way, in 1964 personal plot output had declined to 92 percent of its 1958 value, but by 1967 had reached 105 percent of 1958 value. In addition, between 1965 and 1970, the area of land cultivated for personal plots by kolkhozniki, sovkhozniki, workers, and employees increased from 7.58 to 8.2 million hectares.[103] Between 1965 and 1970, meat production from personal plots increased almost 8 percent, milk production rose 4 percent, egg production went up almost 11 percent, and wool increased 8 percent.[104]

Brezhnev had reason to be pleased with his agricultural policies early in his tenure, but Soviet agricultural performance deteriorated in the early 1970s, marked by poor harvests in 1972, 1974, and 1975; in 1975 in particular the grain harvest was the worst in a decade and necessitated the mass slaughtering of livestock. In addition, the costs of his financial commitment to agriculture were a source of concern. Already by 1976, of the 320 billion rubles that had been invested in agriculture during the Soviet period, 213 billion (nearly 70 percent of the total) had been invested since 1965.[105] Although the CPSU secretary in charge of agriculture, F. D. Kulakov, continued to speak out for additional sums to be invested in agriculture, at the October 1976 plenum Brezhnev admitted that the money to be spent in agriculture was an "enormous sum" and that "it was not easy to find it. We had to reduce somewhat the requirements of other branches of the national economy."[106]

Brezhnev's main "innovation" to the agricultural downturn was to return to his personal plots policy and promote their operation with renewed emphasis. Brezhnev looked to personal plots as a method to increase food production for the purpose of increasing meat and animal supplies in urban centers without an additional massive infusion of new investment monies and did not siphon off investment resources from the socialized sector. During 1977 conditions for conducting trade of privately grown produce were liberalized still further.

Brezhnev was concerned with the continued supply of food to the urban population, and a key component of the urban food supply was the kolkhoz market. In 1977 there were some 6,500 collective farm markets, with more than three-quarters located in urban areas. Far from trying to curtail them and squeeze kolkhoz markets as his predecessors had done, steps were taken to strengthen kolkhoz trade.[107]

To promote the operation of personal plots, in September 1977 a resolution was adopted by the CC of the CPSU and Council of Ministers entitled "On Personal Subsidiary Plots of Collective Farmers, Workers, Employees, and other Citizens and Collective Gardens and Orchards."[108] This resolution had nineteen separate measures and was directed to the Council of Ministers in the union republics, to Gosplan, Gosbank, and Gosstroi, to various ministries, and to state and collective farms. This formal decree (the first since 1964 on personal agriculture) provided for loans to buy young livestock and adopted measures for expanding the production and sale of tools to plot operators.

As Soviet agriculture continued to suffer through years of bad harvests and poor production during the late 1970s and early 1980s, the regime continued to promote personal plots. In January 1981 a resolution was jointly adopted by the CC of the CPSU and the USSR Council of Ministers entitled "On Additional Measures to Increase the Production of Agricultural Products on Personal Subsidiary Plots of Citizens."[109] This January 1981 resolution went beyond the provisions outlined in any previous legislation.[110] An important consideration of this measure stipulated that plot operators who signed contracts with state or collective farms for sale of production were given improved access to feed and fodder supplies and to other needed inputs. In return, if the plot operator sold his livestock to the farm, the farm could then use that production to count toward its own plan fulfillment.[111] The intent was to provide farms with an incentive for assisting personal plot holders. A follow-on resolution was adopted in February 1981 by the CC of the CPSU and the USSR Council of Ministers. This resolution allowed for land to be converted to personal plots, according to established procedure, from "the state land supply, state forest fund, industrial, transport, and other nonagricultural enterprises and organizations, and also unused land of collective farms, state farms, and other agricultural enterprises and also including land that is uneconomically used by poor collective and state farmers."[112]

Even Brezhnev's "liberal" policy toward personal plots did not deregulate their activities. To the contrary, state interventions in the personal sector continued to limit production and income possibilities. First, and most important, personal plots were regulated in size.[113] Local officials were asked to enforce existing restrictions on personal plots and other forms of personal production.[114] In particular, local officials were charged with enforcing land plot sizes to make sure a person did not violate size restrictions, which could lead to increased income potential. For example, in Kostroma Oblast there was a crackdown on violations of land plot sizes in which a "verification" document reported that nineteen individuals in Kostroma raion and the suburbs of Kostroma city, as well as twenty-eight persons in the city of Shar'ye and Shar'inskiy raion had "excessively large" land plots. In the case of Kostroma city and raion, the nineteen persons who were individually named in the document had no less than thirty-eight square meters (the average personal plot was twenty-seven square meters).[115]

Second, the state continued to regulate the number of head of livestock that a personal plot operator could raise. Earlier, the January 1981 resolution had rescinded numerical restrictions. However, this restriction was only lifted if livestock were raised under contract with a state enterprise, in which case the purchase price was regulated by the state. Thus, if a plot operator wanted to raise livestock for free market sales the number of livestock remained restricted.

A third way that the state continued to regulate plot activities was the prohibition on owning horses, as stipulated by the Collective Farm Charter of 1969. For collective farmers, the use of a horse required permission of the respective republic's council of ministers. In August 1982 the Supreme Soviet of the RSFSR passed a law that permitted citizens to own one draft animal, but similar to other republic legislation it stipulated that the citizen who owned the animal could not be a member of the collective farm. There was no restriction, however, on state farmworkers or other citizens owning horses. Thus, after 1982, virtually all rural families except kolkhozniki had the legal right to own a horse.[116] This prohibition was important because horses were used for plowing and cultivating, owing to the chronic shortage of small tractors and other machinery suitable for personal agricultural activities. Thus, with a shortage of mechanized implements, horses were a primary labor-saving device of plot operators.

Despite short-term successes, personal plots were not the answer to the country's agricultural problems, however. By 1980, 83 percent of the labor expended on personal plots was performed by women or the disabled.[117] Plots were labor-intensive and suitable for only a relatively small range of agricultural produce. Thus, personal plots were a conservative default option that allowed the state to increase food supply at the margins relatively easily and cheaply without really changing the structure of rural society.

Gorbachev came to office with an agricultural sector in disarray. During 1981–1985, for the first five-year period since the 1950s, average grain production levels had declined, grain imports had increased to an average of 38 million tons a year, and per capita food consumption had stagnated. In order to improve agricultural performance Gorbachev moved on three different fronts. These initiatives included the liberalization of personal plot operations, the legalization of land leasing, and the establishment of independent peasant farms.

Prior to the adoption of land leasing in 1988–1989 and the development of peasant farms in 1989–1990, Gorbachev turned to personal plots to bolster the nation's food supply much as Brezhnev had done. Similar to Brezhnev's attempt, this was a conservative reform because personal plots did not require that state and collective farms restructure or be broken up, and it is this aspect that led conservatives such as Ligachev to support expanded plot holdings.[118] Under Gorbachev's reforms, personal plots would continue to be tied to state and collective farms and would depend upon them for material assistance. Thus, personal plots represented an inexpensive way to increase food production without requiring structural change in the countryside.

The policy of encouraging personal plot production during Gorbachev's early tenure was suggested by an article in the authoritative Central Committee journal *Kommunist*, which noted that "at the present stage, when individual plots are objectively necessary and fulfill important socioeconomic functions, we should find ways of achieving more efficient development in this sphere."[119] In a 1986 interview the chairman of Gosagroprom, V. S. Murakhovskiy, stressed that "we have to bring about a situation in which every dweller in the countryside . . . takes part in the production of milk, meat, potatoes, vegetables, and fruit. He will provide produce for himself, the surplus he will send to market. This will be a large contribution to the solution of the Food Program."[120]

Personal plots received attention from the Politburo in a July 1987 meeting. This meeting called for "increasing the production and sale of agricultural products" from personal plots among kolkhozy, sovkhozy, and the population. Further, the Politburo recommended to "review the norms for keeping cattle and on the size limits for personal plots."[121] By 1988 personal plots were considered an integral part of socialist agriculture.

The regime's favorable policy disposition toward personal plots was codified in the new draft Collective Farm Charter in March 1988. In a section devoted to personal plots (section 9), for the first time personal plots were referred to as "a component part of socialist agricultural production." This status coincided with momentum to deregulate plot operations, and restrictions on the permitted number of livestock and poultry were conspicuously absent in the statutes.[122] By the time of the important 1989 plenum on agricultural reform, personal plots had gained equal stature with other forms of farming, indicated by the inclusion of personal plots in developing a "multiplicity" of farming operations, and this formula was repeated by Gorbachev himself on many occasions.

The renewed favor with which personal plots were treated was indicated in other ways as well. Over the course of three decades fewer and fewer people operated personal plots, and the number of cattle and other livestock declined as well.[123] Under Gorbachev, however, production from plots and the number of families who operated plots increased significantly. For example, from 1980 to 1985 the number of people operating a personal plot in Kostroma Oblast decreased from 8,800 to 3,600. From that low point, from 1985 to 1991 the number increased from 3,600 to 26,800.[124] With more people operating a personal plot, production naturally increased. From 1985 to 1991 the production of cattle and poultry (live weight) rose from 24.4 to 36 thousand tons, an increase of 50 percent.[125] Other types of production from plots increased as well. Throughout Russia, by 1989 it was clear that a revival in the popularity of personal plots was underway, as from 1985 to 1991 the percentage of total food turnover on the urban collective farm market increased from 3.7 to 5 percent of retail trade turnover.[126]

Nor was Kostroma Oblast an exception. In the Russia Republic, although production in the social sphere was declining it increased for certain products from personal plots. In Russia during 1991, whereas meat, milk, egg, and wool production fell below average levels attained during the twelfth five-year plan (1986–1990), personal plots produced more meat, more eggs, more milk, and more wool than in 1985.[127] Personal plot production was even more important for dwellers in Russia's non–black earth zone, accounting for over 61 percent of potatoes, 50 percent of vegetables, and nearly 90 percent of berry production. Among the twenty-nine oblasts and autonomous republics that comprised the non–black earth zone, in 1991 personal plot production accounted for more than 25 percent of gross food production in twenty oblasts, including more than 30 percent of gross food production in eight oblasts.[128]

By 1990, throughout the USSR there were more head of cattle, more pigs, and more sheep and goats held by personal plot operators than in 1985, a reversal of a decades-long decline. These trends are illustrated in table 6.

More head of livestock translated into increased meat output. After a decade of stagnation in meat output (1971–1980), meat output from personal plots grew by almost 11 percent during 1985–1990.[129] In 1990, output from personal plots accounted for 65 percent of the USSR's potatoes, 33 percent of its vegetables, 30 percent of its meat, 28 percent of its milk, and 27 percent of the country's eggs.[130] All these indices represent gains in comparison with 1985, again a reversal of a decades-long trend.

The liberalization of personal plots failed to address larger systemic problems. The personal plot as a reform default was conservative because the land used was not owned and could not be bought or sold; plots remained small and

Table 6. Livestock Holdings on Personal Plots, USSR, 1980–1989 (million head)

	1980	1985	1989
Cattle	23.0	24.1	24.2
Pigs	14.0	13.9	15.2
Sheep and goats	30.2	33.1	36.3
Poultry	387.7	389.2	410.8

SOURCE: "O lichnykh podsobnykh khozyaystvakh naseleniya," *APK: ekonomika, upravleniye,* no. 3 (March 1991): 22.

thus production was limited; and income flows were finite because of the limited production. Personal plots by their very nature would remain small and labor-intensive. Personal plots could not transform the agricultural sector.

As Gorbachev's thinking evolved, he moved toward liberalizing conditions in which larger agricultural units operated. A new form of personal farming was the individual (or family) peasant farm, legislation for which began to be discussed as early as mid-1989. The general idea was to allow a state or collective farmworker to leave the parent farm, to lease land, and to begin his own independent farming operations, although there were considerable and important differences between national legislation and legislation in various republics concerning landownership and peasant farming.[131] The peasant farmer would conduct independent operations and not be bound by the constraints of the internal lease that required the leaseholder to remain within the parent farm structure.

It was hoped that individual farms would change the face of Soviet agriculture and, over time, address problems of input efficiency and high production costs. However, there were two ways to pursue peasant farming: one way was a radical path based on private property. This path was pursued by the Russian Republic and its leadership, headed by Boris Yel'tsin. (Russian land legislation on peasant farms and land reform will be examined in a subsequent chapter.) There was also a conservative approach to peasant farming, which rejected private property, and this approach was embraced by Gorbachev.[132]

Under Gorbachev, the most important document on land use in the USSR was the Fundamentals of Land Legislation, adopted at the end of February 1990 (published in March). This document indicated that USSR citizens could lease land for peasant farming, but it did not provide for private ownership. The legislation left it to individual republics to define the process for acquiring land—that is, how a farm member could leave the state or collective farm in

Table 7. The Development of Peasant Farms in the RSFSR, 1990–1991

	April 1990	July 1990	January 1991	July 1991	January 1992
Number of farms	240	890	4,433	25,159	49,015

SOURCE: Goskomstat data on peasant farms, 1990–1992.

order to start independent operations, and the size of land plots to be used for peasant farming. The most common method for obtaining land for peasant farming was to lease land from a state or collective farm. After December 1990, Russian legislation allowed members of state and collective farms (and their families) to leave the farm and granted them a land plot free of charge from the parent farm.[133] A dispute thus existed between national and Russian legislation. Under national legislation, land for peasant farming was governed by the Law on Leasing and the Law on Property—which is to say, land could be leased for up to fifty years but could not be owned, bought, or sold. Under Russian legislation, personal property was allowed, although restrictions were placed on the purchase and sale of agricultural land.

During 1990, the highest number of peasant farms developed mainly in those republics with the most liberal land laws.[134] Until the RSFSR adopted more radical land legislation, peasant farms were limited, and thus as of April 1990 only 240 peasant farms were registered in the RSFSR. During 1990 the number of peasant farms in the RSFSR grew slowly. Only after the December 1990 legislation did the number of peasant farms begin to increase rapidly in the RSFSR. Legislation adopted in the RSFSR at the end of December 1990 permitted a farm member to leave the farm (with a land plot) *without* the approval of the farm collective or farm chairman, and to own land privately.[135] Thus, in the earliest stage of land reform it is clear that legislation had a significant impact on individual behaviors. The growth in the number of peasant farms during 1990–1991 is shown in table 7.

Even after the RSFSR adopted legislation to open up state and collective farmland to members who wanted to withdraw, Gorbachev continued to hold out against private property and resisted breaking up state and collective farms. In January 1991, for example, he ordered an inventory of "irrationally used land in kolkhozy, sovkhozy, forestries, and other land users, including land given to ministries and departments of the USSR" during the first half of 1991. After the inventory, some 3–5 million hectares (less than 2 percent of all agricultural land) were to be assigned to the population for use in peasant farms, personal plots, collective gardens, and other uses.[136] The important point is that this land fund contained only land that social sector farms did not want or were not using properly, which implied it was not of good quality and would not affect state and collective farms in any fundamental way.

As the Gorbachev era came to an end, the conditions for personal plots had been liberalized, but their inherent limitations could not be overcome. As Gorbachev moved toward the establishment of a nonstate agricultural sector utilizing larger production enterprises, he chose a conservative option that rejected private ownership of land and that retained the state and collective farm structure.

Social Conditions: The State and Rural Living Standards

The third important state intervention affected social conditions and rural standards of living. The nature of this intervention was important because it influenced the sociopolitical relationship between urban and rural dwellers. The rural standard of living was also an important factor influencing rural out-migration, as rural dwellers sought education and greater economic opportunity. For rural dwellers who stayed, rural social conditions influenced the performance of the agricultural sector by affecting aspects of the production cycle such as storage, transportation, and distribution.

It is commonly agreed that during the Stalin period the state pursued a strategy to protect urban consumers and to pump resources out of the rural sector with little regard for the effect on the rural standard of living. One would point to urban rationing as a strategy to protect urban dwellers from starvation, while later allowing the man-made famine in the Ukraine to claim millions of rural lives.[137] Collectivization is often seen as the epitome of the state's antirural bias as it attempted to transfer resources from agriculture to industry. We should, however, distinguish between collectivization as a strategy and the actual results. Strategically, there is no question that the regime attempted to squeeze resources out of agriculture at the most favorable terms possible, irrespective of the consequences. Undoubtedly the nature of state interventions and the rapid decline of rural living standards were main considerations that motivated different forms of peasant resistance.[138]

When considering the actual transfer of resources from agriculture to industry there is reason to doubt the efficacy of collectivization. For example, in an early debate on the subject with Alec Nove, James Millar argued that net resources moved into agriculture and therefore did not make a contribution to industrialization but rather detracted from it.[139] Similarly, Jerzy Karcz argued that it would be difficult to conclude that collectivization was a solution to the "grain problem" that Stalin had portrayed as threatening Soviet cities and their urban work force. Karcz calculated that, although the marketed output of grain increased during the first stages of collectivization, the increase was accounted for by the decline in the size of livestock herds.[140] A few years later Holland Hunter argued that collectivization led to less grain being produced than without collectivization, and that, despite the death of some 15 million people (thus lowering overall demand), per capita consumption in 1933 was only 43 percent its 1928 level. Hunter concluded:

> What agriculture delivered to other sectors came neither from an increased share of its output at the 1928 level nor from an enlargement of agricultural output. There was a deficit rather than a surplus, an absolute shortfall that cut cruelly into rural living standards and lowered urban living standards as well.[141]

Other Western analyses also questioned whether Stalin's collectivization did in fact facilitate urban bias. Stephen Merl, for instance, argued that although cities had higher consumption norms than rural dwellers and were protected by rationing,

> the food situation in the cities deteriorated in spite of rationing.... The workers suffered more from the shortage of animal products, since these have had a greater significance in their food ration, and the increase in the consumption norms of grain products and potatoes always has been a sign for the deterioration of food supply.[142]

Stalin's antirural bias continued throughout his regime. In the postwar period, for example, state investments into agriculture as a percentage of all investments were less during 1946–1950 than they had been during the first five-year plan. Agriculture received one-fifth the level of government investments as did heavy industry, and most of the investment into agriculture was used for the growing of food, not reconstruction and expansion of infrastructure.[143] Furthermore, the terms of trade between industry and agriculture worsened. In the latter 1940s procurement prices for nonindustrial crops changed little from their 1930s levels. Procurement prices for food were kept low while postwar inflation soared for consumer and industrial goods. State purchase prices for compulsory grain deliveries had not been increased since the mid-1930s and were below production costs. Purchase prices for potatoes and meat even decreased from prewar levels.[144] As a result of this discriminatory pricing, Alec Nove calculated, the price paid for state-ordered potatoes was actually less than the cost of transporting them to the collection point, a cost borne by the kolkhoz. It was, he concludes, "as if Stalin was determined to make the peasants pay for the necessary post-war reconstruction."[145]

In the post-Stalin period there occurred a sharp break with Stalin's antirural policies and attitudes. In his memoir, Khrushchev described Stalin's antipathy toward agriculture: "[Stalin] taught us to think of agriculture as a third-rate branch of our economy. For Stalin, peasants were scum. He had no respect for them or their work. He thought the only way to get farmers to produce was to put pressure on them."[146] Khrushchev's basic orientation was to increase food production for a growing urban population; as a result of higher production and better purchase prices rural incomes would increase and the standard of living would rise. As a by-product of general economic policy, rural dwellers would benefit.

In order to increase food production and raise rural living standards, Khrushchev pursued three main strategies. The first was the relaxation of conditions for operating a personal plot (described above). The second strategy was to increase purchase prices paid to farms. The third strategy was to increase wage rates, to switch from payments in-kind to a monetarized rural wage system, and to offer advance payments to rural workers.

The benchmark to determine improvements in rural living standards was often the relationship between average level of urban and rural wages (not including income from personal plots). Differences concerning collective farm wages make it difficult to ascertain how much the urban-rural gap actually closed under Khrushchev. There is general agreement that income differences narrowed during Khrushchev's rule. Between 1953 and 1965, state farmworkers' annual wages increased from about 54 to 72 percent of the wage level of an industrial worker. Collective farmworkers also closed the urban-rural gap. In 1953 a kolkhoznik earned on average 18 percent of the wage level of an industrial worker, and this increased to 42 percent by 1965.[147]

Brezhnev significantly expanded upon the policy trends begun under Khrushchev, and one could argue that the big winners during the Brezhnev era were rural workers. Whereas Khrushchev improved rural living conditions, Brezhnev was the first to extend the benefits of the Soviet welfare state to rural dwellers. During his tenure, Brezhnev's basic strategy was to pour money into the rural sector. From 1965 to 1981, as a result of Brezhnev's program to industrialize agriculture, the level of state investments to agriculture rose fourfold. Over 500 billion rubles (in 1987 rubles) in capital investments were made from 1966 to 1982. By 1975, state capital investments into agriculture as a percentage of all state investment monies had increased to 28 percent, rising from 20 percent in 1961, and this level remained constant until the late 1980s.[148]

Under Brezhnev urban-rural wage differences also narrowed, although we should point out that average rural wages remained below average industrial wages. There is often disagreement among Soviet authors over exact wage levels and the exact percentage of rural to urban income, but all are agreed that during the Brezhnev period rural wages increased faster than urban wages. For example, one author noted that between 1960 and 1977 the average monthly wage for workers and employees in the national economy increased 180 per-

Table 8. Average Monthly Collective Farm Wage Related to Wage of State Industrial Worker, 1965–1984 (Russia only)

Collective Farm Wage	1965	1970	1975	1980	1984
Average rubles per month	50	79	100	124	159
As percentage of average industrial wage	49	59	60	65	75

SOURCE: Calculated from *Narodnoye khozyaystvo RSFSR v 1981 g.: statisticheskiy ezhegodnik* (Moscow: Finansy i statistika, 1982), pp. 161, 212; *Narodnoye khozyaystvo RSFSR v 1984 g.: statisticheskiy ezhegodnik* (Moscow: Finansy i statistika, 1982), pp. 161, 244.

NOTE: "Wage" does not include state transfer payments, bonuses, or income from personal plot.

cent, the average state farmworker saw his wages rise 240 percent, and the average collective farm wage rose 320 percent.[149] Table 8 shows the narrowing of average monthly wages between a collective farmer and an industrial worker from 1965 to 1984.

By the time Gorbachev came to power, urban-rural income differentials had narrowed significantly, even though in general the rural standard of living was lower than that of a urban dweller. The early Gorbachev shared Brezhnev's view that rural living standards had to be improved in order to increase food production for the urban population. Toward this end the policy of narrowing urban-rural differences was said to be a "component part" of realizing general socioeconomic goals.[150] Gorbachev himself at the Nineteenth Party Conference in 1988 expressed the sentiment that more had to be done for the countryside:

> In particular I want to talk about the social development of the countryside. Here, society has no small debt to rural dwellers. In many raions housing, social-cultural conditions, and medical services in villages are at a low level. To that condition one must add the unsatisfactory construction of rural homes, difficulties with electrical supplies, use of household equipment, and the poor condition of roads.
>
> During the present five-year plan, major measures for improvement in the social sphere of the countryside are being undertaken as never before. A large program of transformation is being implemented, in particular in that most important region of the country, the Russian non–black earth; . . . the essence of agrarian policy at the present stage consists of changing the productive relations in the countryside itself. We should restore the economic balance between town and country, in all ways unleash the potential of state and collective farms through the development of various forms of contracts and leasing. . . . Only on this basis is it possible to sharply increase the effectiveness of agricultural production, to ensure a fundamental improvement in food supplies in the country.[151]

Gorbachev, however, faced a dilemma. The policy of funneling resources to the countryside had often been wasted on weak farms and in general squandered with little positive result. Thus a basic tension existed between using state levers to promote greater equality versus encouraging greater inequality by rewarding the productive. By the late Gorbachev period the idea that urban-rural equality was attainable was quietly dropped as public policy, which in turn gave rural conservatives a plank around which to organize opposition to state policy.

Under Gorbachev urban-rural wages trends may be divided into two time periods. The first time period covers 1985–1990, and although published Soviet data on income levels among professions do not always agree on the exact level of remuneration, they do depict similar trends. According to one set of data for the nation as a whole, from 1985 to 1990 the average monthly wage from state sources for a kolkhoznik increased by 57 percent; that of a sovkhoz worker increased by 52 percent; and the average monthly wage for a state worker and employee increased only 44 percent.[152] Within the RSFSR similar trends were evident: basic monthly wages rose 59 percent for kolkhozniki, 54 percent for sovkhozniki, and 43 percent for workers in industry.[153] In short, despite reform goals to differentiate according to skill, basic wage trends duplicated the trends Gorbachev had inherited from his predecessors.

The second time period covers 1990–1991. During this period the gap between urban and rural wages (which for decades had been closing) began to widen, as state control receded and market forces were introduced into the economy. Data from Kostroma Oblast shows that during 1990 and 1991 average monthly wages for kolkhozniki rose 53 percent, for state farmworkers 60 percent, and for industrial workers 90 percent.[154] Thus, the second period saw the beginning of significant wage differences between urban and rural sectors, a trend that intensified in Russia following the dissolution of the USSR. The urban-rural wage relationship is characterized using data from Kostroma Oblast in figure 1.

Kostroma Oblast was representative of broader trends throughout the country. Nationwide, urban-rural wage differences narrowed through 1990, and in that year data indicated that workers and employees (most of them urban residents) received over 299 rubles a month on average, whereas kolkhozniki received around 265 rubles monthly. By mid-1991 the gap between the urban and rural incomes began to increase, as a worker in industry received an average of 581 rubles a month, and a kolkhoznik received 425 rubles a month.[155] Divergence between urban and rural incomes continued to widen thereafter.[156]

By the late Soviet period, it is important to note, not only did urban-rural

Figure 1. Urban–Rural Wage Relationship, 1985–1992

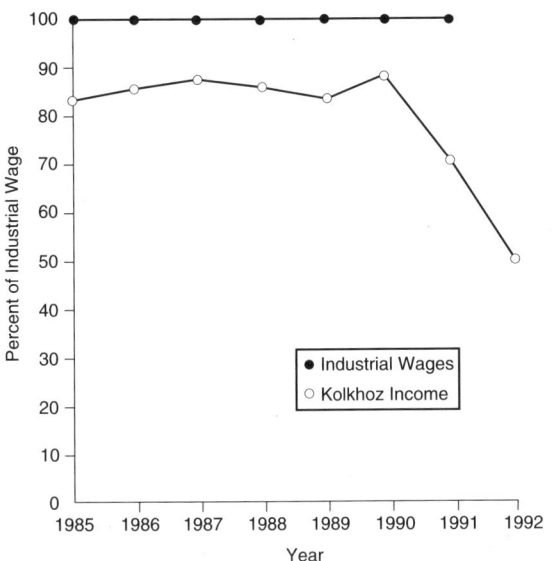

wage differences begin to widen, but other aspects of rural living standards continued to lag, although there is no question that rural living standards had improved since the 1960s.[157] Despite relative gains in rural incomes since Khrushchev, data on rural income levels were misleading. This was so because monetary incomes were important only insofar as they could be used to buy other goods. But opportunities to spend money were more restricted for the rural dweller than for the urban resident. Per capita retail trade turnover in rural localities was about one-half the level found among urban dwellers on a per capita basis. Moreover, the rate of growth of trade turnover per capita in rural localities slowed after the tenth five-year plan (1976–1980).[158] By the late Gorbachev period, the national average for trade turnover was 2.2 times lower in rural localities than in urban centers.[159]

Moreover, Soviet data indicate that since the 1960s urban residents spent a higher percentage of their family income on food, although the percentages declined over time. However, beginning in the mid-1980s, the relationship changed and rural workers spent a greater percentage of their income on food than their worker and employee counterparts.[160] This is explained in part by fewer public catering establishments and state retail food stores in rural areas, which forced rural dwellers to shop at kolkhoz markets and cooperatives where prices were higher.

Other aspects of rural life also lagged behind the urban condition. For example, although rural housing provided more average living space per resident, rural housing amenities were sorely lacking. Published data demonstrated the appalling backwardness of the Soviet countryside. For example, in 1988 almost 93 percent of the housing in urban settlements had piped water. In 1988, however, only 39 percent of the housing in kolkhozy and 42 percent of the housing on sovkhozy had piped water. Nearly 91 percent of the housing in urban centers had sewerage, but only 27 percent of kolkhozy and 30 percent of sovkhozy were similarly equipped. In 1988 more than 75 percent of urban housing was supplied with hot water, but hot water was available to only 10 percent of housing in kolkhozy and 15 percent of sovkhozy. Whereas 90 percent of urban housing had central heating, only 21 percent of kolkhozy and 31 percent of sovkhozy had this luxury.[161] Finally, in 1989 only 12 percent of rural apartments were equipped with gas, whereas more than two-thirds of urban apartments had gas.[162] These data represented significant *progress* resulting from Brezhnev's rural policies.

Social amenities in the countryside remained significantly below those found in urban centers. One Soviet author summarized the situation: "Despite advances in the social infrastructure in the countryside, there remains a considerable gap between it and the urban sector both in quantity and quality."[163] However low Soviet medical standards were in general, they were even lower in rural areas. One study found that the level of medical services for the rural population was "two times as backward as for urbanites," thus explaining why infant mortality rates in rural areas were about twice as high as in urban localities. Many health facilities in rural areas "have limited facilities, are insufficiently staffed with skilled cadres, are poorly equipped with contemporary medical equipment, and many of them are located a significant distance from rural population points."[164] Accessibility of rural doctors was estimated to be two times lower than for urban dwellers.[165] Even when access was available, medicine per capita was 2.4 times less available for rural patients. Furthermore, the material base of rural public health facilities was extremely low: it was reported in 1990 that only 35 percent of raion hospitals had hot water, only 17 percent had piped water, and 27 percent did not have sewerage.[166]

Even in the late 1980s there was a significant gap between the level of education for urban and rural populations. Overall, in 1989 among the urban population 865 persons per thousand had a higher or secondary education, while only 702 per thousand rural persons had a similar education level. The gap was most widely seen in higher education, where 138 per thousand urban persons had any higher education while only 46 per thousand rural persons could make the same claim.[167] This statistic reflected the poor preparation and lack of schools that handicapped rural students. In addition to the poor quality of rural schools, there was a basic shortage of schools. In the mid-1980s, 27 percent of large rural population centers (500 people or more) did not have preschools (affecting some 11 million people), and over 16 percent of rural population centers did not have a middle school.[168]

Finally, there was a marked paucity of cultural and recreational opportunities in rural areas. A survey in 1985 indicated that nearly 50 percent of kolkhozniki enjoyed movies, the theater, or concerts during their free time. At the beginning of 1988, however, some 42 percent of rural settlements did not have any cultural enterprises—libraries, clubs, or movie theaters.[169] Opportunities for recreation deteriorated over time, as the number of clubs and recreational places decreased from 88,500 to 59,800 from 1985 to 1991.[170]

These social trends are important because they affected agricultural performance and, more important, the willingness of the rural young to remain in the countryside. Between 1960 and 1986, the number of agricultural workers in the Russian non–black earth zone decreased by 40 percent.[171] Even after the main wave of rural out-migration ended, between 1979–1988 the Russian Republic as a whole averaged a rural outflow of almost 100,000 persons a year.[172] If we combine rural migration and a falling rural birth rate, the net loss in the Russian Republic alone totaled almost 325,000 rural dwellers a year from 1979 to 1988.[173]

Who migrated from the countryside was as important as how many left. One of the most prominent features of rural out-migration was the flight of the young from the Soviet countryside. A consistent finding of Soviet surveys was that rural dwellers under thirty expressed the greatest desire to migrate out of the countryside, whereas those over fifty expressed the least willingness to leave. In particular, both sexes aged from sixteen to twenty-nine migrated from the countryside in the greatest numbers. Among the reasons cited by young people for their desire to migrate, males expressed dissatisfaction with work, females cited poor supplies of food, and both sexes cited a lack of cultural activities and a shortage of other young people.[174] Using Kostroma Oblast as an example, the age and sex structure of rural migrants is shown in table 9.

The table is significant because it depicts the dilemma of rural demographics. More women left the countryside than men. Peak child-bearing years for rural women were between twenty and twenty-two (for their first child; those with a second child tended to be between twenty-three and twenty-six).[175] The data show that rural women left the countryside before having children or during peak child-bearing years. Moreover, those women who did remain did not have large families. So a number of processes were at work: there were fewer younger women in the countryside to have children, which contributed to a demographic trough; the shortage of rural women affected opportunities of rural men for marriage and a family; and there was a huge loss of rural women

Table 9. Distribution of Rural Migrants to City of Kostroma in 1982, by Age and Sex

	Number of Men	Percentage of Total Male Migration	Number of Women	Percentage of Total Female Migration
15–19	318	19.4	997	40.0
20–24	580	35.5	550	22.1
25–29	256	15.6	330	13.3
30–34	154	9.4	70	2.8
35–39	52	3.2	58	2.3
40–44	86	5.3	42	1.7
45–49	48	2.9	32	1.3
50–59	58	3.5	88	3.5
60–69	38	2.3	112	4.5
70+	42	2.6	208	8.4
Totals	1,632	99.7	2,487	99.9

SOURCE: Calculated from "Migratsiya naseleniya v 1982 godu," KGA, f. R-1951, o. 15, d. 141.
NOTE: These data represent gross rural migrants into the city of Kostroma, not net.

of prime child-bearing age (between sixteen and twenty-nine), as this age group accounted for 75 percent of all female rural migration into the city of Kostroma.

During the Soviet period, as this chapter has shown, the state intervened in ways that affected rural egalitarianism, the rural private sector, and rural social conditions. State interventions were the key variable affecting social relationships among the rural population and between the urban and rural sectors.

For most of the post-Stalin period, the state pursued an egalitarian wage policy that prevented the rise of significant wage differences among categories of rural employees. Only in the last year of Gorbachev's rule did significant inequality in wage rates begin to emerge. State investment and financial policies benefited weak farms and often discriminated against more productive farms. Procurement pricing policies also benefited farms with higher production costs and those with higher resource expenditures. These egalitarian policies were embraced by Khrushchev and Brezhnev, and the Soviet countryside became more egalitarian after 1953. Gorbachev attempted to reverse certain aspects of rural egalitarianism, but he was frustrated by high-level political opposition and the fear of political consequences from below. Ultimately, he failed to reverse significantly the legacy he inherited.

The Soviet state regulated the nonstate (private) sector, embodied most of all by personal plots operated by the population. Although small and inherently limited in its production potential, the state often viewed the personal sector as ideologically incompatible with larger farms, although the nation remained dependent upon plots for their contribution to the food supply. The ambivalence of the regime was reflected in the policies of Khrushchev, who first promoted and then restricted plot operations. Brezhnev reversed the restrictions of the Khrushchev era and in fact used personal plots as a reform default that allowed him to improve food supplies to the urban population while avoiding systemic change in the rural sector. Upon coming to power, Gorbachev also embraced personal plots as a reform option, but quickly moved to land leasing and the establishment of individual farms. Nonetheless, by the end of the 1980s, the popularity of personal plots, as well as production trends from them, reversed their downward trend and began to increase. This trend would become one of the notable features of Russian agrarian reform.

Finally, this chapter has demonstrated that state policies affected rural social conditions. For decades, the low level of rural living standards was ignored or minimized. Although rural living standards improved significantly after 1965, the rural sector continued to lag far behind the urban sector according to almost any living standard that was examined. The primary impact of low rural standards of living was their effect on rural out-migration. The young and the skilled were the most important groups to quit the countryside, leaving behind the old, the unskilled, and the unambitious. The demographic legacy inherited by the post-Soviet regime would have a significant impact upon reform potential.

Summarizing the Soviet Legacy: The Tasks for Russian Agrarian Reform

State interventions defined key aspects of the Soviet agrarian sector. By defining the core issues of rural social policy, the Soviet state was able to influence various aspects of rural dwellers' existence: the incentives they could hope for, their economic environment, and their standard of living. At the end of the Soviet period, as reform initiatives began to unfold, the legacies of the Soviet system required that any future agrarian reform be multifaceted and comprehensive. In order to be successful a number of issues would have to be addressed, including the creation of new rural institutions, new policies, and modifications to existing social relations.

Beyond relatively easy policy changes in retail food and trade policies, a number of other tasks would also be necessary that would have both economic and political importance. The attainment of these tasks would improve agricultural performance, build rural social support for reform, and undercut potential political opposition. In particular, the Soviet legacies in rural social policy would have to be reversed. Using the three variables of rural social policy established at the beginning of this chapter, we may conclude that successful agrarian reform would need to achieve the following tasks (the list below is not intended to be exhaustive):

Incentives
1. To correct investment and financial policies that favored economically weak farms and regions;
2. to move away from individual, farm, and regional egalitarianism;
3. to revise state subsidy, zonal pricing, and price supplement systems;
4. to economically promote and support a competitive rural private sector.

Economic Environment
1. To remove restrictions and regulations on the private sector, including both personal small-scale plots and larger commercial production units;
2. to institutionalize privatization of land and landownership;
3. to create an economic environment in which food producers of all types could prosper;
4. to create economic conditions for a new rural class of private farmers to develop and prosper.

Rural Socioeconomic Conditions
1. To improve rural living standards;
2. to lessen urban-rural wage differences;
3. to create a socioeconomic environment in which the "best and the brightest" would not migrate from the countryside;
4. to improve rural infrastructure and social conditions.

In sum, successful agrarian reform would do more than change policies, facilitate new forms of labor organization, and create new forms of agricultural enterprises. Successful agrarian reform would need to redefine rural political-economic relationships. Successful agrarian reform, therefore, would involve reversing Soviet-era rural social policy.

3

Reform of the Collective Agricultural Sector

In the post-Soviet period, Russian agrarian reform took on two main directions. One reform strategy was the reorganization of state and collective farms. Because the collective sector dominated agricultural production, if state and collective farms could be successfully reformed and transformed the benefits would be substantial: food production should increase; the cost of production would decline; and subsidies could be cut, which would free up funds for more general economic modernization. The goal was to improve agricultural performance by forcing state and collective farms to transform themselves.

In order to analyze state and collective farm reorganization, we shall first identify the problems that necessitated the reorganization of state and collective farms. The next step is to examine the reform legislation that governed farm reorganization, and to analyze the impact of reform on the collective sector and impediments to farm transformation. We shall then investigate the model of farm reorganization that was developed and how it performed. In examining these aspects of reform in the collective sector, we shall address several broad themes relating to the nature of state interventions:

1. The origin of reform of the collective sector, whether from above or from below;

2. whether privatization in the collective sector was sufficient to solve the economic problems plaguing the agricultural sector; and

3. the political motivations for privatization in the collective sector.

The Impetus for Reform

Basic food production in post-Stalinist Russia was not the impetus for Russian agrarian reform. From 1956 to 1990, average grain production (before cleaning and drying) increased during every five-year-plan period except the eleventh five-year plan (1981–1985). The same was true for grain yields. The average production level of high preference items such as beef and pork also increased during every five-year-plan period from 1956 to 1990.[1] Gross agricultural output in the USSR (in 1983 prices) rose from an average 128.3 billion rubles per year during the period 1961–1965, to 220.8 billion rubles per year during the period 1986–1990. During that thirty-year period, the value of output increased 72 percent, as opposed to an approximate 23 percent population increase. Grain output increased during the same period from 124.1 to 196.5 million tons (an increase of 58 percent), meat from 9.3 to 19.4 million tons (108 percent), milk from 64.7 to 105.9 million tons (63 percent), and eggs from 28.7 to 83 billion units (190 percent).[2]

Just as food production increased, so too did food consumption. Specialists have long been aware that average per capita consumption of basic food products such as meat, milk and milk products, eggs, sugar, vegetable oil, and fruits and berries increased during every five-year period from 1966 to 1990.[3] Already by the end of the 1960s the Soviet Union was on a par with the United States, West Germany, and Great Britain in terms of daily consumption of calories and proteins.[4] From 1965 to 1981, the per capita daily consumption of fats increased by 25 percent, whereas per capita daily consumption of carbohydrates decreased. Increased per capita consumption reflected higher production levels and yields. All of this was accomplished despite chronic shortages of feed and protein for livestock, spare parts and fuel oils for farm machinery, and poor rural infrastructure, including the lack of decent roads. Thus, despite popular misperception, the Soviet Union did not suffer from a "food problem" that could be traced to basic production.[5] The leading expert on Soviet agriculture prior to his death a few years ago, Karl-Eugen Wadekin, wrote in the early 1990s, "It has by now become general knowledge that the food crisis in the USSR is more one of distribution than of production and should therefore be primarily approached from the distribution side."[6] Instead, as shortages and imports increased during the 1980s, the problem could be attributed to an irrational price system that facilitated excess demand; inadequate rural infrastructure that contributed to large losses; and an antiquated distribution system.

The "agrarian problem" stemmed from the need for massively increased inputs in order to achieve and sustain higher rates of production. The Stalinist collective farm system was notorious for its inefficient use of inputs, lack of personal responsibility for the final product, and enormous harvest losses and wastage. In short, one crucial impetus for reform was the fact that Soviet agriculture became intolerably expensive in terms of economic and monetary resources.

Why was Soviet agriculture so costly, inefficient, wasteful, and unprofitable? Some have argued that these deficiencies resulted from inherent flaws in collectivized labor organization. Collective labor organization robbed rural workers of responsibility for their labor and concern for the final result.[7] Incentive structures within farms failed to reward hard work. Others argued that the larger economic environment in which state and collective farms operated—including state determination of input costs and output revenues—made sound economic decision making impossible.

It is well to remember that collective and state farms never worked in a normal economic environment. Instead, in the words of Vasilii Vershinin, otherwise one of the earliest proponents of private farming in Russia, state and collective farms experienced the state's "predatory-colonialistic policy toward them, [which] deprived them of economic freedom. Undoubtedly, collective and state farms would be able to work rather effectively and the foodstuff problems would not be so prominent if normal farming conditions existed."[8] To the extent that Soviet agricultural producers operated inefficiently or irrationally, the fact that they operated within an environment that made "rational" economic decisions difficult, if not impossible, must be factored into the equation.

Consider a short list of obstacles and environmental problems that confronted state and collective farms, all of which have been well documented in the Soviet press over the years:

1. They did not control input prices or availability of supplies;
2. They did not control the quality of inputs;
3. They did not control the quality of cadres or specialists assigned to the farm;
4. They did not control procurement prices they received;
5. Farm managers and directors could not fire slackers or drunks;
6. They suffered from shortages of inputs, various types of transport, including refrigerated trucks, as well as shortages of storage and spare parts;
7. They suffered from primitive rural infrastructure, and most capital investments were directed toward "productive" purposes (food production);
8. They suffered from an inability to retain skilled workers, due in part to the wage policies set by the state and the low level of rural amenities which reflected state investment priorities.

In addition to the above list, one must add that farms suffered from poor coordination by planners, duplication of authority, and problems from interministerial competition. In the early 1980s, the agroindustrial complex was managed by no fewer than eleven different ministries. A partial list of ministries that were part of the food production, processing, and distribution process would include the following:

Ministry of Agriculture
Ministry of Trade
Ministry of Cereal and Grain Production
Ministry of Fruit and Vegetable Farming
Ministry of Machine Building for Animal Husbandry and Feed Production
Ministry of Tractors and Farm Machinery
Ministry of Land Reclamation and Water Resources
Ministry of Meat and Dairy Industry
Ministry of Food Industry
Ministry of Rural Construction
Ministry of Fertilizer Production

One may add to this list at least three state committees as well: State Planning, Material and Technical Supply, and Supply of Production Equipment for Agriculture. Given the difficulties of coordination among all these actors—each of whom pursued its own interests—it is no wonder that one of the first changes Gorbachev made in agriculture was to streamline the bureaucracy by combining five ministries into the "super" ministry Gosagroprom in November 1985. The larger point is that given the range of economic factors over which farms had no control, and given the degree of bureaucratic infighting and lack of coordination, it is not surprising that collective farms did not produce "efficiently." The real question is how they produced at all and how they managed to improve food production during the 1960s, 1970s, and 1980s.

Thus, it can be argued that what was needed to reform Soviet agriculture was a better system of incentives and disincentives and an economic environment that rewarded "rational" decisions. This meant, most of all, the introduction of appropriate prices for inputs and outputs, as well as a lessening of central control and the creation of marketing and wholesale channels that farms could pursue without excessive state interference. In other words, making Soviet farms more efficient might have been achieved by "getting prices right" and liberating wholesale trade, distribution, and marketing networks (as well as correcting deficiencies in infrastructure).

A little more than a decade ago, one of the foremost analysts of Soviet agriculture wrote:

All of these sources [of inefficiency and high cost] I assert, start with the inappropriateness of the price system for agricultural outputs and inputs. By inappropriateness of prices I mean that prices do not reflect the underlying supply (or cost) and demand conditions.... *I am now quite fully convinced that the socialized nature of agriculture is not the primary or even an important source of the inefficiency and high cost so prevalent in the Soviet Union.*[9]

In this view, state and collective farms simply responded to the economic environment and set of incentives that confronted them, however irrational the resulting behaviors were. Thus, the Soviet agrarian problem stemmed from the wrong incentives.

The Origins of Farm Reform

Before proceeding with an analysis of farm reorganization in the postcommunist period, it is first necessary to briefly review attempts to reform collective farms in the late Soviet period under Mikhail Gorbachev. By so doing we shall establish a context within which to assess the efforts at farm reform under Yel'tsin.

Despite previous attempts to improve state and collective farm performance (such as the "link" experiments under Khrushchev or the collective contract experiments under Andropov), systemwide reforms that would encompass the operation of all farming enterprises gained momentum only under Gorbachev.[10] During Gorbachev's rule, efforts to reform collective farms opened a debate on the future of these farms. Some specialists argued as early as 1987 that "it might even be possible to disband collective and state farms," but this approach was not embraced by the leadership.[11] The conservative end of the political spectrum envisioned the retention of state and collective farms, but moving them toward the establishment of independent agricultural cooperatives of varying sizes based on self-accounting and self-financing and employing the widespread use of contracts for labor remuneration, sales of food products, and purchases of needed inputs. In short, not only would farm operations be reformed (and liberalized), but the entire environment in which farms operated would be radically transformed along the lines Gorbachev pursued during 1988–1989, emphasizing leasing contracts and economic levers to guide farm activities.

It is interesting that many prominent members of the "radical" wing of agrarian reformers also did not question the continued existence of state and collective farms at this time.[12] True, they believed that state and collective farms would eventually wither away, unable to compete with other forms of farming, but they did not advocate the forcible breakup of farms.[13] Even the radical "five-hundred-day plan," which in general was a plan for widespread privatization and destatization of state enterprises, advocated a two-stage reform of farms, beginning with a "gradual" transformation of state and collective farms into profitable enterprises that used leases and contracts. In the second stage of reform, farms would be transformed into cooperatives.[14] The same plan did, however, advocate a process of bankruptcy and liquidation of unprofitable state and collective farms.[15] During the bankruptcy process, farm assets would be auctioned off by a land bank, along with the local soviet, with the bank acting as the creditor of a liquidated farm. Both farmland and equipment could be acquired or leased. If a farm went bankrupt, the bank could grant loans or advantageous credits to a buyer for the acquisition or leasing of land; and farm machinery could be obtained or leased through loans that would be repaid with future production.[16] As we know, however, the Shatalin plan was never adopted as the economic reform strategy of the USSR.

In general, Gorbachev attempted to reform the operation of state and collective farms but was unwilling to question the continued existence of state and collective farms. For example, during his speech in March 1988 to the Fourth Congress of Collective Farmers Gorbachev stated, "Our collective and state farms remain the basis of socialist agricultural production. We will not succumb to appeals—which are highly questionable and more important, unfounded—to reconsider in general the fate of collective and state farms."[17] Furthermore, the inherent conservatism of Gorbachev was shown in a speech before the Twenty-Eighth Party Congress in 1990:

We reject the demand for "total decollectivization." I am convinced that those collective farms and state farms that are managing their operations skillfully, that are working to improve social amenities, providing peasants with proper living conditions, and producing large amounts of output needed by society deserve every kind of support. Naturally, they will remain an organic part of the Soviet countryside as it undergoes renewal.[18]

Despite his unwillingness to consider the breakup of state and collective farms, Gorbachev freely admitted that changes were necessary and lamented that "the internal structure of collective and state farms does not meet the current needs of farming . . . and most important, does not make people personally responsible for the use of land and other assets and leads to wage leveling."[19]

Thus, Gorbachev attempted to change the operation of the collective sector. An early effort was to change the basic rules and guidelines by which farms operated. Significant changes from the 1969 model charter in the management and operation of collective farms were reflected in the 1988 draft model statutes, published prior to the opening of the Fourth Congress.[20] The draft statutes spelled out greater farm autonomy, internal democracy, and financial responsibility. In particular, the draft granted collective farms more autonomy in their production and marketing operations. For example, within the guidelines of (stable) five-year plans, farms were independently able to work out plans for their economic and social development that were to be confirmed by a general meeting of farm members. Although deliveries to the state remained obligatory, Gorbachev did reveal in his speech that the Politburo had discussed the procurement issue, and the view that emerged was that the system should move to a contractual relationship based on "economic, not administrative" means. In the interim, farms were to sign contracts with state, cooperative, and social organizations for the delivery of food products, but production above planned volumes could be sold on the market or through consumer cooperatives at prices determined by the farm.

With regard to internal democracy on farms, members of farms and their assemblies were given more rights and a greater role in the operation of the farm. Farm chairmen and section heads were to be elected for a limited term of three years, and the draft statutes permitted secret ballot, multicandidate elections. The farm general assembly was given the right to determine the form and level of remuneration for farm chairmen and section heads. In addition, an important clarification was introduced that allowed a farm member to withdraw from the farm, subject to review by a general meeting of farm members not later than three months after the submission of a written notification to resign.[21] During this three-month period, an individual had the right to rescind his announcement to withdraw from the farm. Although the statute did not specify what would occur if the general assembly rejected the announcement (or even if they were required to approve it), the basic right to leave was implied. Two days after the member withdrew from the farm he was to be given his work record book *(trudovaya knizhka)* and any earnings he was due.[22]

Regarding financial responsibility, farms were given greater latitude to define farm needs and to fund the construction of cultural and service projects, as well as housing amenities. Farms were given the right to engage in wholesale trade and international business abroad—implying the right to import or export—with the stipulation that they incurred "full responsibility for the results of their foreign economic ties." Social welfare obligations were expanded for farms, and the farm itself could define wage levels, although the minimum wage principle was not to be violated. In an attempt to link remuneration to productivity, the draft statutes stipulated that collective contracting was "the basic form of labor organization." Family, individual, and other forms of contracting were also looked upon with favor. Residual earnings after wages were intended to be used as an incentive for productivity (although this seldom worked because the "minimum" wage constituted about 80 percent of the total wage—so any incentive was marginal). There was also new emphasis placed on self-accounting *(khozraschet)* and self-financing, with shortages in operating expenses to be covered from farm income. Although this author never obtained a published final version of collective farm statutes, when the final version was approved and adopted in August 1988 press coverage and interviews suggested that the draft version had remained largely intact and that the final version was considerably more liberal than its 1969 predecessor.[23]

Land Leasing

In addition to the rules that guided farm actions, Gorbachev also attempted to reform farm operations. The most notable reform initiative was the introduction of land leasing within the collective sector. By mid-1987 Gorbachev began to speak out more forcefully on the possibility of leasing state land to peasants for independent tillage. At the June 1987 plenum, at which he first unveiled his conception of economic reform, Gorbachev devoted a considerable portion of his speech to agriculture. He chided the Party for the mistakes of the past: "For a long time we tried to lead the economy on the basis of enthusiasm, and sometimes by force of decree. But we forgot Lenin's precepts—that production growth can be ensured on the basis of personal interest and material commitment." At that plenum Gorbachev also explained his idea of widespread leasing, claiming that "the most important task today is to make people full-fledged masters of the land again."[24] He further urged that kolkhozy and sovkhozy lease abandoned houses and land plots on a contract basis to urban dwellers who wanted them for productive purposes.

The basic decision to adopt leasing as the main direction of reform was made by mid-1988, although formal approval would not come for another nine months. The decision-in-principle was evidenced by remarks at a Central Committee meeting Gorbachev made prior to the Nineteenth Party Conference, at which he argued that, with new forms of economic organization, the country had the resources to "make considerable progress in resolving the food question within 2–3 years." He argued that the persons who "block the transfer to these new forms of work cannot be trusted with the leadership of a kolkhoz or sovkhoz."[25]

At the end of August 1988, Gosagroprom published its recommendations on leasing agreements.[26] A number of important provisions were proposed, including that leases could be contracted by teams, cooperatives within farms, individual families, kolkhoz members, or sovkhoz workers. A leaseholder need not be a member of a kolkhoz or a sovkhoz in order to lease land. Leases were to be long-term, up to fifty years, and were conditioned on the lessee's interest in improving soil quality and ensuring protection of the environment. Lease payments would be reduced by the amount invested by the lessee in land improvement. Estimated profit per hectare was to be used to determine the size of lease payments, taking into account fertility of land, pedigree of animals, and productivity based on specific farm conditions. All payments were to be paid to farms and counted as part of farm income. Land on loss-making farms could be leased for the first two years without charge. Output prescribed by contract was to be sold by the lessor (or on his behalf) to procurement agencies at state purchasing prices; all output over and above that could be sold at the lessee's discretion. Last, leaseholders were entitled to all types of social welfare on an equal basis with kolkhoz or sovkhoz workers, including pensions, vacations, and unemployment benefits.

The March 1989 plenum officially codified land leasing as the Party's reform package, although Gorbachev actually opted for a conservative reform when he argued that it was "scientifically and practically ungrounded" to disband kolkhozy and sovkhozy in order to hand their land over to leaseholders. The plenum approved long-term leases of up to fifty years and permitted leases to be inherited by family members. Leases did not, however, transfer ownership of land or any leased object—the state remained the owner.[27] The conservative nature of the leasing reform is apparent in that the most common form of leasing was the "internal lease," which bound the lessee to the parent farm. Leasing allowed an individual or a small group of persons to work under contract on a particular plot of land, with a defined quantity of livestock, or other productive objects, and retain any residual profits after paying "rent" and costs incurred during production.[28]

Peasant Indifference to Leasing

On paper the number of farming operations and collectives working under leasing conditions was impressive. For example, it was reported that in mid-1989 more than twenty-three thousand sovkhozy and kolkhozy, or about 45 percent of all farms in the USSR, had signed at least one lease contract.[29] As of January 1, 1991, in the RSFSR, 63 percent of all state and collective farms used leasing contracts.[30]

Despite relatively high numbers of farms with at least one leasing contract, to a large extent leasing was conducted as a campaign, similar to other political campaigns in a mobilized society.[31] Of the farms that used lease contracts, very few incorporated leases into all branches of production. In other words, farms would conclude one or two contracts in order to report that they had "adopted" this progressive form of new farming. As of January 1, 1991, in the RSFSR, for example, only 9 percent of all farms used leasing contracts in all their productive branches. As a local example, in Kostroma Oblast only one farm (out of more than three hundred) used leasing contracts in all branches of production.

The main reason that rural workers did not embrace leasing more enthusiastically was that "independent" lessees were regulated by a set of burdensome restrictions. The lease reform must be understood as an inherently conservative reform that continued to tie the "independent" farmer to the parent farm. There was continued state control over the lessee, as well as detailed involvement by the director of the parent farm. Internal leasing, the most common form of lease contracting, held much less freedom than was commonly acknowledged.[32] In an internal lease, the lessee performed only one function in the production cycle and therefore remained tied to the parent farm. During an interview with the farm director at the Borovikskoye sovkhoz in Kostroma Oblast in 1990, I was told that he decided what land on his farm would be used for leasing, and how much. The freedom to choose what products to grow was restricted, as the lessee had to grow products that "corresponded" to the farm's infrastructure, and the director had to approve these decisions. The lessee was obligated by contract to sell his output to the farm, and prices for this output were set by the farm. These contract prices were not allowed to exceed state procurement prices, and the lessee was not allowed to sell his output to sources other than the farm. The lessee was also dependent on the farm for feed, fodder, seed, technical support, and mechanized assistance. Oftentimes the lessee was the claimant of last resort, which meant that his input supplies were not guaranteed, despite the presence of a contract. Nonetheless, he was responsible for the delivery of his output according to contract.

Because many aspects of production remain controlled even under leasing conditions, in essence the burden of risk shifted from the farm to the lessee. The burden of risk did not weigh favorably against the possible benefits. One survey conducted by Goskomstat in 1989 revealed that 54 percent of lessees were not satisfied with the conditions of internal leasing. The basic reasons cited by lessees were "the absence of any kind of independence, violations of lease obligations for the supply of material-technical resources, low prices for products along with high payments for land, cattle, and machinery."[33]

The intention behind leasing reforms was to provide incentives for increased performance by offering the opportunity to work outside the collective setting. Beyond the formalism with which leasing was implemented, however, the real story was its limited impact. Relatively few farms used leasing in all their operations, and the inherent disincentives to undertake leasing outweighed potential gains from increased income.

In conclusion, we see that early in the reform process farm transformation was quite moderate, although Russia had already taken the forefront by mid-1991.[34] During this early period, a consensus existed that state and collective farms would continue to be the bulwark of the agricultural sector, and even the most "radical" plans advocated the breakup only of unprofitable farms. With the demise of the Soviet Union, however, reform in Russia became more far-reaching and assertive.

Stage 1: Reform Legislation and Farm Reorganization, 1992–1993

The first significant step in the reform of state and collective farms in Russia can be dated to December 1991. Legislation adopted in that month evolved from previous policy positions and did not "solve" the question of the ultimate fate of state and collective farms. It did, however, define the course of farm transformation for the next few years and indicated a political willingness to use state power to define and implement reform legislation. On December 29, 1991, President Yel'tsin signed a government resolution entitled "On the Procedure for the Reorganization of Collective and State Farms."[35] The purpose of the resolution was to destatize (privatize) collective enterprises and farmland. To promote the creation of private farms the resolution repeated the legal provision that farm members had the right to receive a share of farmland upon withdrawing from the farm. In order to privatize state and collective farms, the resolution instructed all farms to reregister themselves by March 1992 and, if required, to reorganize (a later decree in March 1992 extended the time frame to January 1, 1993). To oversee farm reorganization and to facilitate privatization, "local commissions on land privatization" were created within "every sovkhoz and kolkhoz."[36] Managers of collective farms and directors of state farms had "personal responsibility" for the fulfillment of the 1991 resolution.

Similar to previous directions of reform, the resolution began with an attack on unprofitable state and collective farms. Although figures varied, chronically unprofitable farms numbered between twenty-five and thirty-five hundred throughout all of Russia, out of a total of some twenty-seven thousand state and collective farms. According to the resolution, "chronically unprofitable" farms were obligated to reorganize and to adopt a new form of labor organization. During the bankruptcy process, the first preference was that a profitable enterprise would take over a bankrupt farm. If another enterprise did not assume responsibility for the debts of the bankrupt farm, debts were to be settled through the selling of farm assets in an auction, and participation in the auction was "mainly for workers of the farm."[37]

But the original resolution also went beyond unprofitable farms. Point 14 of the resolution, if taken literally, could have meant the closure of large numbers of state and collective farms, for it required that farms that were unable to pay off their debts and meet their payrolls were to be declared bankrupt by February 1, 1992, and to be liquidated during the first quarter of 1992. Because a vast majority of farms had debt (the Soviet practice was to periodically write off farm debt), this point of the resolution could have forced the closure of large numbers of farms. Rural conservatives and opponents to Yel'tsin's reforms argued that the resolution was a thinly veiled attempt to destroy state and collective farms for political reasons.

According to conservatives, most of the financial hardships that farms endured were due to state nonfulfillment of promises and financial obligations. Conservatives charged that the state's economic policies were responsible for rising farm debt and farms' inability to pay loans and wages. They pointed out that only about one-half of grain sold to the state during 1990–1991 harvests had been paid for, and they argued that other state debt to farms hurt farms' ability to meet their bills. In 1990 when original land reform legislation was adopted, rural conservative interests supported planks that emphasized land and enterprise privatization in exchange for government promises to rebuild and revitalize the countryside by devoting no less than 15 percent of national income to rural needs.[38]

Rural conservatives took the promise seriously and argued that the government breached its legal commitments during the second half of 1991 when prices and terms of trade turned against the entire agrarian sector and promised investments were not forthcoming.[39] Thus, conservatives argued that the provision forcing indebted farms to be liquidated was a deliberate trick by reformers and a violation of the principle that all types of farming—private and socialized—had an equal right to exist, a provision that had existed since the adoption of the 1990 Law on Land Reform.

Whatever the political intent, the December 1991 reorganization resolution was constrained by the necessity of feeding the country, and (contrary to some sensationalist Western press reports) the resolution did not mean that Yel'tsin was altogether abandoning socialized farming. As one local Russian editorial commented shortly after the resolution appeared, the issuance of the resolution "does not deprive kolkhozy and sovkhozy of the right to exist."[40]

In fact, having issued a radical decree to force farm reorganization, Yel'tsin almost immediately began to backtrack. During the next month, both he and government spokesmen attempted to clarify the intended targets of the resolution. The most explicit statement in this regard came at the end of January 1992 when Yel'tsin explained:

The reorganization of collective and state farms does not mean they are going to be disbanded.... If collective and state farms are in good condition and work for a profit, let them go on doing so. But we do have 2,600 loss-making farms. A decision has to be taken on their future during the first quarter. This land should be given to peasants or anyone else who wants to work the soil. There will be no further subsidies.[41]

Less than three months after the December resolution, and after enduring a barrage of criticism, the Russian government adopted another resolution in March 1992 that allowed state and collective farms to retain their previous status, that is, to remain state and collective farms if a meeting of farm members so voted.[42] The right to retain a collective form of labor organization was legalized, and a farm retaining its previous status was not violating policy or obstructing reform. Thus, during stage 1 of farm reform, state institutions introduced radical measures that soon thereafter were moderated. Following the March 1992 resolution, farms were given an "out" that allowed them to escape fundamental restructuring. The importance of this escape can hardly be overstated, for the overwhelming number of farms retained a collective form of labor organization.

The Farm Reorganization Process

During the reorganization process, state and collective farms were able to select from a variety of forms of labor organization. No matter what option was ultimately chosen, the reorganization process shared certain similarities. The first step in the reorganization process was for the farm chairman to create a "commission on privatization of land and reorganization" within each farm. These commissions were often staffed by the "leading" members of the farm. The farm's commission would brief the farm members on what each option entailed. This was an important aspect because differences among the various reform options were not always clear to farmworkers, and even Western analysts sometimes did not understand the nuances. Press reports indicate that often the chairman "suggested" a preferred form. After briefings had concluded, the chairman of the commission—usually the farm chairman—convened a general meeting, at which every member of the farm had equal juridical status, and a vote was taken by show of hands regarding the farm's future form of labor organization, the form of landownership, and other questions concerning farmland and property.[43]

During the reorganization process, a farm member had a onetime opportunity to receive land in-kind. If he decided to remain a member of the farm collective, then his land shares transferred to the farm and became general farm property. If he later decided to leave the farm, he could not receive land but would be compensated the cash value of his land.[44] As is well known, an overwhelming percentage of reorganized farms either retained their collective farm status or became joint stock farms—about 84 percent of former state and collective farms. In these cases the farm and farm property remained intact, which meant that very little land was actually distributed to individuals. At the end of the farm reorganization process, former state and collective farms and their successors still held about 90 percent of rural farmland in Russia, although technically this land was no longer considered state property.

After the vote by farm members as to the farm's future, an inventory of farmland and property was taken by the farm accountant and head engineer. If the farm was to remain intact, farm property was divided into "movable" and "immovable" funds. Things like water resources, roads, schools, and medical facilities were included in the immovable fund, whereas trucks, tractors, various machinery, equipment, and automobiles were placed in the movable fund. Land was also inventoried. Assuming that farm members voted to keep the farm intact, a farm would distribute land shares to all those who were eligible to receive them. However, the way the reorganization system worked, the land shares were conditional, and only those persons who actually left the farm received their land in-kind and were assigned specific plots of land.[45] In the case of land distribution in-kind, the oblast Committee on Land Reform (later changed to Committees on Land Resources and Land Surveying) designated specific land plots to each applicant using a standardized procedure that numbered land plots and then assigned them to persons in the same order as the applications were made.

A certain quantity of land could be retained by a reorganized farm or agricultural enterprise. This quantity was based upon raion land norms. "Surplus land"—defined as the land above the land norm multiplied by the number of farm members—had to be made available for farm members who wished to receive their land in-kind, and another portion of surplus land had to be made available for purchase or lease to those who wished to undertake private farming and wanted to increase farm size above the designated raion norm. The farm meeting that decided the future of the farm also decided where the free land and land for purchase or lease would be located.[46]

One problem was that only a small amount of land was made available for private farming, because former state and collective farms remained intact by forming joint stock farms, limited partnerships, or retaining their previous status. Kostroma, Moscow, and Rostov Oblasts showed that often private farmers had less than 5 percent of an oblast's arable agricultural land.[47] Nationwide, private farmers held about 6 percent of agricultural land at the beginning of 1997.[48] Thus, a significant problem of land distribution was that often very little of the land possessed by former state and collective farms changed hands during farm reorganization, and this in turn would have repercussions for the development of a rural land market.[49]

Following the outcome of the farm vote, all reorganized farms were required to register their name and their new status according to special forms, after which they were permitted to open bank accounts for receipt of credits and payments. If a farm voted to disband completely, a "liquidation commission" was formed and all farm assets, including land and property, were distributed to members.

Farm Reform Options

State and collective farms had a number of options to choose from when undergoing the reorganization process. Flowing from the March 1992 resolution, one option was for state and collective farms to retain their original status. This option entailed holding a meeting, taking a vote of members and, following the vote, filing new registration forms with local agricultural and tax officials. One agricultural expert explained that primarily the strongest farms decided to retain their previous status, and for this reason there was a fair amount of regional diversity based on climatic and geographical factors that affected farm profitability.[50] Other reform options included forms of farm organization that changed the legal status of farms but basically kept the farm and its operations intact. Farms could disband altogether and create either associations of private farms or simply individual private farms.

Joint Stock Farms

As the farm reform process unfolded, the most common choice of collective organization was the joint stock farm, with origins dating back to 1988 when a state farm in Stupinskiy raion in the southern part of Moscow Oblast formed into a joint stock farm. Based on this experience, in mid-1991, experimental joint stock farms were reported in Yaroslavl' and Pskov Oblasts in the Russian Republic. Within the joint stock variant there were open and closed joint stock farms. "Open" meant that shares could be traded and owned by non–farm members. "Closed" joint stock farms (also referred to as communal property with limited responsibility) meant that shares could not be obtained by persons outside the farm, and farm members could not trade or cash in their shares without the approval of a farm's general meeting.[51] An overwhelming majority of farms undergoing reorganization chose the joint stock option, and within this variant the most popular type was the closed farm.

The "highest organ" of farm management in joint stock farms was supposedly the general assembly of farm members, which met twice a year or, under urgent conditions, with two-thirds agreement of the assembly. In between assembly sessions, the farm was run by a management committee comprised of between five and fifteen individuals who were elected for a three-year term by open (nonsecret) majority voice vote. This committee was responsible for production and financial decisions on the farm. The chairman of this committee also served as the chairman of the farm.[52]

A joint stock farm—open or closed—meant that the farm and its property remained intact and labor organization remained collective, that is, based on the brigade system. Workers of reorganized farms remained eligible for state pensions (and reorganized farms continued to contribute to the state pension fund). Even in its reformed form, a joint stock farm received state subsidies and compensations; the farm had to meet delivery obligations to regional and federal food funds; and its workers were officially considered to be employed in the state sector, not the private.

Share Distribution in Joint Stock Farms

During the process of reorganization into a joint stock farm, farm property and land was divided among members, with each member receiving a certain quantity of both land and property "shares" free of charge. The value of land shares was determined by the number of claimants who had a right to shares, and this list of persons was decided upon by a general meeting of the farmworkers' collective. In general, this list included farm employees; pensioners who resided on farm or agricultural enterprise territory; and persons employed in the social sphere on the farm. In addition, those who were temporarily away from the farm (for military service, for instance), those who had a right to return to their place of work (who were temporarily living or working in a city, for example), and persons who were fired or released from the enterprise after January 1, 1992, also had a right to a land share. Following a September 1992 resolution all persons received equal land shares.[53]

Once the list of eligible persons was established, the actual division of land could take one of two forms. In the first variant, land plots were given a monetary value and persons eligible for land shares received an equal value of land shares. The second variant divided land in-kind, and each person received equal land shares based on size, quality of land, and so on, which could not exceed raion norms.[54] The first variant was the option most often chosen.

Calculation of the value of property shares varied from farm to farm according to "concrete conditions" of the farm. In general, once all farm property had been inventoried and valued, a common method of division was for each farm member to receive property shares based on the length of service to the farm, professional status in the farm (which often meant type of work), and level of income during the past three to five years. Other farms calculated labor input by taking the number of days worked, subtracting the number of days not worked when the person did not have "a good reason," and multiplying that sum by a ruble value per workday—in other words a continuation of the Soviet workday (trudoden') system.[55] Based on the number of land and property shares held, each farm member received a "dividend," which could be paid quarterly, every six months, or once a year depending on the farm. Many farms opted for annual payments.

Because joint stock farms transferred land and property shares into the hands of farm members, and because remuneration was better tied to output results, some argued for the "progressive" nature of joint stock farms. This statement was only partly true and applied mainly to dividends. For most reorganized joint stock farms, basic farm wages continued to comprise the bulk of earnings. Dividends were the analog to Soviet "bonuses" that were supposedly tied to farm output. The dividend system worked as follows: The farm's management committee created a "share fund," expressed as a percentage of net income obtained from food production and other productive endeavors. For example, in one published account the members of a joint stock farm in Pskov Oblast contributed 20 percent of the net value of receipts from production to the creation of a share fund for the payment of dividends. Once a fund was formed, the farm's management committee decided how much to distribute to farmworkers.

There were two options for the payment of dividends. The first option was to distribute as dividends a flat percentage of the farm's profits. For example, taking the case in Pskov Oblast again, 27 percent of farm "profitability" appeared to be the base below which dividends were not paid. But with a 27 percent farm profitability, 4 percent of the profits were paid as dividends; and with each 1 percent growth in profitability, dividends increased by 0.1 percent of the profits.[56] Using a somewhat different system, the joint stock farm "Russia" in Chelyabinsk Oblast distributed 70 percent of profits above the base share fund as dividends, a method that was intended to create incentives for increasing production.[57] Even using a flat percentage rate not all farm members received equal payments. Coefficients were calculated and graduated on the basis of farm seniority, or how long a person had worked on the farm, with a person having twenty years of seniority receiving about twice as much as a person who had worked between two and five years. This system therefore benefited workers with seniority and not necessarily the most productive farm members.

The second option was to pay a differentiated dividend rate according to categories of farmwork. This model was geared toward rewarding the more difficult professions and skills and was recommended for farms with high profitability. In order to prevent abuses, farm leaders, white-collar employees, and specialists could not set higher rates of dividends for themselves.[58]

In both variants certain conditions had to be met. For example, a certain minimum number of days had to be worked in order to be eligible to receive a dividend. Dividends were treated as ordinary income and were subject to income taxes. If the value of production decreased, so too did the value of the share fund for distribution, the precise amount to be determined by a meeting of farm members. Thus the size of dividends varied yearly and in some unfavorable years might not be paid at all. If a person left the farm he lost his right to receive a dividend. Finally, farm pensioners could receive dividends if their pension was lower than the dividend.[59]

Farm Disbandment

Another option in the course of farm reorganization was for the state or collective farm to disband. During the general meeting at which the farm's future was decided, members had the right to vote to disband the farm and create either individual private peasant farms *(krest'yanskiye khozyaystva)*, or an association of peasant farms in which individual farmers would work cooperatively together. These forms of farm organization required that a state or collective farm physically divide and distribute its land and property, in short, to cease to exist as a unified entity. Of course, the Law on Peasant Farming allowed peasant farms and farm associations to form independently, that is, not as a result of state or collective farm dismemberment. But during the farm reorganization process the disbandment of former state and collective farms was the most radical step that could be taken.

The method of valuing and dividing land and property was the same as described above: the agricultural land of the farm was valued and assigned equally to each member of the farm, irrespective of the length of service to the farm or size of personal income. Property was also divided based on variables chosen by the farm, usually including length of service, professional status, and size of past income. The main difference was that, after members voted, these assets were distributed to farm members.

Once a former state or collective farm was disbanded and its assets divided, either individual farming operations could be conducted or private farmers could form collective operations in the form of farmers' associations or small cooperatives. Both farmer associations and small cooperatives were cooperative endeavors that encompassed productive, marketing, and service operations. The purpose of these organizations was to "increase the income of Association members and strengthen the material-technical and social base of its members by coordinating activities and organizing joint production of agricultural products." Both associations and cooperatives had management organs, and these organs usually defined the agricultural activities that would be conducted. New membership was approved by a general meeting of existing members or by the management committee. Members were free to resign from the association at any time. Finally, these organizations were based on full self-financing and were entitled to engage in foreign trade.[60]

It is not necessary to spend a great deal of time examining cooperative forms of private farming because data on farm reorganization showed that such forms comprised an extremely small percentage of postreorganization farm operations. By January 1, 1994, with 95 percent of farm reorganization completed, only 936 associations of private farms existed and 1,861 agricultural cooperatives had been formed in the entire country. In comparison, over 82,000 individual private peasant farms had been created out of the farm reorganization process.[61]

There are several reasons why associations of private farms were not more popular. Interviews with private farmers in Kostroma Oblast indicated that part of the relative unpopularity of farmer associations and small cooperatives was due to the dispersed nature of private farmland holdings. One farmer related that his closest private farmer neighbor was many kilometers away. Although a general shortage of farm equipment forced private farmers to share resources, at the same time significant distances between farms made such cooperation difficult. Private farmers often incurred marketing difficulties (finding a buyer for their produce), and so private farmers did have the sense of competing for market share with other private farmers.

The Political Impact of Reform Institutions

Among Russian leaders there was little disagreement about the need to reform state and collective farms. There was also a variety of choices. So there was nothing deterministic about the course chosen by the Yel'tsin regime. During stage 1 of reform, Yel'tsin hardly chose the most radical of reform options. Although the first stage of farm reorganization was rather moderate, in the mind of Russian conservatives the December 1991 resolution violated the compromises worked out in 1990 and the principles established in various laws that guaranteed the equality of different forms of farming—private, collective, and cooperative. Therefore, there was an immediate political impact from the attempt to reorganize state and collective farms, which polarized leaders in high echelons of power.

One political casualty was the relationship between Russian President Yel'tsin and Vice President Aleksandr Rutskoy, who was named to oversee agrarian reform in late February 1992.[62] Yel'tsin and Rutskoy differed on the desired fate of state and collective farms and on how reform in this area should be conducted. These differences became apparent almost immediately. On March 16, 1992, Rutskoy sent a telegram to the heads of autonomous republics, krays, oblasts, and okrugs in Russia, in which he argued that the "central task" was to create acceptable conditions of work for the entire agro-industrial complex. He argued that agricultural production would continue to depend on the "stable and effective work of profitable state and collective farms." Therefore, he instructed the heads of territories that they were personally responsible for guaranteeing "normal work on these farms, abolishing all instructions and administrative decisions that hinder effective work on collective and state farms," which might be taken to imply disregard for farm reorganization initiatives.[63]

Rutskoy continued to emphasize the need for reliable food supplies and to criticize agrarian reform because it was leading to serious declines in food production. About two weeks after the telegram, Rutskoy was interviewed in a "liberal" agricultural paper, in which he denounced the fact that

> in a number of regions reorganization of farms is proceeding as a "campaign." . . . Under the slogan of privatization, joint stock farms and peasant farms are encouraged to plunder profitable farms and to distribute them into the hands of people who seldom have a direct relation to the land. Such a practice brings losses not only to agroindustrial enterprises but to the entire economy of Russia.[64]

Other differences existed as well. Rutskoy felt that private farms were too expensive to establish and that state funds should be devoted to state and collective farms to help them operate better.[65] He favored free land sales via coupons, which liberals felt would threaten the existence of newly emergent private farms. Owing to divergent views on the course of agrarian reform and other policy differences, Rutskoy was stripped of his responsibility for agriculture in April 1993.

Furthermore, as agrarian reform became radicalized, policy divisions arose over the course of reform pursued by the regime. One example concerned the moderate Viktor Khlystun, who served as minister of agriculture from 1992 to October 1994 (he was then reappointed in May 1996).[66] During his first term as minister of agriculture, Khlystun tried to walk a fine line between the people who surrounded him, who were much more conservative on agrarian reform, and the Yel'tsin regime, which was pushing radical reform with more urgency after the events of October 1993.[67]

Khlystun favored reforming state and collective farms, but he did not advocate their destruction and in general he operated in a "pragmatic" manner. As political and financial pressures continued to be exerted on the collective sector, Khlystun finally broke with the regime. He expressed his disagreement in an interview just prior to his resignation in October 1994. In this interview he criticized the state's financial policies toward the agricultural sector. In a key passage he remarked:

> The development of a market infrastructure is much more important than further experiments on regulating ownership relations. Any form of ownership by rural dwellers will

exist in difficult conditions and will not be able to sell its production if it does not possess market information, if there is no market structure and infrastructure.[68]

Thus, Khlystun indicated a basic disagreement with the priorities of reform, and he resigned at the end of October 1994. Although he cited health reasons for his resignation, it was reported that the motivation was in fact displeasure over the course of the agrarian reform policy, and at one point Khlystun commented, "it would be beneficial for us to exchange Boris Yel'tsin."[69] After speaking out against the financial policies toward the collective sector during his absence from the ministry, in 1996 he agreed to return as minister in order to "make a difference" and to moderate some of the government's policies.[70]

Another political impact was the loss of rural support. By attacking state and collective farms the Yel'tsin government alienated rural leaders. Rural leaders feared the loss of authority and perhaps even their jobs. If market reforms were successful, the rural planning and management bureaucracies would no longer be necessary. Reduced procurement targets, direct links between producer and processor or retail outlet, and the obsolescence of farm planning, all threatened agricultural officials at the raion, oblast, and national levels. Furthermore, by not fulfilling state promises to revive the countryside through increased financial support, the government alienated the rural population at large. As has been consistently demonstrated, rural dwellers expressed less support and more uncertainty about farm reform, land privatization, and economic reform in general than did the urban population.[71] Part of the reason for these types of responses had to do with the natural conservativeness of the rural population (a universal phenomenon) and its age structure. But an important aspect was also the impact that reform had on rural dwellers' standard of living.

The consequences of state reform policies were significant. The mainstream rural sector perceived the government as an intractable enemy (a view fueled by rural conservatives), a government dedicated to the destruction of the collective agricultural sector. The loss of most rural support left the government dependent on the Association of Peasant Farms and Cooperatives of Russia (AKKOR), which represented private farming interests but was not popular outside those interests. Thus, the government from the beginning had weak rural support, which complicated and frustrated reform efforts.

The Impact of Farm Reorganization

Many analysts, both Russian and Western, argued that the results of farm reorganization during stage 1 were mostly cosmetic, that is, farms changed their legal status through reregistration but farm operations remained essentially the same. Indeed, a rural survey of 507 farmworkers in Bryansk Oblast in

1994 confirmed that farm members themselves felt few positive changes as a result of the farm reorganization. When asked whether reform allowed them to work better, less than 7 percent of farm leaders and head specialists and less than 2 percent of skilled workers responded that it had. Less than 50 percent of the farmworkers in all categories felt that "their labor was affected by any positive changes or their work was more stimulating," and less than 9 percent said that reform had increased independence or responsibility in worker collectives. The conclusion was that, "because of the formal reorganization of farms, farmworkers at present do not feel themselves owners."[72]

There is little disagreement that farm reorganization accomplished less than intended. Specifically during stage 1 of farm transformation, reform initiatives failed to achieve several important things. Among the most important, reform failed to create sufficient incentives for farm personnel to leave the parent farm. Instead, an overwhelming majority of farm members remained within a parent farm, and these farms continued to operate essentially as they always had.

The lack of sufficient incentives could be measured in two ways. First, the macroeconomic environment for leaving the farm was hostile: private farm start-up costs were high; machinery, equipment, and other inputs were either expensive or difficult to find; inflation was rampant; rural infrastructure was substandard; consumer markets were unreliable; and revenue from output seldom covered production costs. One Rostov Oblast farm chairman enumerated the "objective conditions" of leaving a parent farm: land was difficult to obtain, workdays were long and hard, profits were meager, the level of mechanization was low, and credits were often nonexistent. "Why would anyone want to leave for that?" he asked rhetorically when interviewed in 1993. This question—Why would anyone want to leave when the options were so unattractive?—was repeated many times to this author in various regions.

Second, the basic conceptualization of reform did not reflect an understanding of rural societal fabric. Rural services, such as they were, came from state and collective farms. The fact that state and collective farms were much more than a production unit—but instead comprised a rural community—appeared to be forgotten by those who designed reform or who advised the Russian government. Medical care, housing, day care, schooling, and subsidized food, all came from state and collective farms. Alternative sources of these services were never identified. As this author toured several former collective farms in Rostov Oblast during July 1993, he was told by farm chairmen that even after "reorganization" their farms would continue to provide housing, day care, and medical care to their members. One joint stock farm, the Don, in Salsk raion, had its own bakery where it baked bread and bread products for its members.

The main meal was available for next to nothing for farmworkers. All this would be lost if a farmworker became a private farmer. Remaining in a collective, then, provided both psychological and material security.

So rural preferences were important in explaining patterns of farm reorganization. To understand the impact of reform on state and collective farms, we shall examine two aspects of farm reorganization, using oblast- and raion-level data. The first, oblast-level data on patterns of farm reorganization represent the choices that farm members made about the future of their farm. The second, raion-level data give spatial patterns of farm reorganization.

Patterns of Farm Reorganization

During the reorganization process it became increasingly clear that collective forms of labor organization were more popular than private farms. Reformers anxious to find the silver lining noted a steady decrease in the percentage of farms that retained their previous status at the national level. In July 1992, almost 44 percent of state and collective farms had retained their previous status. By October 1992, this percentage had dropped to less than 43 percent; by January 1, 1993, to 35 percent; and by October 1, 1993, to 34 percent.[73] In mid-1993, about one-third of state and collective farms had retained their previous status. By January 1994, farm reorganization was virtually complete; 95 percent of state and collective farms had reregistered, and 34 percent of state and collective farms that had reorganized retained their previous status, that is, remained state and collective farms.

Emphasis on the percentage of farms retaining their previous status is misleading, however. The real story of reform is that, even after reorganization, most farms chose some collective form of labor organization. Overall, about 84 percent of farms that had reorganized adopted some form of collective labor organization, the most common of which were joint stock farms of the closed type.[74] Trends for state and collective farm reorganization in European Russia during 1992–1994 are illustrated in table 10.

By January 1994, 47 percent of reorganized farms had transformed into joint stock farms of the closed type or farms with limited responsibility, and 11 percent formed a cooperative agricultural enterprise.[75] There were, of course, regional variations. These are illustrated in table 11, showing reorganization results in European Russia.

The table depicts a general pattern in which southern regions showed a higher retention rate of state and collective farms. The table also clearly indicates that the greatest number of private peasant farms were created in southern regions. On average for European Russia, three families per state or collective farm withdrew from the farm, with an average of three or four members

Table 10. State and Collective Farm Reorganization in Russia, 1992–1994

	July 1, 1992	January 1, 1993	July 1, 1993	January 1, 1994
Farms registered				
Number	8,391	19,719	23,365	24,344
Percentage	33	77	91	95
Farms retaining previous status				
Number	3,664	6,990	7,862	8,373
Percentage	44	35	34	34
Open joint stock farms				
Number	324	328	307	272
Percentage	4	2	1	1
Closed joint stock farms				
Number	2,803	8,551	11,153	11,493
Percentage	33	43	48	47
Number of associations of peasant farmers	612	748	931	936
Number of agricultural cooperatives	411	1,662	1,973	1,861
Number turning into subsidiary farms of enterprises	245	347	364	424
Number of sovkhozy converting into kolkhozy	—	251	217	237
Number converting into other forms[a]	1,099	2,062	2,151	2,273
Number of private peasant farms created	23,337	43,590	62,000	82,000

SOURCES: *Reorganizatsiya kolkhozov i sovkhozov Rossiyskoy Federatsii (po sostoyaniyu na 1.07.1992 g.)* (Moscow: Goskomstat, 1992); *Reorganizatsiya kolkhozov i sovkhozov Rossiyskoy Federatsii (po sostoyaniyu na 1.01.1993 g.)* (Moskow: Goskomstat, 1993); *Reorganizatsiya kolkhozov i sovkhozov Rossiyskoy Federatsii (po sostoyaniyu na 1.07.1993 g.)* (Moskow: Goskomstat, 1993); *Reorganizatsiya kolkhozov i sovkhozov Rossiyskoy Federatsii (po sostoyaniyu na 1.01.1994 g.)* (Moskow: Goskomstat, 1994).

a. Unidentified in primary source.

Table 11. The Reorganization of State and Collective Farms in European Russia as of January 1, 1994

	Number of Farms That Reregistered	Number of Farms Retaining Previous Status	Percentage of Farms Retaining Previous Status	Number of Peasant Farms Created	Ratio of Private Farms Created to Reorganized Farms
Russian Federation	24,344	8,373	34	81,628	3.35
European Russia	19,806	6,998	35	61,411	3.10
Northern region	638	124	19	1,284	2.01
Northwestern region	734	155	21	3,257	4.43
Central region	4,549	1,112	24	5,879	1.29
Volga-Vyatka region	2,247	997	44	1,314	0.58
Central black earth region	2,371	704	30	3,231	1.36
Volga region	3,404	1,570	46	9,933	2.92
Northern Caucasus region	2,615	1,099	42	23,403	8.95
Urals region	3,248	1,237	38	13,110	4.03

SOURCE: Author's calculations from *Reorganizatsiya kolkhozov i sovkhozov Rossiyskoy Federatsii (po sostoyaniyu na 1.01.1994 g.)* (Moscow: Goskomstat, 1994).

per private peasant farm.[76] A 1994 World Bank–sponsored survey in five Russian oblasts found similar results: about 30 percent of farms surveyed had experienced zero exits by farm members; about 35 percent of farms experienced between one and nine exits; and 35 percent experienced more than ten exits.[77] Thus, about two-thirds of state and collective farms experienced less than ten exits (an "exit" is a person; the average rural family size is 3.2 persons, so three families would equal just over nine persons).

Spatial Patterns of Farm Reorganization

Both tables 10 and 11 depict a large number of private farms created out of the reorganization process—precisely what reform was intended to accomplish. But there is more to the story than mere numbers of private farms. Raion-level data within oblasts are useful for gaining insight into spatial patterns of private farm creation resulting from the farm reorganization process. Although results in individual oblasts may or may not be replicated nationally, they are useful in illustrating some of the basic problems that accompanied farm reorganization.

Private farmers, more so than state and collective farmworkers, suffered from substandard rural infrastructure and the lack of adequate rural roads. Therefore, an ideal situation for private farmers would be a location with good proximity to major urban centers so that they might take advantage of better roads, an existing food market, skilled labor for hire, lower transportation costs, and perhaps an agricultural institute for technical advice. Flowing from the way in which development was conceptualized during the Soviet period, a common land-use pattern for Russian cities was for state and collective farms to surround the cities, thereby providing easy access to food markets. The use of land around cities for agricultural purposes had high-level support among key liberal reformers owing to the importance of food supplies from "suburban" farms.[78]

Thus, regarding land reform and spatial patterns of private farming there were two types of "good" land—land that was inherently productive due to its mineral and organic content, and land that was "good" because of its location. Farms with good location would have lower transportation costs and easier marketing that could at least partially offset advantages of farms that had higher production and higher yields but more remote locations. Given the tenuous financial position of most private farms, therefore, good location was of central importance. The question is, How much "good location" land did private farmers receive during farm reorganization? Raion-level data from two central region oblasts help to clarify this question.

The first oblast for which farm reorganization data are presented is Kostroma Oblast. At the oblast level, farm reorganization patterns within Kostroma Oblast were in some ways different from national trends. According to unpublished statistical data obtained by this author, as of January 1, 1994, 97 percent of kolkhozy and 90 percent of sovkhozy had reregistered. Of those farms that had reregistered, only 9 percent of kolkhozy and 16 percent of sovkhozy had retained their previous status, much lower than national averages. That so few farms retained their previous status suggests the weak financial position of most farms in the oblast.

On the other hand, the most popular form of labor organization for reorganized farms was a joint stock farm of the closed type, chosen by nearly 76 percent of the farms. In this sense Kostroma duplicated national trends. Adding all forms of collective labor organization showed that 86 percent of reorganized farms chose some form of collective labor organization, including 5 percent who chose an undefined "other" form of collective agriculture.

An important factor in the creation of private farms was what happened during the farm reorganization process in Kostroma Oblast. By January 1, 1994, 397 private peasant farms had been created out of the farm reorganization process—or about 35 percent of the total number of private farms in the oblast at that time. At first glance it appears that the farm reorganization process was an important contributor to private farm creation. However, such a conclusion would be misleading. Unpublished data showing farm reorganization results by raion demonstrate that private farmers lost the battle for land in the key areas of the oblast. In fact, in Kostromskoy raion and the four other raions that comprise the southwestern corner of the oblast, no private farms were created during farm reorganization. In general, Kostromskoy raion and the four other raions not only have the best location relative to the main food market, the city of Kostroma, but also are among the most productive and profitable raions in agricultural production.[79] In short, state and collective farms in those five raions with the best proximity to the city of Kostroma, the oblast center and the largest city in the oblast, did not lose their advantageous position to private farms.

Furthermore, even within these most advantageously located raions, state and collective farms (and their successors) did not lose their dominant position. On January 1, 1994, the five raions with the best location relative to the city of Kostroma accounted for 256 private farms, or 23 percent of the total number in the oblast. However, total land assigned to private farms was less significant: private farmers had 2.45 percent of all agricultural land in Kostromskoy raion; 2.26 percent in Krasnosel'skiy raion; 2.8 percent in Nerekhtskiy raion; 2.38 percent in Sudislavskiy raion; and 2.03 percent in Susaninskiy raion.[80] The conclusion, then, is that, at least in Kostroma Oblast, farm reorganization did not put much "good location" land into private farm hands during the reorganization process.[81]

The same general pattern is found in Kaluga Oblast, also located in the Central region but to the southwest of Moscow. According to unpublished data obtained by the author, as of January 1, 1994, virtually all collective and state farms had reregistered—99 percent of collective farms and 97 percent of state farms, of which 28 percent of collective farms retained their previous status, as did 13 percent of state farms. Of the 335 farms that reregistered, 328 adopted some form of collective labor organization, whether it was retaining the previous status, forming an open or closed joint stock farm, or creating some "other" form of collective enterprise. Raion-level data show that in the entire oblast only twenty-seven private peasant farms were created out of the reorganization process; and these farms were located in just two of the oblast's twenty-four raions. Three farms were created in Iznoskovksiy raion and the other twenty-four in Meshchovskiy raion, both of which are located far from the oblast center and largest city, Kaluga.[82] Iznoskovksiy raion is a sparsely populated (8,500 permanent residents) rural raion located in the far northwest part of the oblast, near Smolensk Oblast; whereas Meshchovskiy raion is also a sparsely populated (15,100 permanent residents) rural raion that is located in the center of the oblast.[83]

Impediments to Farm Transformation

How do we explain these results? The question really has two parts. One part concerns the creation of private farms from state and collective farms and asks, Why were they so disadvantageously positioned? The second part concerns the operation of state and collective farms and their successors and asks, Why were operations not substantially changed as a result of farm reorganization?

Reorganization and Private Farms

With regard to the first part, often farmworkers did not want to leave parent farms in order to create private farms. Why? There are a number of reasons. A well-documented reason has to do with the difficulty of obtaining land early in the reform process and the opposition from state and collective farm managers. There is no question that this was a key factor, and had land been more easily obtainable more farm members might have taken up private farming. Yet the difficulty of obtaining land is not the only explanation.

A second factor is that the rural demographic situation was quite unfavorable and that relatively few young people were located in rural localities who were willing to assume the risks inherent to private farming. The demographic problem was particularly true in non–black earth areas, although even black earth areas were also affected. Whereas the most common age to undertake private farming was between thirty and forty, and men usually headed private farms, in most non–black earth oblasts from one-quarter to one-third of the rural population was too old to work and the number of rural men was lower than the number of rural women. Furthermore, although it is impossible to know the exact number, many collective farms had at least as many—if not more—pensioners as working-age members; and this, combined with decades-long out-migration of the young and the skilled, meant that relatively small percentages of the collective farm population were potential private farmers.

The inherent difficulties of private farming also discouraged many prospective farmers, especially as reform matured. Interviews with farm and agricultural officials consistently yielded the view that it was hard to find enough people who wanted to become farmers. Stories in the press depicted similar situations. One paper noted that in Moscow Oblast few collective farmers became private farmers. The article recounted the experiences of three families who left the collective to begin their own farm. "They immediately encountered massive problems—no machinery, nowhere to obtain gas, nowhere to sell produce, and so on. People saw this and did not want to follow their example."[84] The article revealed how collective farmers were able to buy food products at Soviet-era prices (milk at 2.50 rubles per liter, butter at 2 rubles per kilogram, sugar at 2 rubles per kilogram, meat at 25 rubles per kilogram), which reinforces my observations and conversations on farms in Rostov. In addition, in Kostroma Oblast the farm directors who were interviewed indicated that, even though rural wages were low and often were not paid for many months, farmworkers were "taken care of" by payments in-kind, so they did not starve.

Finally, farm reform measures did not create incentives for farmworkers to want to leave. The economic environment was hostile and uncertain, with rural services of poor quality and available primarily through existing parent farms. Private farming was very high risk, as inputs were prohibitively expensive, credit was hard to obtain, and most of the farms were even more primitive than weak collective farms. Further, the infrastructural base for private farming was primitive and did not improve as time progressed.

With respect to the poor location of private farms, one reason was that farm managers and local agricultural officials often assigned the least desirable land to would-be farmers. But again there is more to the story than conservative opposition. First, well-positioned state and collective farms were likely to be more prosperous than remote farms, and farm prosperity correlated with the desire to remain on the farm. Thus the fact that raions close to major urban centers did not give rise to a large number of private farms may be explained by farm location. Farms with better locations could provide for their workers better, which translated into less desire on the part of workers to leave or disband the farm.[85] Second, we must remember that private farmers competed for land with other private agricultural users. Private farmers competed for land from raion land funds with persons who wanted land for other small-scale agricultural activities such as collective fruit and vegetable gardens, dacha plots, and private plots. These activities were less capital intensive, more risk averse, and hence more desirable. As a consequence, among possible privatization options, private farming was the least preferred among rural dwellers, which meant that would-be private farmers often lost the competition for good agricultural land.

Reorganization and Farm Operations

The second aspect of farm transformation concerned farm operations, and here we ask, Why were farm operations not substantially changed? The most fundamental reason is the nature of the incentives and the economic environment that farms confronted. The economic intent to reform farm operations was undermined by reform strategy itself.

The reform of state and collective farms were affected in a fundamental manner by the consequences of price liberalization in January 1992. Whatever else one wants to say about other aspects of price reform (how it led to fuller shelves through decreased demand; how it closed the gap between production costs and retail prices; how it brought consumer demand and supply into better balance; how it lessened the need for consumer subsidies), the architect of price liberalization—Yegor Gaydar—very likely did not anticipate the significant shift that turned the terms of trade against the agricultural sector as a result of price liberalization. Prior to price liberalization, officials and experts within Russia were estimating that prices would not increase by more than a factor of three or four. No one even imagined increases of twentyfold and more during the first year. As a consequence of price liberalization, agriculture suffered because price relationships between agriculture and industry turned drastically against food products. We shall examine in more detail the deterioration in the terms of trade faced by food producers in the following chapter, but for now we are mainly interested in the state responses to these unforeseen consequences.[86]

One important consequence of price reform was an immediate deemphasis on rural efficiency. Already by spring 1992 the Yel'tsin government showed that it placed priority on food supplies over farm efficiency, and presidential decrees indicated that old financial policies were slow to fade. With promises of subsidies and supplementary credit resources, the state pursued an egalitarian policy toward farms so as to ensure sufficient state reserves to meet urban demand for food.[87] One example of the continued egalitarian nature of rural-rural relations was that special subsidies for animal husbandry products were introduced in May 1992, which divided the country into two large price zones, one for the far north and regions equal to it, and the rest of the country. Producers were paid a subsidy in rubles per ton of produce delivered to state reserves. The far north price zone received subsidies for milk, cattle, and poultry that were twice as high as the rest of the country, and five times as high per thousand eggs.[88] In other words, these subsidies were basically a continuation of the Soviet-era "cost-plus" system, which had benefited high-cost producers. By the end of 1994, state subsidies to the animal husbandry sector equaled 18 percent of the purchase price for cattle and poultry, 27 percent for milk, and 8 percent for eggs.[89]

Although subsidies continued to animal husbandry, and in fact accounted for 85 percent of all farm subsidies, the animal husbandry sector remained unprofitable.[90] The sector responded to low purchase prices, high input prices, chronic nonpayments, and higher borrowing costs by slashing livestock herds drastically. From 1991 to 1995 on state and collective farms and their successors, the number of cattle declined from 47.1 to 31.1 million; the number of pigs declined from 31.2 to 16.7 million; and the number of poultry decreased from 465 to 318 million.[91] (During the same period, holdings by the population—livestock raised on personal plots—increased significantly; and these holdings were substantial, as personally owned livestock alone accounted for one-third of all hogs and close to 40 percent of all cattle.)

Similarly, a government resolution adopted in June 1992 stipulated that all farms were eligible to receive compensation for fuel, machinery, and fertilizers per ton of various food products produced.[92] Once the details of the resolution were published, it was clear that compensations were offered irrespective of how efficiently those resources were used.[93] The purpose was to compensate farms for the fact that prices for inputs were rising faster than prices for production. However, the policy on input compensation meant that economic policy clashed with the political requisites of food policy, with the result that neither policy's requirements were satisfied.

Periodically, the cost-plus subsidy policy came under attack from agrarian specialists, for instance from Viktor Khlystun at the end of 1992 and the beginning of 1993. In response, the government promised a "new approach" to financial support for agriculture in 1993. When the "new" subsidy policy was revealed for 1993, however, there was very little to distinguish it from the 1992 policy.[94]

At the end of January 1993, the new prime minister, Viktor Chernomyrdin, stated that "it is very important that all agricultural producers will be subsidized," and that the government would "provide additional subsidies to kolkhozy, sovkhozy, and farmers." Shortly thereafter, in early 1993, the government announced a whole series of production subsidies and compensations to collective and private farmers for 1993; and these subsidies for agricultural production would be indexed to the rate of inflation.[95] For example, from the federal budget, compensation up to 30 percent of the cost of mineral and chemical fertilizer would be paid monthly to state, collective, peasant, and other farm organizations. Up to 50 percent of the cost of fuels used in agricultural production would be compensated.[96] The state would also compensate farms and farmers for 50 percent of the cost of drought insurance from the federal budget. The state would compensate farms and farmers up to 50 percent of the cost of farm equipment and machinery. During the first half of 1993, farms would receive compensation for 50 percent of the cost of transporting agricultural products.[97] State declarations of "urgent" financial support became annual occurrences, often coming early in the year prior to spring sowing and then again later in the year. Thus, similar types of state-funded programs were announced in November 1993, February 1994, March 1995, July 1995, February 1996, April 1996, June 1996, and February 1997.[98]

Subsidies to agriculture are necessary; they are found in every country in the world. In fact, Russian agricultural officials often argue that higher production corresponds to higher subsidies, something that may be true given the high-cost nature of plant growing due to climatic reasons and high-cost animal husbandry due to shortages of mechanization and infrastructure. In the Russian case, however, the problem was that agricultural subsidies continued the Soviet cost-plus system, which carried no penalties for waste. Moreover, weak farms often continued to receive preferential treatment in resource allocation. Interviews with agricultural officials in the agricultural administration for Kostroma Oblast and data received from them revealed that, during 1993, preferential subsidies continued to be distributed to weak economic farms and regions, just as in the past.[99] Thus, reform institutions changed forms of ownership and labor organization but did little to create incentives to change farm production operations or the treatment of weak farms.

The last factor that contributed to the lack of significant changes in farm operations concerned the economic environment that was created by agricultural trade policy. Flowing from political pressure exerted by agricultural interests, state trade policy created an environment to protect domestic food producers. Competing against more efficient food producers likely would have damaged domestic food producers who remained less efficient and less price competitive; in short, free trade would have allowed market mechanisms to function. Instead, a protectionist policy—a policy that was supported by both the liberal AKKOR and the conservative Agrarian Party—sheltered food producers and forced domestic producers to compete mainly with other inefficient domestic producers.

We shall examine trade protectionism in more detail in the following chapter, but for now we should merely note that a protectionist policy entailed tradeoffs. On the one hand, a protectionist policy sheltered domestic producers. The head of the administration on price policy within the RSFSR Ministry of Agriculture openly admitted that the intent of import tariffs was to make domestically produced food competitive with (cheaper) imported foodstuffs—even though such a move would increase food prices for the urban consumer.[100] On the other hand, a protectionist trade policy also slowed the impetus for domestic food producers to transform their operations. That is to say, the importance of food trade policy was that, although it helped domestic producers financially, it also removed a key stimulus of foreign competition that would facilitate the transformation of farm operations.

Summary

By mid-1993 many people both in and out of government in Russia argued that little about farm operations had changed as a result of farm reorganization. The attempt to transform farm operations suffered from the failure of reform legislation to create proper incentives. The problems were twofold. Farm managers themselves indicated that farm reorganization had not led to increased responsibility, labor discipline, or productivity.[101] Westerners blamed these results on the lack of reform implementation, Russian reformers blamed the "agrarian lobby" and other conservative forces within Russia. Neither Russian nor Western analysts understood why attempts at farm reorganization failed to achieve their objectives; and both failed to consider the incentives that reform initiatives created or failed to create. Ignored was the fact that price liberalization undermined efforts to transform farm operations. Production subsidies failed to stimulate efficient use of inputs. The trade policy environment removed the urgency to lower production costs.

Second, farm reform and transformation failed to create incentives that would motivate substantial numbers of farm members to create private farms. For a variety of reasons private farms that were created out of the farm reorganization process had poor locations, which affected their productive potential. Because social-economic conditions in the countryside were misunderstood by urban elites, reform legislation failed to create proper incentives that would bring results according to the small family farm model being advocated by Western advisors.

By the end of 1993, both Western advisors and Russian reformers agreed that if real farm reform was to be realized, something more drastic would have to occur. Within this context, following the forcible disbandment of the national parliament in October 1993, agrarian reform took a sharp turn and became radicalized. Following the defeat of Soviet-era conservatives, collective agriculture enterprises were targeted. The view emerged that collectivized agriculture was not an acceptable reform option—even if that form of labor organization was preferred by farm members. Western intervention in the form of advisors and organizations sided with radical reformers and also advocated drastic reductions in financial support to agriculture.[102] Such views argued for a "sink or swim" approach to food producers, asserting that farms would either adapt or fail.

Stage 2: The Radicalization of Farm Reorganization

Renewed attempts to privatize Russia's collectivized sector coincided with the publication of radical reform attitudes toward the collective sector in the spring of 1994. The publication of "analytical notes" from Yel'tsin's analytical center were important in that they reflected the type of analysis Yel'tsin was receiving and revealed the true intentions of the most radical core within his administration. In late March 1994, a document signed by the leader of the analytical center, P. Filippov, was leaked and published in a specialized, mostly proreform agricultural newspaper. The document was highly ideological and fanatical in its opposition to collective farms. The content of these notes was amazing for its mendacity and complete misunderstanding of reality in the Russian countryside. If these notes were in any way representative of the information being given to Yel'tsin, then it is no wonder that rural and farm officials, when they met with him at the end of June 1994, concluded that Yel'tsin was grossly misled and unaware of actual rural conditions.[103]

The document flowed from the "model" of the Russian countryside that was being advocated by Western institutions. This view favored a social organization of the Russian countryside that was based on small private farms. The March 1994 document was explicit in its use of state financial levers as a political weapon to undermine the existence of all farms with a collective labor organization. Even though farm reorganization had only recently concluded, the document suggested that all joint stock farms convene a general meeting during February–March 1994, the purpose of which would be the dissolution and distribution of all land and property assets to small collectives of private farmers (up to thirty individuals). Responsibility for the dissolution of farms would be given to raion officials (the document called for presidential decrees to "stimulate pressures for reform from below"). Farms that dissolved by the fall of 1994 would be granted credits for six months; all social-service objects would become municipal property. Agricultural enterprises that did not dissolve and disperse their assets would have all credits stopped.[104]

Following the publication of this document, Viktor Khlystun, who supported land reform and private farming, termed this document "absurd." Former Deputy Prime Minister Aleksandr Zaveryukha, in charge of agriculture, spoke out against the document, claiming it was not an official document and carried no executive authority.[105] Nonetheless, in early April 1994 it was reported that a draft decree had been prepared for Yel'tsin's signature that would "require the immediate division of joint stock farms and collective farms into small collectives of peasant private farms before the spring sowing."[106] Later, the regime tried to distance itself from this document, with one Roskomzem official saying that the "government is categorically opposed to deciding the question of farm reorganization by forcible means."[107] At about the same time another report indicated that Yel'tsin would not sign the decree because if the decree were implemented it would lead to "undesired social-economic consequences."[108] The leaking of this document caused such a reaction that in mid-Apriil 1994 a letter from the head of the Russian government was released which clarified that the document reflected the views of the preparers and not the government. The letter further stated that the government had no intention of forcibly breaking up agricultural enterprises into smaller production units.[109]

Views such as those reflected in the document prepared for the president's signature are important because they constituted a drastic departure from previous positions about the future of state and collective farms. For years Yel'tsin himself consistently maintained that only "unprofitable" state and collective farms be transformed, amalgamated, or closed. He did not target state and collective farms as a class but, rather, made farm performance the primary criterion. While there always was a fringe element among agricultural and political elites that opposed state and collective farms in general, a number of commentators, including both liberals and conservatives, did not think that state and collective farms should be destroyed.

In fact, administration views reflected in the March document represented a departure not just from its own past views, but from other "liberal" rural views. Supporters of private farming have often been critical of the state-collective farm system. But those supporters of private farming also have not advocated the rapid abolition of state and collective farms and in fact are not opposed to all forms of collectivism. For example, at the second congress of AKKOR in February 1991 it was noted that although newly created private farms were not receiving much support from state and collective farms, "the farmer movement is by no means aimed at confrontation with the kolkhoz system, at its collapse. Mutual understanding and mutually advantageous cooperation are more important."[110]

In late 1994, V. Bashmachnikov, the president of AKKOR, was quoted as saying that "only large collective farms can supply the country with food . . . and therefore the state should support collective farms."[111] Bashmachnikov later denied making this comment, but he never did explain where it came from. Another liberal commentator known in the West, Zhores Medvedev, published a long article in *Sel'skaya zhizn'* in which he canvassed the entire agricultural spectrum, including the fate of state and collective farms. He, too, concluded that it would be dangerous to believe private farmers could feed the country more cheaply and more quickly than former state and collective farms.[112] In fact, by the end of 1996 private farms contributed 2–3 percent of the nation's food supply (measured in ruble value). For that reason, government agricultural officials such as Khlystun and his successor, A. Nazarchuk (1994–1996), as well as former Deputy Prime Minister Zaveryukha believed that state and collective farms should continue to exist but urged the transformation of their operations.

Other voices expressed concern over decollectivization without an economic purpose. For example, the vice president of the Russian Academy of Agricultural Sciences noted that "many collective and state farms have achieved excellent results.... Surely it would be unwise to break up everything that has been created."[113] An American economist warned that "the naive promise of privatization and decollectivization of agriculture, without careful thought given to the accompanying institutional changes beyond the farm gate, will retard rather than revitalize the Russian economy."[114] Another analyst noted that "almost all of the society-wrecking advice given to the Eastern European countries points them in the opposite direction of successfully developing capitalist countries. An extreme and naive application of free-market ideologies has produced structure-determining policy decisions which do not engage product or factor market realities."[115]

Thus, a much more assertive stance vis-à-vis state and collective farms and their successors began to emerge in the post-October 1993 period. These views represented not only a departure from past policies toward collective farms but also a departure from the views rural liberals and proreform elites had held for years. The Russian government, with the help of Western agencies, drafted and began to implement a farm privatization program that would move farms toward radical reform. It is in that context that farm privatization experiments in Nizhegorodskaya Oblast are noteworthy.

The Nizhniy Novgorod Model of Farm Privatization

Russia's land privatization program was worked out between January and June 1993, and then in November 1993 the first privatization experiment to facilitate the transfer of land to farm members was undertaken in Nizhegorodskaya Oblast (hereafter referred to as Nizhniy Novgorod Oblast, for short). The so-called Nizhniy model was a procedure to transfer land and property into the hands of farm members by privatizing farms.[116] The development of this model was based on the help of the International Finance Corporation (IFC), an affiliate of the World Bank, the British Know-How Fund, and support from the Canadian government. This help was solicited by the USSR government as early as November 1991. Although some Russian input was offered, primary credit for introducing the program goes to foreigners. According to the governor of Nizhniy Novgorod Oblast, Boris Nemtsov, "foreigners did what Russians could not, they took the idea [of farm reform] to its logical end."[117]

The initial farm to be privatized was the former Pravdinskiy state farm in Balakhna raion, located sixty kilometers to the north from the city of Nizhniy Novgorod. This farm had previously reorganized itself into a joint stock farm. Two closed auctions—that is, open only to members of the association Pravdinskaya—were held. The first was the land auction, held November 9, and the second was the property auction, held November 12, 1993. During the privatization process the farm's 3,600 hectares were divided among 634 persons who had legal claim to a land share (being members of the farm). The land available for "purchase" was divided into 210 lots, and during the auction 172 of the lots had only one bidder, but there were a few cases in which competitive bidding led to a doubling of the starting price. After the auction a total of 151 hectares remained unallocated. Following the land auction, an auction of the farm's property was held.

As a result of the distribution process by the auction, twelve new farm businesses were formed. The size of each enterprise varied according to the number of persons who chose to participate in the business and the size of their entitlements. The largest new agro-enterprise acquired 915 hectares and was to become a mixed farm for milk, meat, and grain. The smallest new enterprise consisted of only 5 hectares and was to be a small family farm.[118]

Following the Pravdinskiy experience, other farms in the oblast underwent farm privatization and the auction process as part of a model program to test the process. In March 1994 Prime Minister Chernomyrdin, deputy prime ministers A. Zaveryukha and A. Chubays, minister of agriculture V. Khlystun, personnel from other ministries, and administrative heads from thirty-seven regions in Russia arrived in Nizhniy Novgorod Oblast to witness land and property auctions in the Sixtieth Anniversary of October collective farm, located in Gorodetskiy raion. It was reported that a smiling Chernomyrdin watched the land and property auctions.[119]

After Prime Minister Chernomyrdin returned from Nizhniy Novgorod Oblast, he announced that the pilot scheme used in Nizhniy would be the basis for nationwide farm privatization, and on April 15, 1994, he signed a resolution approving the Nizhniy variant for the rest of Russia.[120] The IFC was asked to help draft the legislation that would establish the national program using the model documents.[121] At the end of July 1994, government resolution no. 874 was signed, approving the Nizhniy model of farm privatization and making it the official policy of the Russian government. The processes of farm privatization flowed from the model documents drafted by the IFC, with help from other Russian contributors.[122] To facilitate the introduction of the Nizhniy model to other areas, in February 1995 Prime Minister Chernomyrdin signed a government resolution that codified the process for dividing land and property shares among farm members along the lines of the model program.[123]

In fact, the model was adopted elsewhere. Whereas only six farms were privatized during 1993 (all in Nizhniy Novgorod Oblast), by the end of 1996, 179 farms in six oblasts had privatized, including 127 farms privatized in Nizhniy Novgorod Oblast, 30 in Orel Oblast, nine in Riazan, six in Rostov Oblast, five in

Kirov, and two in Volgograd Oblast. In addition to the six oblasts where farm privatization had been completed, another five oblasts had a list of farms that had formally applied for privatization based on the Nizhniy model. These oblasts included Krasnodar, Voronezh, Tula, Samara, and Moscow. In addition, by winter 1996 the second edition of the privatization program manual had been distributed to all 26,000 collective farms in Russia.[124] Also, by late 1995 the IFC and USAID-funded training seminars had been given in four oblasts in order to transfer the skills and expertise to local raion administrations needed to implement the farm privatization model.

How the Nizhniy Model Works: Farm Privatization and Auctions

The process for farm privatization proceeds in several distinct stages. We should note that the decision to adopt a farm privatization program is separate from the decision to reorganize a state and collective farm. Farm reregistration and reorganization precedes privatization, a procedure we examined in previous sections.

To review briefly, during the first step in the reorganization process, a commission on land privatization and reorganization was created in each farm, in which the following people would serve: the director or manager of the farm, other leading members of the farm, a representative of the raion land reform committee, a representative of the raion agricultural administration *(upravleniye sel'skogo khozyaystva)*, creditors to the farm (if any), and representatives of the local government. These commissions were responsible for all tasks relating to the reorganization of the farm, including establishing lists of entitlement (share) holders and distributing land and assets to entitlement holders upon request. During farm reorganization, the farm commission convened a meeting of farm members in order to present the reorganization plan, a timetable for reorganization, and to present the list of shareholders. After the general meeting, a land and property inventory was conducted. Following an inventory and any corrections in the entitlement list, the farm commission calculated the size of land shares, which were then distributed to those who desired to leave the farm.

Farm privatization takes the reorganization process further. A farm that has decided to privatize must first undertake preparatory work before land and property can be divided. Once a farm has adopted the privatization plan, the farm commission on land privatization composes a list of who is eligible to receive land and property shares.[125] Legislation from September 1992 regulated how land was to divided and assigned to farm members, and according to this government resolution land shares were to be distributed equally among all members of the farm. Those who qualify for land entitlements include employees of the farm, pensioners residing on the farm's territory, employees in nonproductive activities who reside on the farm's territory, and permanent employees who lost their jobs after January 1, 1992. The list of those who qualify to receive property shares is the same with the exception of those employed in nonproductive activities.[126]

The value of a land share is expressed in an amount of land. Three units of measurement are used to calculate land shares, and each farm member receives a certificate that reflects these three units of measurement. The first unit of measurement is in hectares, and this is calculated by dividing the total land area of the farm, according to the inventory, by the number of land entitlement holders. The second measurement is in rubles. The ruble value is expressed as the normative land price (fifty times the normative land tax) multiplied by the size of the land share.[127] The third measurement is in "ballohectares." The term "ballohectare" has no direct translation but in essence means land points. This measurement of land quality is based on a hundred-point scale, taking into account land quality, the distance from the road, and from population points.[128] The intent of land distribution was to give equal shares of land to all farm members using these three measures.

After land shares are determined, a property inventory is taken of all farm property, including fixed, current, and cash assets. Property is divided into "lots." Property lots are formed on the basis of independently functioning units, and assets that cannot operate independently should be listed as one lot. For example, the law states that all tractors, automobiles, combines, and so on, are placed into separate lots. If a piece of machinery can only be used in combination with another piece, then the two are grouped together in a single lot. Garages for automobiles and tractors are placed in separate lots. Equipment, buildings, and materials for processing and repair projects are also put in separate categories. Grain dryers and their equipment are placed in one lot.[129] Compound lots list each component and its value, as well as the value of the whole. So-called current asset lots—such as seed, fertilizers, feed—are listed separately. For example, a warehouse with equipment and seed would be listed as two lots: one for the warehouse and equipment, and one for seed. Standing crops are listed separately but not sold. Instead, they are obtained by whoever obtains the land on which they are located. In this case payment is made in property shares at the value equal to the value of the standing crop.

In order to "purchase" property lots, farm members use property shares. Property shares are calculated by the farm commission. Property shares are determined by applying a coefficient to the work contribution of an individual, taking into account length of service and size of income. The coefficient is derived by dividing the total property fund by the total work contribution of all property entitlement holders. Each property holder's work contribution is multiplied by the coefficient and is expressed in "share rubles."[130] Share rubles are then used to bid on property lots at auctions.

The methodology for determining work contributions is chosen by the farm commission and needs to be approved by a farm general meeting. Although there are several methods for valuation of work, some examples include:

1. averaging the salary for the past five years and multiplying that by the number of years worked;
2. using the current salary and multiplying that by the number of years worked; or
3. summing the salaries earned by an individual for all years worked.[131]

Each of these methods yields a "work contribution" measured in rubles. As one article noted, pensioners often ended up with large property shares since they had served the farm the longest.[132]

All of the above represents the conclusion of the preparatory stage for farms undergoing privatization. Once preparatory work is completed, a general meeting of farm members is convened. This meeting is considered the supreme administrative body of the farm and will settle such issues as amendments to the list of entitlement holders, approval of the method for calculating property shares, confirmation of the right of the farm commission to implement reorganization, approval of the privatization program of the farm, and setting a timetable for share distribution and the date of the auction.[133]

Preparation of Certificates

The second stage is to prepare land and property certificates for those who are eligible. Before land is distributed, the farm must receive a Certificate of Landownership, which is a legal document serving as a land deed, that is, specifying land that the farm owns. In order for a farm to receive this Certificate of Landownership, the minutes from the general meeting of farm members, the list of entitlement holders, and an application from the chairman of the farm commission must be delivered to the raion Committee on Land Resources.

The next step is to ensure that each individual receives his or her certificate of ownership rights. This may be done by the raion Committee on Land Resources, or the farm commission may prepare temporary individual certificates. Whichever method is used, the farm commission is charged with making sure that all entitlement holders on the list receive their certificates. Further, the certificates must contain the "land-point" valuation of land, expressed in ballohectares, as described above.

Unlike land certificates, certificates for property shares can only be prepared by the farm commission. After determining the size of each individual's shares, property certificates contain the following information: the certificate number, the certificate holder's name, the name of the former enterprise, the size of the property share in entitlement rubles, the date of the general meeting at which the method of property valuation was approved, and a stamp and signature from the chairman of the farm commission.[134]

Share Distribution

The third stage is to distribute land and property shares. A meeting is convened, the purpose of which is to distribute each qualified person's land and property certificates. The meeting at which land and property shares (or entitlements) are distributed is to be well publicized; announcements should indicate the date, time, and place of the meeting. All shareholders are to be invited. In order to record the receipt of certificates, one register for property and one for land shares should be present at the meeting, with each showing the certificate number, the name of the shareholder, and a space for the recipient to sign for the receipt of his or her shares. Thus, during the farm privatization process the question of who will receive what is worked out in advance, and the actual process for the distribution of shares is simply the culmination of all work completed up to this stage.[135]

The Auction

Preceding the auction, all land and property to be "sold" at the auction should be organized into lots. A lot is defined as a grouping of property or a particular piece of land. To each lot should be assigned a value. For land, the value is expressed in ballohectares. For property, the value is expressed in share rubles. The value placed on each lot will determine how many ballohectares or share rubles will be necessary for a person or group of persons to acquire that lot at the auction. The chief economist and chief accountant are responsible for formulating property lots. The chief agronomist is responsible for land lot preparation, including arable and pasture lands. Land lots may vary in size. Each land lot is numbered, and this number serves as the lot number to be bid upon during the auction. Every land lot should contain the following information: the location of the land, the crop rotation, the area in hectares, the value expressed in ballohectares, and the starting price, again expressed in ballohectares. The farm commission has authority over the final version of land and property lots. Following the preparation of lots, lots for auction should be posted one week prior to the share distribution meeting.

The word "auction" is actually a misleading label because, as we have seen, farmland and property is divided up among farm members prior to any auction.[136] Auctions are really a means to distribute collectively held assets. Prior to an auction, shareholders know the value of their entitlements, expressed in ballohectares or share rubles.[137] However, they do not know the exact location of their land, or exactly what property will be possessed. Auctions therefore finalize the distribution of assets among the shareholders. Prior to an auction each participant is to receive information about the place and time of the auction, the rules of the auction, a list of auction participants that indicates the size of land and property shares, and a list of land and property lots with the starting price. An auction proceeds in three rounds.

Once land and property shares have been distributed, but prior to an auction, share recipients have the right to decide what to do with their shares, and a number of options exist. For example, shareholders may combine their shares in order to organize a new agricultural business, collective in nature but smaller than previous collective farms. This option requires that shares be combined with other individuals' shares in order to raise sufficient "capital" to equip and run the enterprise. Share recipients may form individual peasant farms—usually, though not necessarily, based on family relations. Or an individual may start his own enterprise and work alone, using his own shares to acquire land and property. The main point is that after the distribution of shares, individuals have the freedom to choose what they want to do with their shares.

After this decision (whether to join a collective group or work individually), an application for specific land and property must be filed with the farm commission.[138] This step requires that the new group or individual examine the list of land and property lots in order to ascertain what they wish to obtain. In the case of a group, they will need to determine if their combined entitlements are sufficient to purchase desired assets. One may only bid (that is, submit an application) for a lot if one possesses sufficient shares to do so. Only by submitting an application does the group or individual retain the right to bid on a specific lot of land or property during the first two rounds of the auction. Applications, therefore, serve as official claims for specific land or property. Only those shareholders who have applied for specific lots may participate in the bidding for those lots. For this reason, auctions serve essentially as a means of conflict resolution among farm members, and an auction actually only has limited competition.

Both land and property auctions consist of three rounds. The first round of the land auction puts up for bidding those land lots that were applied for by more than one applicant. Bidding for land is conducted in ballohectares, beginning with a starting price determined by the farm commission. The winner is the applicant who bids the most ballohectares, or land points, per hectare. If the winner has enough ballohectares to obtain the lot, the entire lot must be purchased. In "exceptional" cases, an applicant may obtain only a portion of a lot, corresponding to the amount of ballohectares he or she possesses.[139]

The second round of the land auction is conducted for those land lots with only one applicant who may purchase the lot at the posted value (again using ballohectares). The third round of the land auction is for land lots that were not applied for by any applicant. These lots are offered for bidding to any applicant, and starting prices are adjusted downward.

The same general formula is followed in the property auction, which is held after the land auction. In the first round specific property lots with more than one applicant are auctioned, this time using share rubles. The winner is the highest bidder. The second round is for property lots with only one applicant. The third round is for property lots for which there were no applicants, and again the starting price is adjusted downward.[140]

Following the auction, there is a two-week period during which exchanges may be made. For the most part, however, all that remains to be done is to oversee the transfer of land and property to the new enterprises. The farm commission has responsibility for this task as well. An Act of Transfer and Acceptance document records the transfer of property from the balance sheet of the old enterprise to the balance sheet of the new enterprise. All property should be accounted for using this document, and all transfers should occur "as soon as possible" after the auction.[141]

An Assessment of the Nizhniy Model

The Nizhniy model was intended to introduce a "bottom-up" variant of land reform by allowing farm members to choose what activities to engage in and how to organize themselves.[142] In point of fact, the bottom-up model was introduced from above, but nonetheless the model attained widespread international attention, domestic support, and domestic attention. During a trip to Washington, D.C., in March 1995, former minister of agriculture Nazarchuk expressed support for the Nizhniy model. Early in 1995, IFC and British Know-How advisors worked hard with the Chernomyrdin government on an agricultural strategy that would broaden the adoption of the Nizhniy model. Coverage of privatized farms appeared on both Russian TV and Ostankino; and four auctions in Rostov, Orel, and Ryazan were given coverage on the newscast Vremya. Moreover, newspaper reports on privatization results in the oblasts noted above suggest that at one level the model has done what was intended. Several of those reports indicated improved discipline and decreased theft, absenteeism, and alcoholism.[143] These benefits undoubtedly flow from a system that makes an individual more responsible for his actions.

Early results from farm privatization in Nizhniy indicate that many of the new agricultural enterprises were more specialized than their state and collective farm predecessors had been. Because of orders from Moscow during the Soviet period, farms sometimes attempted to grow or raise a variety of plant and animal produce that may not have been adapted to particular natural conditions.[144] Specialization increases the chances for greater productivity and efficiency but does not in and of itself guarantee it. Although all food producers suffered from a hostile economic environment, privatized farms in Nizhniy Novgorod as a group had higher cereal, potato, and milk yields than the oblast average. Privatized farms also took less land out of production than the average farm in the oblast, and livestock holdings decreased by less than the oblast average.[145]

To shed further light on the changes ushered in by the Nizhniy model of farm privatization, a survey of thirty-nine reorganized and twenty-seven non-reorganized farms in Orel, Nizhniy, and Rostov Oblasts was conducted by the Agrarian Institute (Moscow) during the summer of 1996. According to this survey, the quality of farm leadership had improved on reorganized farms, and farm leaders took more interest in the opinions of workers when making decisions. There was better work discipline and less drunkenness when compared to nonreorganized farms, and in general nonreorganized farms reported more problems in all aspects of their operation.

Reorganized farms also received somewhat better economic results: profits were somewhat higher in Rostov Oblast on reorganized farms. Workers in reorganized farms derived less income from personal plots and more income from farm labor in Rostov Oblast.

However, in Nizhniy Novgorod Oblast workers on third-wave reorganized farms spent more time working their personal plots than workers in nonreorganized farms. Also, income differences between reorganized and nonreorganized farms were insignificant—less than 10,000 rubles a month—and in Orel Oblast farmworkers in nonreorganized farms actually earned more on average than those on reorganized farms during the second wave of privatization.

Workers in reorganized farms were more likely to take advantage of their right to lease their landshares, whereas nearly no one did so in nonreorganized farms. The most common practice was to lease the landshare back to the farm. However, respondents to the survey indicated that there was little competition among potential leasees, a fact that often allowed the farm manager to dictate the terms of the lease. For this reason, more respondents from first- and second-wave privatized farms in Nizhniy Novogord were not satisfied with the terms of their lease than were satisfied. In both reorganized and nonreorganized farms, less than 5 percent of respondents in Rostov Oblast expressed a desire to sell their landshares.

When considering these results, we should note that significant differences were evident among the "waves" of reorganized farms, with pilot farms performing much better and achieving all-around better results than farms that privatized during the subsequent first, second, or third waves, with third-wave farms in some respects similar to nonreorganized farms.[146]

There are also significant obstacles to overcome if the model is to be effective in transforming Russian agriculture. First, we should note the very small number of farms that have expressed an interest in this form of privatization. Assuming that the 1995–1996 rate remains constant, it would take more than 270 years to privatize all collective and joint stock farms in Russia. Second, it remains unclear if the privatization model will be adopted by Russian farms on their own—and will spread—without the active intervention of American, British, and Canadian governments. The IFC and its advisors have no plan to be involved in the implementation of the model nationwide, so farm receptivity and the training of domestic advisors are crucial variables if the model is to have appeal in areas the IFC has no presence. In certain places where local officials did try to implement reform, IFC documents indicate that "farm privatization has been carried out with more enthusiasm than careful attention to legal requirements."[147]

The most crucial question for farm privatization is whether the Nizhniy model facilitated more effective use of land and made agricultural enterprises more productive, more efficient, and less costly—the basic economic goals of agrarian reform. One of the criticisms of agrarian reform had been that land was transferred to new users and forms of ownership without considering who was able to use the land most effectively.[148] Thus we are justified in asking whether the Nizhniy model holds long-term promise for Russian agriculture.

When considering the efficacy of the Nizhniy model, we should note that it was introduced in conditions common to past Soviet economic experiments: special resources were made available for the experimental farms, advantages that were not available to nonparticipants in the model. Nizhniy farm privatization occurred as a result of significant Western and state intervention in order to create favorable conditions, and the IFC spent millions of dollars in an effort to help several farms in Nizhegorodskaya Oblast privatize.[149] Although many of the costs associated with implementing the Nizhniy model were onetime costs such as legal costs, development of documents, and land titling and mapping, IFC officials have acknowledged that the model would be cost prohibitive if it were to be applied throughout the nation. Because neither the IFC nor its British or Canadian counterparts have the financial resources to fund nationwide farm privatization, the goal is to institutionalize the privatization process, which leads to our next question.

Boris Nemtsov, the former governor of Nizhniy Novgorod Oblast, argued that the "leaders of these farms [that privatized] themselves asked for the experiment."[150] The question is, Why? To what can we attribute the interest in farm privatization when a similar inclination was not shown during the earlier stages of farm reorganization? The answer is that the oblast government itself provided a strong motivation for farms to privatize and that special financial advantages were offered to farms who chose to privatize.

The first farms to privatize were scheduled to be declared bankrupt and so farm members' only choice was between privatizing or closing the farm. Although financial data is not available for all farms that privatized, there is strong evidence that privatized farms in other oblasts also were extremely weak and most likely would have closed. For example, the Forty Years of October farm in Ryazan Oblast, which undertook privatization during 1994, had not paid wages in more than a year prior to privatization and had not received any credits for three years; all its property had a lien on it for past debts; the number of cattle had declined by one-half; and the milk output per cow was less than one-half of the raion average. Under these conditions the farm manager "asked" that the farm privatize along the lines of the Nizhniy model.[151]

In Nizhniy, to provide further incentive for the "right" decision, during the spring of 1994 new agricultural enterprises that formed as a result of farm privatization were given subsidized credits from the oblast budget, to the sum of 200,000 rubles per hectare. Oblast support dropped to 50,000 rubles per hectare by winter 1994 but increased again to 150,000 rubles per hectare by spring 1995.[152] The term was for three years at 13 percent interest, at a time when other agricultural borrowers—including private farmers—had to borrow at rates of 210 percent per annum from Rossel'khoz bank.[153] Various tax advantages were also given to agricultural service and processing enterprises that adopted the privatization model described herein. Thus, farms that undertook privatization were given strong incentives to do so in the form of economic advantages that other farms did not have.

Furthermore, it was reported that in 1995 Nizhniy Novgorod Oblast intended to spend 21.8 billion rubles "in the creation of conditions for effective activities" of privatized farms—this sum equaled 12.5 percent of the entire oblast budget.[154] As a result of these financial advantages, there are significant questions as to whether the somewhat better economic results of privatized farms were simply a result of high levels of targeted financial assistance for each farm, in addition to the priority supply of inputs and material resources that inevitably flow to experimental farms. Thus it is not clear what is really being demonstrated on Nizhniy farms—the superiority of privatized farms or the production capacity of farms when given high levels of targeted financial assistance and sufficient input supplies.

Although financial "selective incentives" are precisely the type of levers the state can use to elicit desired behaviors, it is also necessary to note that it is the cost of these advantages that makes the Nizhniy model difficult to duplicate on a wide scale. Nemtsov revealed that the privatization experiment for six farms cost the oblast budget 46 million rubles (about $46,000) in 1993.[155] Costs have subsequently declined and now average about $3,500 per farm. In addition, 2 billion rubles were allocated in 1994 from central funds to the oblast to help conduct further privatization (about $1 million).[156] Thus, questions remain about the ability to duplicate the Nizhniy model on a large scale throughout the country, given the financial concessions that are necessary to jump-start the process. Indeed, a primary restraint for widespread implementation of the Nizhniy model is the lack of local resources to support farm privatization, and it was admitted that the Nizhniy model was limited to areas where the West had a presence and was able to financially support privatization.[157]

The Nizhniy model also has come under criticism from Russian agricultural economists. In particular, Nikolay Radugin criticized the model for breaking up agricultural enterprises. He argued that "the whole world understands and views as an axiom that the larger the scale of production the more profitable it is and the smaller the plot, the higher the production expenditures, and, it would appear, the products themselves are more expensive." The "smaller producer can never withstand competition in the cultivation of cereal crops."[158]

There is no question that the Nizhniy model went the farthest in implementing a system of property rights and farm privatization. However, it is unrealistic to think that property rights and farm privatization can improve agricultural performance without competitive market channels. Even a system of property rights does not ensure the creation of a market environment. Privatized, small-scale independent farming operations also suffer from economies of scale disadvantages. The success of the Nizhniy model depends ultimately on the macroeconomic environment and the nature of financial policy toward agriculture adopted by the state. Privatized, independent farming operations are vulnerable to input and output monopolies, as are collective agricultural enterprises. Privatized, independent farming operations are also vulnerable to disadvantageous terms of trade, as are collective agricultural enterprises. The importance of macroeconomic policy on farm performance was expressed by Boris Nemtsov in 1994, who stated that, "All farms, independent of whether they are reorganized or not reorganized, now are situated in extraordinarily complex conditions."[159] Any benefits that may accrue from farm privatization are mostly lost until and unless competitive market structures are implemented.

Conclusion

The impetus for reform of the collective agricultural sector came from above, with state-defined institutions governing the nature and form of farm reorganization. This meant that, similar to the Soviet system, the state was able to define the range of reform options. Furthermore, through the definition of legislation, the state exerted a primary and direct influence on the operations of farms.

The purpose of farm reorganization within the collective sector was to facilitate cost effectiveness. Although farms reorganized and many changed their legal status, reform legislation introduced in December 1991 failed to transform farm operations. One explanation for this result was that the regime was hampered by the need to meet current food requirements, and therefore reform institutions were undercut by retreats that allowed farms to "reform" without really changing anything. Farm responses to mandatory reorganization exploited loopholes in reform legislation.

More important, farm reorganization failed because the proper incentives were not created for farms to change the way they operated. Macroeconomic and rural social policies needed to support both new and existing agricultural enterprises, giving them the opportunity to transform their operations and become more efficient. Instead, the economic environment outside the farms was hostile, and the few rural services that existed came only from collective sector farms. For these reasons relatively few families chose to depart the parent farm.

Radical attitudes about the collective sector came to the forefront in response to failed farm reorganization, signifying an important evolution in the approach to the collective sector. Emanating from a combination of Western advice and an internal political impulse to de-Stalinize Russian society, reform policy evolved from closing unprofitable state and collective farms to closing all farms with collective labor organization, even those that had undergone reorganization and legally were no longer state enterprises. As a result of the change in basic attitudes, farm reform was used to address political—not economic—problems.

Despite the political campaign to privatize collective enterprises, it is not clear that making agriculture more efficient required that farms be broken up into smaller units. World trends are witnessing the dying out of the small family farm, yet that is the model that has driven Russian agrarian reform. It is unlikely that privatization and property rights alone can improve Russian agriculture. It was necessary to create an economic environment that would stimulate and facilitate farm transformation, even if that meant higher financial expenditures in the short term (as with defense conversion).

The most important economic necessity for Russian agriculture was to create market structures. Market structures create incentives that allow and motivate farms to adapt, to become more efficient, and to make decisions that control costs. However, state influence hindered market competition and thus farm adaptation. Marketization of agricultural processes, therefore, did not occur. Instead, Russian farms operated in a monopolist, nonmarket environment with an inadequate infrastructure, which undermined the potential benefits of privatization. Although marketization was not accomplished, farm privatization was waged as a political campaign to undermine the economic and political power of the old rural elite, the farm managers.

4

Financial Levers and the End of the Social Contract

Reform in the collective sector was pursued as a two-part strategy. On the one hand, policies and rural institutions were changed (see chapter 3). Second, state financial levers were used to enforce reform initiatives and to change the rural economic environment. The importance of state financial levers was that they affected not only enterprises but food producers and rural dwellers in general. State financial interventions, therefore, had both economic and political effects. This chapter will address four main questions in investigating the nature of state financial interventions:

1. Did reform legislation create the "right" incentives that would address the problems in the collective sector bequeathed by the Soviet system?
2. How did state financial interventions affect the rural standard of living?
3. What were the economic effects of state financial interventions?
4. What were the political effects of state financial interventions?

Ending the Rural Social Contract

The use of financial levers to enforce the reorganization and privatization of the collective sector implied the end of the social contract toward the rural sector. The social contract in the countryside had benefited state and collective farms and their workers. Because the social contract had been linked to rural egalitarianism and high production costs, it was economically necessary to move away from those policies. The task for the postcommunist regime, therefore, was to try to reap benefits from agrarian policies that moved beyond the social contract, but to do so in a way that would not engender rural opposition or have a deleterious effect on farms' production capacity or on rural living standards.

The social contract approach argues that an implicit exchange occurred between regime and society, in which each side delivered something the other desired, an approach used to apply to the whole of the regime-society relationship.[1] According to the social contract approach, the Soviet regime pursued policies that improved the availability of consumer goods, increased wages, provided job security, and increased the standard of living for the population. In turn, the population delivered political compliance and quiescence.

The social contract is normally applied to the Brezhnev era, but one could argue that such a policy was first begun under Khrushchev and expanded under Brezhnev. There were problems with the social contract approach, which failed to distinguish between urban and rural populations, and ignored the fact that the urban-rural relationship continued to be characterized by inequality.[2] Nonetheless, there is no question that the rural standard of living improved dramatically from 1960 to 1985,[3] a period of political stability and rural quiescence that stands in sharp contrast to the rural upheaval and protest experienced during the Stalin years.[4]

The political implications of the social contract approach were twofold: first, political support for the regime was contingent upon the regime's providing economic and social security. Second, the leadership was constrained in its resource allocation by the need to provide social welfare, income equality, and a host of other aspects of economic security. If the state did not deliver these "goods" it risked discontent and the loss of legitimacy.[5] In essence the Soviet state evolved from one in which political support was compulsory and compelled to one in which political support was conditional and performance-based, a basic belief of John Locke although none of the analysts who used the social contract as a framework were quite so explicit in making this connection.

The Russian Social Contract

Russian agrarian reform began with government promises to revive the countryside. With a view toward correcting social deficiencies in the rural sector that remained from the Brezhnev period, the Second Congress of People's Deputies of the RSFSR in December 1990 adopted the "Law on the Social Development of the Countryside" and a resolution, "On the Program for the Revival of the Russian Countryside and the Development of the Agroindustrial Complex," both of which were signed into force by then chairman of the

RSFSR Supreme Soviet, Boris Yel'tsin. The resolution on the revival of the countryside pledged the "priority" development of the agroindustrial complex. Budgetary appropriations would be devoted to the construction of rural housing, health, school, and cultural buildings; and for the construction of enterprises of trade, roads, and intrafarm gasification, electrification, and communications. To fulfill these goals the Russian government promised to allocate not less than 15 percent of the national income of the RSFSR to the agroindustrial complex, beginning in 1991.[6]

The Law on Social Development of the Countryside spelled out in more detail steps that the government would take. The law repeated previous pledges for the "priority" distribution of state capital investments for rural social development and provided for annual financing for a range of rural projects. These projects included the development of private farms; funding for various aspects of rural infrastructure; construction and repair of food-processing plants, trade facilities, and retail food outlets; measures for land improvements; the preparation and training of skilled cadres to manage a reformed rural sector; subsidies for unprofitable products grown on experimental farms and enterprises; and the provision of material-technical resources to agriculture in order to meet state orders during the transition period to market relations.[7]

In addition, the social development law provided for price parity between agricultural and industrial prices and for compensation to the agricultural sector to cover the increases in wholesale prices and services that were to go into effect January 1, 1991. It also extended certain tax advantages to private and collective farms. Many of the points in this law were later repeated in a resolution that was signed by Yel'tsin in April 1991, entitled "On the Political and Social Situation in the RSFSR and Measures for Exiting the Crisis," which was adopted at the Third Congress of People's Deputies in late March–early April 1991.[8] These early documents, which constituted the legislative basis of the regime's approach to rural social development, indicated that the regime hoped to cushion the blow of the market and provide for some degree of social protection for the rural sector.

Even after market reforms were launched, the government continued to promise to fund the revitalization of the countryside. For example, in July 1992 a "general agreement" was published between the Russian government, the Agrarian Union of Russia, and the Russian Council of Collective Farms and Other Forms of Farming. This agreement began by pledging the Russian government to fulfill the legislation and decisions of the Second and Third Congresses of People's Deputies of the RSFSR on land reform, peasant farming, social development of the countryside, and financial support for food producers. In addition, the agreement indicated that the government would provide subsidized credits to private farmers as well as state and collective farms; promised to pay previous state debts for products purchased during 1990–1991 as well as all previously owed compensations; promised future compensations for fuels used to grow food products sold to the state; freed all food producers, including those using leased land, from land taxes on unclaimed land; and promised a number of forms of support.[9]

Were these governmental promises and policies sincere? Allegations of purposeful deceit by reformers do exist, but we have no way of determining the government's sincerity with any degree of certainty.[10] There is no question, however, that a substantial gap existed between state promises and reality, and for that reason the real story is one of unfulfilled state promises, laws, and policy statements. As a result of unfulfilled promises, rural living conditions deteriorated, farm finances suffered, and this in turn affected reform potential.

The Politics of Ending the Social Contract

Yel'tsin moved forcefully to end the social contract in a way that Gorbachev never did. There are several possible reasons to explain the rapidity with which social contract policies were abandoned. First, rural dwellers were perceived to be politically nonthreatening.[11] When "new" officials came to power in Moscow in late 1991–early 1992, they inherited a tradition in which rural interests were poorly organized, compliant, and malleable. There had been no rural resistance to state interventions since the early years of collectivization. Various reforms—from the Virgin Lands to leasing and privatization—had been implemented from the top down without rural protest, so there was little precedent to believe that the future would be different. The extent to which financial conditions turned against the rural sector suggests state leaders believed rural voters not to be a potent political force. Extracting resources from the countryside, procuring food but not paying for it, not remunerating farmworkers, allocating only a fraction of budgetary allotments to agriculture, and other forms of state discrimination were attractive for the obvious economic advantages that accrued. That is, because the state was able to impose unequal terms on the agricultural sector, it was able to divert scarce resources—monetary and otherwise—to other preferred sectors of the economy, in effect propping up other sectors at the expense of agriculture. For example, during the first six months of 1992, subsidies to industrial enterprises increased eightyfold, so that by spring 1993 only one industrial enterprise had been declared bankrupt.[12]

A second reason for the sharp turn against the rural sector is that Russian reformers misunderstood the nature of politics and underestimated the necessity of building rural support. Quite simply, they forgot it is impossible to reform the rural sector with only urban constituents. The lesson from the Third

World—specifically Latin America and Africa—is that successful governments build rural political support. One popular method in Africa was to build rural support by differentiating among farms and farmers. Some regions, farms, and farmers benefited more from reform, while others benefited much less, or perhaps not at all. Through selective incentives—whether it be in taxation policy, investment projects, or subsidies for inputs—the state was able to build a base of rural political support. In effect, a patronage system was created, in which the continued receipt of preferential treatment was contingent upon political support. Robert Bates found this pattern of government intervention in Ghana and other African states, whereby "the agricultural programs of African governments thus become basic units of rural political organization."[13] However, by pursuing financial policies that discriminated against all food producers, the Russian government undermined its bases of rural support and in fact united proreform rural groups with antireform rural groups on financial issues.

Third, and most important, the end of the social contract was inherently linked to the campaign to privatize the collective sector. The social contract had been inherently linked to benefits extended to the rural collective sector. Movement away from the social contract, however, became an ideological campaign rather than a search for a more effective agricultural sector. This was indicated by the publication of "analytical notes" from Yel'tsin's analytical center (see chapter 3). The March 1994 document was highly ideological and was explicit in its use of state financial levers as a political weapon to undermine the existence of all farms with a collective labor organization. According to this document all farms with collective forms of labor organization were to dissolve by the fall of 1994. Those that did would be granted credits for an additional six months, after which all state financing would cease.[14]

The origins of farm privatization as a political campaign had two sources. The first source was domestic in nature: the drive to de-Stalinize Russian society and all institutions therein. The radicalization of farm reform policy coincided with the need for macroeconomic stability, and officials such as Yegor Gaydar and Boris Fedorov argued for less state spending, fewer subsidies, and reduced state support. There is no question that these steps were necessary in order to control inflation, get control of the deficit, and achieve macroeconomic stabilization, which would facilitate the transition to a free market economy. The pretext for farm privatization was greater economic efficiency, but the "efficiency argument" became a cover for the ideologically driven goal of privatization for political purposes.[15] As time passed the post-Soviet regime in Russia moved away from the question, How do we make agriculture more productive and efficient? to the question, How do we destroy rural institutions that are ideologically anathema? Because state and collective farms were relics of the

Stalin era, they were ideologically anathema. Only in the realization of the "moral imperative" to destroy those farms could Russian agriculture throw off the chains of its Soviet history.[16]

The second source came from Western governmental and institutional advice—such as that from the World Bank and its affiliates and the International Monetary Fund—that "collective is bad and private is good."[17] This advice attempted to duplicate American agricultural institutions in Russia. Western advice, which pushed farm and land privatization, was tied to much needed credits, loans, investments, and technical assistance, even to the extent that Russian budget makers built into the budget the loans from international organizations in an attempt to achieve macroeconomic stability and deficit control.[18] Concerns regarding the replicability of the American experience were ignored. For example, when the head of the international economics section at the Academy of National Economy in Moscow, Vladimir Popov, argued that "it is unlikely . . . that Russia will or should adopt the 'American market model' in the near future," his words fell on deaf ears.[19]

The correctness of farm privatization at the expense of larger production units was dubious even using the documents of Western organizations that were instrumental in guiding the Russian government along its privatization path.[20] Nonetheless, doubts about the utility of farm privatization became fewer and later disappeared altogether. Privatization gained steam and took on a campaignlike fervor. To oppose land privatization, or even to raise questions about its efficacy, was heretical. In the hyperbolic political atmosphere that characterized Russian politics, only communists and their supporters could support the continued existence of collective farms.

Farm privatization plans failed to provide alternative access to rural social services—such things as dwellings, schools, medical care, social security, and subsidized or free meals. Instead it was simply assumed that individuals would be willing to leave the parent farm and lose their security net. The absurdity of asking people to leave a parent farm, undertake enormous risk and debt, and lose access to rural social services was made clear to me as I toured farms in Rostov Oblast in 1993. As I visited collective and state farms, I saw how many farms had their own bakery to make bread for their workers. Farmworkers were provided their lunch *(obed)* for a pittance—one ruble when a comparable meal cost about 1,000 rubles in a dining hall or buffet at the time. Although schools, kindergartens, and medical care were of low quality, they were the only services in existence. Only in the third year after farm privatization did USAID begin to fund projects that would privatize rural services, finally realizing that social services and infrastructure were as important to the outcome of reform as privatized production units.

From an economic standpoint, the ending of the social contract was necessary. At the same time, the strategy was adopted for political purposes. As a consequence, rural political support for reform eroded. One would expect rural conservatives to be opposed to reform for obvious reasons, both philosophical and practical. Yet the erosion of political support by rural liberals is perhaps the most telling indictment of the Yel'tsin agrarian reform strategy.

In tracing the evolution of support for the government's reform program, it is evident that something happened to cause the erosion of rural interests' support for reform. For example, during 1992 the minister of agriculture, Viktor Khlystun, was a strong supporter of agrarian reform and the direction it was going. His proreform positions are known from published interviews and from Craig Infanger, a Western scholar who worked for two years inside the Russian Ministry of Agriculture.[21] By late 1993, however, Khlystun—while still supporting the ideas of reform in general—spent a lot of time defending agricultural interests and the need for financial support, and he worried about the effects of the government's financial policies on the farming sector.[22] By 1995 Khlystun was openly criticizing state financial policies. He argued that the losses resulting from the state's policies toward agriculture exceeded the cost of maintaining subsidized credits and financial support.[23] Former Deputy Prime Minister Aleksandr Zaveryukha, in published articles and speeches, also spoke mainly about the crisis in agriculture and the financial troubles of food producers.[24] Beginning in 1994 leaders of AKKOR, the Agrarian Party, and officials in the Federation Council complained often about the government's financial policies and the destruction of the countryside, private farmers, and the collective sector's productive capacity.

Criticism of the government's rural reform strategy continued during the next several years, as many observers argued not only that the course of reform had been the weakest aspect of postcommunist reform, but also that reform brought a "crisis" situation to the agricultural sector.[25] The situation in agriculture led the State Duma of the Russian Federation in early 1997 to accuse the government of pursuing a "mistaken course of reform in the agroindustrial complex" leading to a "catastrophic situation" in the countryside and nation as a whole.[26]

Thus, it is significant that the government's reform policies had alienated nearly every person and rural group concerned with agriculture and rural interests. This list covered the entire spectrum of political ideologies, from conservatives such as Rutskoy, Zaveryukha, and Lapshin to liberals such as Khlystun and Bashmachnikov. Whereas rural interests in general were ideologically dichotomized, on financial issues state reform policies had galvanized rural interests and led them to speak with a relatively united voice in opposition to state policy. What were the state policies and the state financial levers that led to the loss of rural political support for Yel'tsin's brand of agrarian reform?

The State and Financial Levers

Lost in the conventional wisdom of an "agrarian lobby" that exerted predominant influence in Russia is the fact that the rural sector received fewer financial resources as reform progressed, despite widespread acknowledgment that the countryside needed massive infusions of funds in order to be revitalized. Contrary to the laws and promises made by the Russian government in 1990 and 1991, agriculture did not receive the resources it needed (or was promised) in order to rebuild. This result suggests that rural interests were politically weak.

Financial Levers and Budget Allocations

Given the backward economic conditions that prevail in the Russian countryside, an obvious conclusion should have been that the countryside needed massive infusions of funds for rural infrastructure, and only after this infusion was completed would it make sense to think about privatizing production units. But this approach was politically untenable because reformers began with the premise that subsidies to agriculture supported the continued existence of state and collective farms. What reformers did not see was that such a policy undermined absolutely all agricultural producers; and misplaced priorities and inadequate infrastructure (as they found out) undercut any gains achieved by land and farm privatization. Thus, a "less is better" strategy was pursued.

The collective sector of Russian agriculture was adversely affected by twin processes. On the one hand, farm incomes suffered because consumer demand fell dramatically after price liberalization in 1992 and subsequent inflation. With reduced consumer demand, farms had trouble selling their produce, which forced them to curtail production, which further contributed to lower farm income. Food-processing plants were affected by falling consumer demand, and during the summer of 1995, I visited many plants in Kostroma where only one-quarter or so of the production lines were operating.[27] Plant managers, when asked what their biggest problem was, nearly always answered "to sell our products and to increase our income."

However, falling consumer demand was not the main reason for the financial difficulties faced by food producers. After an initial drop, consumer demand rebounded, just not for Russian food products. As agrarian reform progressed, food imports increased dramatically, particularly for meat and meat products. During 1994 meat imports rose by more than 450 percent, and imports of poultry more than 550 percent in comparison with 1993.[28] By late 1995, it was reported that imports accounted for about 40 percent of total food consumption, and the largest cities such as Moscow were said to receive 70–80 percent of their food from imports. By early 1996, Russia was spending $1 billion every month on imported food, which meant that every three months Russia spent as much on food imports as was allocated to agriculture for the year.[29]

The second, and most important, reason for the financial difficulties of food producers resulted from the use of state financial levers. After decades of steadily increasing resource flows to the rural sector, Russian agrarian reform witnessed lower levels of rural investment, budget allocations, credits, subsidies, and so on to the agricultural sector. Russian reformers—with advice from Western organizations—argued that curtailing resource flows into the countryside would force farms to adapt and become more efficient. Flowing from a "less is better" strategy, agriculture did receive fewer resources. In constant rubles financial support to agriculture declined by a factor of six from 1991 to 1994.[30] Budgetary allocations to agriculture were more than halved in 1995 in comparison with 1994. As a consequence, allocations to agriculture as a percentage of the national budget continued to decrease yearly, falling to 3.8 percent for 1995, down from 8.3 percent in 1994.[31] In 1996 the sum allocated to the agroindustrial complex declined even further, to 3.16 percent of the national budget.[32]

To compound the financial problems of farms, the state promised credits and monies but did not distribute them. At midyear 1994, for example, it was reported that less than 6 percent of the monies budgeted to agriculture had been dispersed, and by October 1, 1994, less than 30 percent.[33] According to the former chairman of the Committee on Agrarian Policy in the Council of Federation, Vyacheslav Zvolinskiy, the pattern in which not all budget allocations were distributed had occurred for several years during reform. Overall, Zvolinskiy estimated that "during the years of reform the state failed to give the APK 55–60 trillion rubles," which meant that instead of receiving 15 percent of domestic income, as stipulated in legislation, the agricultural sector received just over 2 percent.[34]

Original budget outlays for 1994 had promised agriculture 18.1 trillion rubles, but eventually only 13.1 trillion were actually dispersed (about $5.9 billion).[35] The distributed sum included 4.5 trillion rubles for the purchase of agricultural products.[36] The pattern continued thereafter as well. In 1995, despite explicit assurances from Prime Minister Chernomyrdin that at least 60 percent of budget allocations to agriculture would be dispersed during the first half of the year in order to help with sowing, an open letter to President Yel'tsin claimed that in fact only about 35 percent were distributed.[37]

On December 1, 1995, agriculture and the fish industry had received less than 57 percent of their budget allocations for the year (5.8 trillion rubles out of 10.2 trillion in budgetary allotments).[38] The situation was much the same in 1996. Agriculture and the fish industry were allocated 14.5 trillion rubles for the year (which included sums to be used for state purchases of foodstuffs for its food funds), but on December 1, 1996, had received 53 percent (7.7 trillion out of a budgeted 14.5 trillion.[39] During the first half of 1997 the situation deteriorated further, as agriculture received only 11 percent of its annual budget allocations, despite the need to front load dispersements in order to meet sowing needs. On May 1, 1997, agriculture and the fish industry had received only 1.9 trillion rubles out of a budgeted 16.1 trillion.[40]

Fewer financial resources to agriculture hurt the collective sector since it had received about 90 percent of state allocations since reform began. Furthermore, state and collective farms received an average of 86 percent of their income from food production in 1994 (the other 14 percent from subsidies), whereas reformed farms (successors to state and collective farms) received 87 percent of their income from food production and private farms received 90 percent.[41] As a consequence of fewer financial resources, farms had to curtail the amount of land under cultivation, which reduced production. This cycle in turn helps to explain why, through 1996, agricultural production continued to fall faster than other branches of the economy. Falling output from the collective sector allowed liberals to argue that the collective sector was in decline and should be cut further, while the private sector was ascendant and should receive budgetary priority.[42]

To conclude, could one argue that there was something to the idea that "less is better"? I would argue yes, but with conditions: if "less is better" had been accompanied by competitive market channels, by the creation of incentives that would motivate farms to cut resource use and production costs, and by the repair and construction of rural infrastructure. Under these conditions, then, one could argue that "less is better" was part of a coherent strategy.

Capital Investments into Agriculture

Because rural infrastructure was often primitive and incapable of supporting a modern food system, capital investments into agriculture were particularly important. Capital investments were necessary to repair and construct modern storage facilities and food-processing plants, to improve land quality and construct rural roads, to establish modern communications, as well as to address a large number of other deficiencies. In addition, state capital investments were needed in rural infrastructure to take advantage of the increased production that was expected to result from the creation of private farms. However, since agrarian reform was begun, several financial trends undercut any potential gains.

The first trend was a decrease in federal investment credits and monies to agriculture. During the first five years of reform, 1991–1995, capital investments budgeted to the agroindustrial complex (measured in constant rubles) declined by a factor of 10, from 65.8 billion rubles in 1991 to 6.3 billion rubles in 1995. Investments specifically devoted to agriculture fared even worse, declining by a factor of 17 (from 37.4 billion rubles to 2.2 billion rubles).[43] In constant 1991 rubles, the volume of investments into agriculture declined by more than 50 percent during 1992 in comparison with 1991. Included in this decline during 1991 was a 47 percent drop in capital investments for agricultural processing enterprises, such as the construction of storage facilities for potatoes, fruits, and vegetables. Reclaimed land in 1992 equaled 25 percent of the area improved in 1991. In addition, the construction of rural infrastructure declined by 50 percent in comparison with 1991.[44]

The situation did not change for the better thereafter. In nominal terms it appeared that budgetary outlays were increasing; in real terms, however, the sum of capital investments declined by a factor of 3.5.[45] In 1993, in constant rubles, capital investments into the agroindustrial complex reached only 67 percent of the level in 1992.[46] In 1994 the situation was much the same, as federal capital investments to agriculture declined again in real terms, and only about 27 percent of budgeted capital investments were distributed. With fewer funds overall and more allocations targeted for food production, construction of rural infrastructure suffered. Overall, from 1991 to 1994, the percentage of all federal capital investments devoted to agriculture totaled 31 percent in 1991, 20 percent in 1992, 17 percent in 1993, and 10 percent in 1994 and less than 5 percent in 1995.[47] During 1996, capital investments decined a further 31 percent in comparison to 1995.[48]

The second trend was a change in budgetary sources of investment monies. The federal government shifted more responsibility for capital investments, rural construction, land reclamation, and construction of processing facilities to local budgets and agricultural enterprises, a move that complemented the decentralization of the financial system and the development of local autonomy.[49] During the first half of 1994, for example, capital investments from federal funds comprised about 16 percent of all rural capital investments, a decline from the 18 percent of federal capital investments during 1993. The "basic source" of capital investments during 1994 was the agricultural enterprise itself.[50] This trend continued and by the end of 1996 agricultural enterprises were, on average, providing 65 percent of their own investment capital.[51]

The changing structure of capital investments meant that—because of disadvantageous terms of trade, problems with nonpayments, skyrocketing production and construction costs, and a sharp decline in farm "profitability"—agricultural enterprises had funds for little more than sowing and harvesting, and therefore capital investment projects simply were not undertaken.[52] One notable consequence was that rural capital construction ground to a halt. For example, during 1994 the construction of rural roads was three times less than in 1991, and 50 percent fewer gas lines were laid than in 1991.[53] Deficiencies in rural infrastructure in turn contributed to the continuation of the Soviet legacy: massive harvest losses (some 21 percent of the 1994 harvest).[54]

Terms of Trade

In addition to reduced volumes of capital investments, food producers also suffered from a deteriorating price relationship between agricultural and industrial products. Despite the alleged strength of the "agrarian lobby," the agricultural sector continued to suffer from the twin problems of reduced state support on one hand, and sharply higher production, construction costs, and input and fuel prices on the other. Russian agrarian reform began with promises of price parity. The principle of price parity between agricultural and industrial goods, established by 1990 law, was violated already in 1991.

With the liberalization of prices in 1992, the terms of trade turned even more sharply against agricultural products. During the first year of price liberalization, purchase prices for food products increased by a factor of 10 during 1992 (in comparison with 1991 prices), while industrial prices rose by a factor of 26.[55] In 1993, purchase prices for agricultural products increased by a factor of 8.5. The index of industrial prices increased by a factor of 10.5, including tractors by a factor of 11, combines by a factor of 12.5, mineral fertilizer by a factor of 11, and fuel and lubricating oils on average by a factor of 11.[56] During the first nine months of 1994, the index of purchase prices for agricultural products rose by a factor of 3.3. Industrial prices increased by a factor of 5.2, including electricity by a factor of 11.2, fuel by a factor of 8, mineral fertilizers by a factor of 7, combines and livestock feed by a factor of 5, and tractors by a factor of 4.[57]

Disparities in prices and unequal terms of trade between agricultural and industrial products extracted significant volumes of monetary resources from the agricultural sector. According to data revealed at a national agricultural conference attended by deputy heads of oblasts and krays, agricultural officials, scholars, and Deputy Prime Minister Aleksandr Zaveryukha, disadvantageous terms of trade cost the agricultural sector 24 trillion rubles in 1994, and over 40 trillion rubles since 1991 (about $18 billion).[58] Overall, price disparities were seen in the following magnitude. In mid-1995, the chairman of the Committee on Agrarian Questions in the State Duma reported that the index of agricultural prices had increased by 334 percent since 1991, but during the same time prices for industrial products had increased by a factor of 1,383, fuels had increased 4,435 percent, and electricity had increased over 9,300 percent.[59]

Higher prices for industrial goods are not uncommon in economies around the world, so this was not a unique feature to Russia. What set Russia apart were differences in degree, not kind, and the rapidity with which the terms of trade deteriorated. The roots of industrial-agricultural price disparities in Russia are explained by several factors. First, the state exacerbated terms of trade problems through its continued influence over agricultural wholesale—and even retail—food prices. Whereas many wholesale and retail prices were freed on January 2, 1992, agricultural purchase prices continued to be defined by the state during 1992–1993. Thus, food producers depended upon administrative price increases, while farm input prices rose without restraint. Purchase price increases were exceeded by even higher price hikes for needed inputs, and thus, even with new purchase prices, farms were unable to cover production costs.[60] Even thereafter, as the food market was liberalized and more trade gravitated to "privatized" wholesale agents, state "recommended" purchase prices had a large influence on "the market," particularly for perishable products.

Another factor explaining industrial-agricultural price disparities was that food producers continued to confront monopoly purchasers who were able to dictate purchase prices.[61] Although the state attempted to privatize the collective sector, it failed to create competitive market channels of trade. Actual purchase prices were significantly below state-set prices. State "recommended" purchase prices, which were introduced in 1994 in an effort to raise farm incomes and address the growing scissors problem, did not correct price disparities. Interviews with farm managers in Kostroma Oblast indicated that "recommended" prices were commonly ignored, and thus food producers found it difficult to cover production costs.[62]

A related problem that stemmed from the lack of demonopolization was the vulnerability of food producers to input prices. With underdeveloped private trading, farms had little choice but to pay what input suppliers demanded. As a consequence, monopoly buyers of farm products underpaid producers, and monopoly sellers of inputs to farms overcharged. This problem was especially serious for the animal husbandry sector, a sector of the agricultural economy where state purchase prices were low and trading alternatives few, because of the lack of private transport and the perishable nature of the food product. As prices for feed increased many times faster than purchase prices, sectoral profitability fell dramatically, output per animal fell, and collective sector farms curtailed livestock herds from 47.1 to 25.3 million head of cattle from 1990 to late 1996.[63]

Unequal terms of trade were further compounded by other factors. Because of inflation, higher federal taxes were levied on "increased" farm revenues—in essence tax bracket creep. As a result, although purchase prices for grain alone were raised twice in 1992 and three times in 1993, real farm income (in constant rubles) fell. However, income taxes paid by farms increased during 1993. Because of the effects of inflation and an increase in the tax rate on farm "profits," farms paid 2.3 times more in taxes during 1994 than during 1993, even though gross agricultural production declined 9 percent and nearly 60 percent of all farms were unprofitable.[64]

Even though retail prices were rapidly increasing, farms could not benefit because of state-defined purchase prices. Food processors were often limited in the prices they could pass on. Limitations on price markups were adopted to protect urban consumers (policies that were adopted at the oblast level). These limitations applied to how much a food-processing plant could mark up prices to retail stores, and how much retail stores could mark up prices to consumers. Since food processors were limited in their price markups, this put downward pressure on the market prices they offered to farms and private farmers who attempted to sell their production. In Kostroma Oblast such limits were introduced at the end of July 1992 and remained in effect until they were removed in the summer of 1993. In July 1994 norms on price markups were reintroduced for food-processing plants in the oblast, such as 10 percent for milk and meat products, and 15 percent for bread and bread products.[65] Those price markups were removed at the beginning of 1995, but interviews with food processors and retail store managers indicated that the limits continued to be observed, and some managers were not even aware the limits had been abolished.[66]

Financial Levers and Rural Living Standards

Just as state financial levers were central to defining the economic environment for food producers, financial levers also affected rural living standards. The desire to improve rural living standards was originally reflected in legislation adopted in December 1990 that envisioned the revival of the countryside. Nonetheless, already in 1991 it was apparent that those promises of rural social development would not be fulfilled. In August 1991 the RSFSR Supreme Soviet Committee on Social Development of the Countryside, Agrarian Questions, and Food published a resolution, which concluded that the RSFSR Law on the Social Development of the Countryside was not being fulfilled. The committee noted that, during the first half of 1991, there were shortages in deliveries of all types of building materials, including "extreme shortcomings" in pipe used for gasification, rural housing, and electrical equipment. From January to June 1991, only 21 percent of the planned gas-line network was completed, and only 8 percent of the gas main lines. During 1991, in comparison with 1990, 31 percent fewer rural dwellings were built, 34 percent fewer general schools, 32 percent fewer preschools, 42 percent fewer outpatient clinics, and 30 percent fewer clubs and recreational facilities were brought into operation.[67]

Commenting on the government's social policy in the countryside, then chief specialist of the Agrarian Committee of the Russian Supreme Soviet, N. Radugin, argued that

the government of the Russian Federation in 1991 did not fulfill the resolutions of the Second Extraordinary and Third Congress of People's Deputies ... did not create a real financial and material priority for the development of the APK, and did not compensate farms in full measure for the increase in wholesale prices for material-technical resources.[68]

Overall, only 7.5 percent of Russia's national income was allocated to the agroindustrial complex (APK) for the social sphere of the countryside in 1991, not the 15 percent envisioned in 1990 legislation. As a result of underfunding, goals for social development of the countryside remained unfulfilled.

Nor did rural living conditions improve as investments in the rural social sphere continued to decrease. In his speech before the Sixth Congress of People's Deputies in April 1992, Minister of Agriculture Viktor Khlystun revealed that the percentage of national income allocated to the agroindustrial complex and the social sphere of the countryside would be "somewhat lower" in 1992 than in 1991. Rural construction, as a result of decreased investment, suffered further declines during 1992 in comparison with 1991: 31 percent fewer rural dwellings, 30 percent fewer general schools, 58 percent fewer preschools, 46 percent fewer clubs and recreational facilities, and 44 percent fewer intrafarm roads were constructed and operational.[69]

During 1993, capital investments directed specifically to the social development of the rural sector declined another 21 percent in comparison with 1992 and 52 percent from investment levels in 1991.[70] By late 1993, the construction of new schools, children's gardens, and clubs was 50–60 percent of the level attained in 1992.[71] Overall, at the end of 1993, only 20 percent of the rural housing, 13 percent of the preschools, 11 percent of the clubs and houses of culture, 17 percent of the gas lines, and 27 percent of the hard-paved roads that had been envisioned in 1991 programs were actually in operation or had

Table 12. Construction of Rural Infrastructure, 1990–1994 (brought into operation annually)

	1990	1994	1994 as Percentage of 1990
Rural housing (mil. sq.m)	16.4	6.1	37
Number of kindergarten seats (thous.)	71.9	9.4	13
Number of school seats (thous.)	92.0	32.7	36
Intrafarm roads, hard surfaced (thous. km)	28.3	2.4	8
Water lines (thous. km)	5.4	1.2	22
Sewerage lines (thous. km)	263.2	23.4	9
Hot water lines (thous. km)	0.6	0.1	17

SOURCE: N. Radugin, "Sotsial'naya infrastruktura Rossiyskoy derevni," *APK: ekonomia, upravleniye,* no. 1 (January 1996): 28.
NOTE: Includes all sources of financing.

been constructed.[72] In constant rubles the volume of capital investments into agriculture for all purposes had declined by a factor of 6 in comparison with 1990.[73] It should be noted that, in real terms, the nation's GDP had declined, as had budgets, and thus when we consider the amount of money in real terms it is clear that agriculture received even less investment than the smaller percentages of investments to agriculture might otherwise suggest.

In 1994, 1.12 trillion rubles were allocated for rural construction projects from the federal budget (about 5 percent of the sum allocated to agriculture as a whole), but during the first half of 1994 not a single ruble was actually distributed. Only at the end of the year were 150 billion rubles (about $41 million) distributed, but this equaled only 13 percent of the original budget allocation.[74] As a result, by the end of 1994, the level of construction of rural housing, schools, children's gardens, and clubs were two or three times less the level obtained in 1991.[75] Trends in rural infrastructure are indicated in table 12.

Decreased funding for rural social development affected not only the number of new construction projects, but also the ability to repair existing facilities. Again, rural infrastructure continued to crumble because of underfunding: it was estimated that 50 percent of rural clubs and recreational facilities required major repairs, 33 percent of rural housing was decrepit, 50 percent of electrical lines had worn out and needed replacement, and (as anyone knows who has traveled rural roads either within or between farms) it appears that repairs had been ignored for years, if not decades. Therefore resurfacing was a primary

Table 13. A Comparison of Urban-Rural Housing Amenities, Russia, 1993 (percentage of housing with such amenities)

	Rural Localities	Urban Localities
Gas	77	71
Indoor running water	51	94
Indoor plumbing	39	93
Central heat	40	92
Hot water	21	81

SOURCE: V. Mashenkov and Ye. Lysenko, "Sotsialnoye razvitiye sela," *APK: ekonomika, upravleniye,* no. 9 (September 1994): 3.

need, and yet the construction and repair of intrafarm roads continued to decrease. Three years after resolutions were adopted to revitalize the countryside less than one-quarter of rural roads were hard-paved.[76]

In conclusion, when agrarian reform was begun in postcommunist Russia, rural living conditions were below those of urban dwellers, and thus, to be fair, we should be clear that rural social deficiencies did not just suddenly appear in the postcommunist period but rather were long-term problems. However, we should also be clear that rural living standards were exacerbated during the postcommunist period due to underfunding and a series of broken governmental promises.[77] Broken state promises for rural revitalization affected the level and quality of amenities in rural housing. Very significant differences in amenities continued to exist between urban and rural residents as shown in table 13.

As the agricultural sector continues to receive less investment monies and a lower percentage of the national budget, it is difficult to envision how urban-rural differences will be narrowed in the future. In certain regions such as Moscow Oblast, the construction of suburban subdivisions based on Western home models for urban dwellers makes it likely that the gap between urban-rural living conditions and amenities is likely to grow.[78]

Urban-Rural Wage Differences

With the introduction of radical reform in 1992, nearly forty years of social policy began to reverse as urban-rural wage levels diverged, leading to a growing gap between the income levels of rural and urban workers.[79] In 1990, for example, according to official statistics the average monthly wage for state farmworkers constituted 105 percent that of an industrial worker, while the wage of a collective farmworker constituted 88 percent (not including income from personal plots). In 1991, however, state farmworkers earned 80 percent, and collective farmworkers 70 percent of the average monthly wage of an industrial worker. Beginning in 1992, the income gap widened further, as state farmworkers earned 63 percent and collective farmworkers received 49 percent the average monthly wage of an industrial worker.[80]

By the end of 1995, available evidence for aggregated agricultural incomes showed that the deterioration of rural wages continued. Agricultural workers had the lowest income of any branch in the economy, 222,400 rubles a month (not including income from personal plots). In comparison, a person in transportation made an average of 745,000 rubles a month, a person in construction made 678,100 rubles a month, and an industrial worker made 553,700 rubles monthly. Thus rural wages comprised about one-third of the average monthly wage of these urban professions (which meant that ordinary rural workers earned an even smaller percentage).[81] Trends in the industrial-agricultural wage relationship are illustrated by the chart below (figure 2).

As the chart demonstrates, a significant erosion occurred in the relationship between urban and rural wages beginning with economic reform. What the chart does not—and could not—show are the months when rural wages were not paid at all. Nonpayment of wages affected the entire Russian economy, not only farms and farmworkers, so no one escaped this cruel fate. However, one could argue that workers and employees in the agroindustrial complex suffered most of all because their wages were the lowest in the economy. Total wage debt to the industrial sector was higher than in agriculture throughout Rus-

Figure 2. Urban-Rural Wage Relationship, 1985–1995

sia—owing to larger salaries. However, because agricultural personnel had significantly lower average monthly salaries, more people were affected. By the end of 1996, throughout Russia, more than twenty-five thousand agricultural enterprises were in arrears on wage payments, more than any other branch of the economy, and nearly twice the rate of industrial firms.[82]

The wage nonpayment problem, which began in 1992, had not been solved, despite government promises, even by the end of 1996. In fact, as wage indebtedness increased, labor strikes became more common. For the economy as a whole, the number of enterprises experiencing labor strikes decreased from 6,273 to 264 from 1992 to 1993. But they increased to 514 in 1994; to 8,856 in 1995; and while the number of strikes during 1996 declined somewhat to 8,278, the number of people participating in strikes increased nearly 36 percent in 1996 when compared to 1995.[83]

In the rural sector, lower rural wage levels and chronic nonpayment are also important because they likely acted as a stimulant to rural out-migration. Of course, rural out-migration was a decades-long process. But by the 1970s already many demographers considered the rural population to be largely "exhausted," which meant that the age structure was not conducive to further migration. In the past the number of farmworkers declined as part of the general decrease in the rural population. But a more recent trend runs contrary to general demographic patterns. During 1992–1994 the net rural population increased for Russia as a whole (and for all regions except the Northwest during 1991–1993). In a reversal of a pattern that had endured for several decades, the rural population increased as a result of rural in-migration, even as the rural birth-to-death ratio deteriorated.

However, during this period farm populations continued to decline. In Kostroma Oblast, farm data show that from 1986 to 1990 (a five-year period), the number of farmworkers involved in farm production declined by 15 percent. During 1991–1995, however, the number of farmworkers involved in farm production declined by 26 percent.[84] This latter statistic is notable because it occurred after decades of rural out-migration, leaving the farm population older, less mobile, and with less migration potential.

Second, these data from Kostroma Oblast suggest that farm personnel with high migration potential—the young—are even more rapidly departing agriculture. In the words of one journalist, rural out-migration reached "threatening proportions" particularly among the rural young.[85] Third, there is suggestive evidence that past patterns of farm out-migration continue to hold true, that is, the most skilled are the most likely to depart. From 1990 to 1993, the number of kolkhozniki in Kostroma Oblast declined by just over 8 percent. During the same time period, managerial staff *(rukovoditeli)* declined by 20 percent, specialists by 11 percent, and white-collar employees by 15 percent. Thus, rural out-migration resulting from deteriorating rural social conditions would have an impact on production potential and the demographic structure of the countryside.

Financial Levers and Their Economic Impact

Although one of the overall goals was to make agricultural production less costly, by pursuing a "less is better" strategy state economic levers resulted in a number of economic consequences, including a precipitous drop in agricultural production, per capita food consumption, a deterioration in farms' financial condition, and farms' use of the "market." We shall survey each in turn below.

Agricultural Production

The economic impact of contemporary agricultural reform on agriculture contributed to significant declines in the output of basic foodstuffs as well as productive inputs used by food producers of all types. Food production declined as a result of fewer hectares being sown, and yields per hectare fell.[86] The decline of agricultural equipment and the fact that the rate of depreciation exceeded the replacement rate meant reductions in the gross number of agricultural machinery. The production of mineral fertilizers also fell drastically even as exports increased. As a result, the application of fertilizer declined to among the lowest in Europe; soil fertility worsened and this affected land reclamation, which virtually ceased. At the beginning of 1995, the Russian Ministry of Agriculture estimated gross food output had fallen by 33 percent in comparison with 1990.[87] Gross output declined by 8 percent during 1995 in comparison with 1994, and during 1996 agricultural production fell by another 7 percent.[88] Various production trends in the agroindustrial sector are indicated in table 14.

From the table it is clear that Russian agriculture in 1995 had produced less of nearly everything in comparison to 1990.[89] With regard to inputs one might be tempted to argue that less is indeed better. In fact, when agricultural reform was begun in 1990 one of the primary goals was to improve agricultural performance by using fewer inputs per unit of output. However, crop yields, labor productivity, and food production have declined during agricultural reform. As the table shows, grain yields in 1995 were 63 percent of their 1990 level; in 1996 grain yields rebounded somewhat, but still were more than 30 percent below their 1990 level.[90] In addition, by the end of 1995, milk output per cow had declined, eggs per chicken had decreased, and slaughtered cattle yielded less meat in comparison with 1990.[91] Owing to the decline in capital equipment available to mechanized workers, labor productivity fell.

Food production declined, but not only in areas where one would logically

Table 14. Selected Production Indices of the Russian Agricultural Sector, 1990–1995 (all categories of farms)

	1990	1995	1995 as Percentage of 1990
Area sown for grains (mil. hect.)	63.1	55.0	87
Grain harvest (mil. tons)	116.7	63.4	54
Grain yields (centners/hect.)	18.5	11.6	63
Meat production (all kinds—mil. tons)	10.1	5.9	58
Cattle herds (mil. head)	58.8	43.3	74
Pig herds (mil. head)	40.0	24.9	62
Milk production (mil. tons)	55.7	39.3	71
Egg production (million)	47.4	33.7	71
Potato production (mil. tons)	30.8	39.9	130
Vegetable production (mil. tons)	10.3	11.3	110
Sugar beet production (mil. tons)	31.1	19.1	61

SOURCES: Jaclyn Y. Shend, *Agricultural Statistics of the Former USSR Republics and the Baltic States,* statistical bulletin no. 863 (Washington, D.C.: U.S. Department of Agriculture, 1993); *Former USSR Situation and Outlook Series, WRS-96-1* (Washington, D.C.: U.S. Department of Agriculture, 1996); *Rossiyskiy statisticheskiy ezhegodnik 1996* (Moscow: Goskomstat, 1996); and author's calculations.

expect. The evidence suggests that all agricultural areas were significantly affected by state financial levers. Even the best agricultural regions did not escape unharmed. The Northern Caucasus region—the single largest producer of grains in Russia—in 1994 produced just 73 percent of its 1991 volume of grains, and the two most productive wheat-growing regions—the Volga and Northern Caucasus—produced 60 and 81 percent of their 1993 levels of wheat during 1994.[92] Similarly, yields and area under cultivation for a range of other plant products decreased in other black earth regions, showing that productive food-growing regions were hurt as much as were the less productive regions.[93] In 1995, for example, Krasnodar kray—an important grain region in the Northern Caucasus region—suffered a 30 percent drop in grain production in comparison with the average levels during 1986–1990; likewise, rice production declined 43 percent and sugar beets and vegetables both declined by 30 percent in the kray, using the same comparative time frame.[94]

The Volga region is one of the country's most important agricultural regions, producing annually about 20 percent of the nation's grain, 15 percent of its meat, and 13 percent of its milk from 1991 to 1993. However, due to the effects of reform, and in particular the disparity between agricultural and industrial prices, the amount of land used to grow grain declined about twenty-one thousand hectares between 1991 and 1993 (about 3 percent of the total area used for grains in the Volga region). Overall, by the end of 1993, the region produced 92 percent of the amount of grain produced in 1990, 89 percent of the meat, only 50 percent of the milk, and 89 percent of the eggs.[95]

Decreased gross production was not the only problem. Farms had difficulty obtaining farm machinery, fuels, and fertilizers, with the result that land erosion increased and soil fertility decreased. Farm productivity fell also because of decreased use of inputs and level of equipment. For example, the Volga region as a whole was using about 10 percent of the mineral fertilizer in 1993 compared to the average levels of 1986–1990. Furthermore, according to at least one academician, there was between 21 and 35 percent less farm equipment and machinery in use during 1993 in comparison with 1991.[96] Thus even relatively productive agricultural regions did not escape the consequences of financial policy.

Food Consumption

For years the Russian consumer could expect slow but steady increases in per capita consumption in addition to retail price protection in the form of retail price subsidies. The shortages and the long lines that Soviet consumers experienced resulted from low fixed prices at state stores, not from shortages in production. Despite an irrational price system and well-documented problems in distribution, average per capita consumption of basic food products such as meat, milk and milk products, eggs, sugar, vegetable oil, and fruits and berries increased during every five-year period from 1966 to 1990.[97] From 1965 to 1981, the per capita daily consumption of fats increased by 25 percent, while per capita daily consumption of carbohydrates decreased. By the late 1980s, Western specialists concluded that "the per capita level of food energy (calories) nearly matches that in the United States" and that "findings . . . suggest that the impact of shortages of quality foods—items high on the Soviet consumer's scale of preferences—may be primarily upon consumer satisfaction rather than on physiological need."[98]

As Russia moved toward a market economy, food production experienced a precipitous drop, and since 1991 significant retail price increases occurred. These two processes led to a rather sharp decline in the demand for high-preference items such as meat and dairy products—items that were heavily subsidized under the Soviet regime.[99] Per capita consumption changes are indicated in table 15.

Among those most affected by lower consumption were women and pensioners who consumed significantly less than males aged from sixteen to fifty-nine, which is to be expected. Pensioners ate the least of all age groups, with

Table 15. Selected Food Consumption Indices in Russia, 1990–1995 (in kilograms per person)

	1990	1995	1995 as Percentage of 1990
Meat and meat products (all types)	75	55	73
Milk and dairy products	386	253	66
Eggs (number consumed)	297	214	72
Bread and cereal products	119	121	102
Potatoes	106	124	117
Vegetables and melons	89	76	85

SOURCES: Jaclyn Y. Shend, *Agricultural Statistics of the Former USSR Republics and the Baltic States,* statistical bulletin no. 863 (Washington, D.C.: U.S. Department of Agriculture, 1993); *Potrebleniye osnovnykh produktov pitaniya naseleniem Rossiyskoy Federatskii* (Moscow: Goskomstat Rossii, 1996); and author's calculations.

the exception of children under six. Children aged from seven to fifteen had average consumption levels that approached those of women aged from sixteen to fifty-four.[100] There was evidence of hunger among the most vulnerable segments of the population in Moscow and other large cities.[101] Furthermore, in late 1994 and during 1995 in certain locales, rationing was introduced for meat, butter, and sugar for the first time since 1991.[102]

Not only did overall food consumption decline in comparison to the late Soviet period, but the structure of the Russian diet changed. Although food production continues to be marginally subsidized at the farm level, the removal of subsidies at the retail level meant that meat and dairy products increased in price more rapidly than plant products. Russian consumers adapted by substituting more starch and carbohydrate calories for animal husbandry and dairy calories. As a result, by the end of 1995 Russians were eating more bread, bread products, and potatoes; they consumed less meat and milk compared to 1991 and substituted eggs for their protein source (see table 15). Based on the average diet, Mikhail Lapshin of the Agrarian Party claimed in 1995 that Russians ate 40 percent less protein than they needed and had a vitamin deficit estimated at 60 percent.[103] One may argue that the reduction in high-preference items such as meat and milk products was "good" in the sense that it was more healthy than the typical Russian diet, which is heavy with fat and cholesterol.

Financial Conditions on Farms

Price discrimination and unequal terms of trade also affected farm operations. To get a sense of the financial impact on farms, this author spent a week on the Novosel'skoye state farm in Kaluga Oblast during July 1994.[104] The farm was an experimental farm, established under Khrushchev in 1959, and had always received priority supplies of inputs. For that reason the farm's grain yields were about 50 percent higher than the raion average and more than double the oblast average. Conversations with farm personnel and the acquisition of unpublished farm data revealed the extent to which even a productive farm had been affected by economic reforms.

The head of the trade union of the farm explained how the farm suffered from multifaceted financial pressures. First, there was no wholesale market for farm inputs, no competition from sellers, and so even while agroindustrial plants that supplied farm inputs had been privatized, they remained monopolies that could dictate prices. Second, there were few "real" options for the sale of agricultural produce. Food processors remained monopolies that could dictate prices; "market" prices were below state prices; and state prices, while seemingly attractive, suffered from chronic nonpayments. To meet spring sowing costs the farm had to borrow money at 270 percent interest, which meant that for every 1 million rubles borrowed 3 million had to be repaid after the harvest.

Because of these financial pressures even this progressive, productive farm had experienced a decline in its output and overall performance. These trends are indicated in table 16. The table indicates a number of interesting trends that affected not just this farm, but all farms. First, it is clear that farm performance suffered, both in terms of production potential (hectares sown to grain, number of cattle) as well as actual performance (yields per hectare, output per cow). Throughout Russia, earnings from food output accounted for the largest percentage of farm income, but because of disadvantageous terms of trade overall production and animal stocks declined.[105] Second, owing to the cost of farm machinery, the farm used fewer pieces of machinery and was unable to afford new equipment or make repairs on the old. Third, a notable increase occurred in the cost of production in real terms, which reflected the growth in prices of fuels and spare parts. Fourth, although the rural population as a whole increased in Kaluga Oblast, the farm population itself continued to decline—and, similar to previous migration patterns, the young and the skilled were the most likely to depart. Last, due to financial pressures the farm did change its production mix, but it did so in a way that would not lead directly to increased food output. Instead, the farm reduced the area of land that was sown with grain and increased the area devoted to livestock feed—a practice that essentially allows the field to grow wild with grasses.

Similar financial pressures existed for agricultural enterprises throughout Russia and resulted in an increased number of "unprofitable" farms.[106] For ex-

Table 16. Economic Indicators of Sovkhoz "Novosel'skoye," Kaluga Oblast, 1993

	1990	1993	1993 as Percentage of 1990
Grain yield (cent./ha)	34.1	25.1	74
Milk output per cow (kg.)	3,495	2,655	76
Grain production costs (rubles/centner)	12.46	2,049.41	16,448
Potato production costs (rubles/centner)	13.08	1,810.18	13,839
Number of workers	746	666	89
Number of tractors	183	154	84
Number of cattle, total	5,332	4,431	83
Agricultural land (ha.)	12,556	10,899	87
Area sown (ha.)	9,077	9,795	108
Area sown with *korm* (ha.)	3,997	5,082	127
Area sown with grain (ha.)	4,800	4,433	92

SOURCE: "Analiz khozyaystvennoy deyatel'nosti Novoel'skogo sovkhoza, 1990–1993," unpublished data from Novosel'skoye sovkhoz, Kaluga Oblast, 1993.

ample, in 1990, about 2 percent of all state and collective farms were officially unprofitable (the percentage rises to 18 percent if "low" profitable farms are included, defined as those with profitability rates of less than 15 percent). In 1991, when Yel'tsin unleashed his farm reorganization and land privatization decrees, a core of 5 percent were said to be unprofitable.[107]

During the reform period, the interest rate question was of central importance to reform, and a key aspect was state policy. Throughout most of 1993, collective agricultural enterprises were given subsidized credits, although at a somewhat higher interest rate than private farms. For example, in 1992, when private farms received credits at 8 percent per annum, state and collective farms received credits at 25 percent; in 1993, when private farmers received credits at 25 percent, state and collective farms paid 170 percent.[108] Higher interest rates and the ending of subsidized credits to farms in September 1993 adversely affected collective agricultural enterprises, and the percentage of unprofitable farms in 1993 increased to about 14 percent of all agricultural enterprises. By 1994 the percentage of unprofitable collective enterprises increased dramatically, with estimates varying between 47 and 57 percent of all collective agricultural enterprises.[109]

During 1995, an estimated 70 percent of all collective agricultural enterprises were unprofitable, and this figure rose to 75 percent in 1996.[110] State financial policies influenced the interest rates that farms faced, and interest rates in turn affected farm profitability and production capacity. When subsidized interest rates were available (which in real terms were negative rates), farms had little incentive to increase efficiency or productivity. Higher interest rates were necessary to force farms to think hard about their expenses and their production costs. After October 1993, the stated intent of a tight money policy was to squeeze out high-cost unprofitable farms and to reduce state financial support. However, what the pursuit of this policy has accomplished since late 1993 is instead the near destruction of the entire agricultural sector, both strong and weak farms. After October 1993 and during 1994, similar to private farms, collective agricultural enterprises had access to either the 210 percent interest rate established by the Central Bank, or rates from commercial banks that were often double those of the Central Bank. Even as interest rates dropped to around 150 percent, interviews with farm managers suggested that they were unable to afford loans for anything other than basic production.

Although higher interest rates were necessary, higher interest rates had other consequences. Farms responded to lower income, high interest rates, and higher production costs by curtailing acreage sown and food production. Lower food production in turn resulted in lower farm income, compounded in part by decreased consumer demand. Further, losses and wastage costs were still borne by farms because they lacked funds to improve (or even maintain) those aspects of farm operations.

Use of the Market

Farms responded to their economic environment by changing their product mix and decreasing production. Another important response by farms was to use "the market" against the state. In this case the term "the market" refers to greater marketing options and less state control over food trade.[111] As a consequence, the amount of grain sold to the state was less than in the 1920s. Part of the reason was deliberate state policy to promote private channels of trade, but even reduced state demand went unfulfilled.[112]

Since 1992 in Russia a slow evolution of the domestic food procurement system has occurred. A domestic food trade "market" has arisen, characterized by the reduction of state control in defining the terms, volumes, and outlets of food trade, and the establishment of alternative channels of food trade. The food trade system has responded to the reduction of state regulation and control. As a result, the development of a market-based food trade system is one of the successes of reform, although trading freedom has economic consequences.

The Evolution of the Procurement System

Initially the Russian procurement system resembled the Soviet system, a system that emphasized obligatory deliveries to the state, although Russian procurements were established at reduced levels. In 1992, the unsuccessful food tax was eliminated, and in its place a different system for state deliveries arose. In early January 1992, Yel'tsin signed a presidential decree that stipulated that all agricultural producers were required to deliver up to 25 percent of their sugar beets, potatoes, vegetables, and fruits; up to 35 percent of their grain; and up to 45 percent of their milk, cattle, and poultry.[113] The percentage of output for obligatory deliveries for individual regions would be based on the average annual production attained during 1986–1990. Failure to conclude or fulfill contracts signed with procurement organizations would result in a fine, to be paid to procurement organizations, equal to the local market value of the undelivered production.

This "new" Russian procurement system in 1992 was not much more effective than the old Soviet one. Sharp price disparities between industrial products and grain resulting from price liberalization and rampant inflation led to a situation in mid-August 1992 in which the state had obtained only about one-half the volume of grain as at the same time a year earlier, and only about one-half of the volume specified in contracts. In response, Yel'tsin forbade farms from selling their grain until their state delivery quota had been met. If a farm violated this decree, the state could confiscate the proceeds from that illegal sale.[114] In order to motivate farms to sell more to state procurement organizations, in mid-August 1992 the government increased grain procurement prices; and in November 1992 state grain prices (in rubles) were doubled from the August 1992 level.[115] Despite these steps, state grain procurement targets in 1992 were not fulfilled, reaching about 26 million tons (24 percent of gross grain harvest), or about 10 percent more than in 1991 even though the 1992 harvest (after thrashing) was 20 percent higher than in 1991.[116]

Perhaps because the Russian procurement system was not much more effective in 1992 than the Soviet system had been, at the end of 1992 and beginning of 1993 numerous changes were introduced in domestic food procurement policy that affected all food producers. As a result of those changes, beginning in 1993 the Russian domestic food procurement system became less regulated. At the end of January 1993 the Russian government announced that obligatory food deliveries to the state had officially ended.[117] To replace obligatory deliveries, in February 1993 a presidential decree formally established federal and regional food funds. The federal food fund henceforth constituted the state's central reserve to meet various social needs.[118] The intent of the federal food fund was "to supply food products to the military, health organizations, to support territories that had limited means of their own production, to Moscow and St. Petersburg, regions of the far north, and also for the creation and maintenance of state reserves and reserves for the liquidation of extraordinary situations."[119] The establishment of food funds represented the lessening of state intrusion into the food market, especially for grains. The state moved from trying to direct all food trade to a system with minimum levels of reserves to meet targeted state needs.

The federal food fund defined purchase targets that were below the amount procured by the state under the Soviet system. To meet food fund targets, deliveries to both federal and regional funds were characterized as not "obligatory" but contractual, and purchases were to be at prevailing market prices. Both collective agricultural enterprises and private farmers were to sign contracts with state procurement organizations for delivery of food to these food funds. The government would issue credits in order to purchase foodstuffs for the federal food fund. In the case of grain, Roskhleboprodukt (the agency in charge of food purchases for food funds—later to be privatized) was responsible for delivering the required volume of grain to the federal fund and received state credits to purchase grain. Farms were paid for their grain in state credits. Although the system was contract-based, federal fund obligations were disaggregated to the oblast level, whose officials assigned raion targets, which in turn were translated into farm quotas. As an added inducement, food producers who sold their output to the state were eligible to receive various production subsidies as well as compensation for equipment, fuels, and fertilizers used in the production of produce sold to the federal fund.[120]

The Law on Grain, adopted in May 1993, tried to liberalize food trade further by introducing competition into the grain purchasing process by stipulating that the state would select, on a competitive basis, purchasers who, acting as state representatives, would buy grain from producers at "negotiated market prices."[121] Reality was somewhat different as Roskhleboprodukt remained the monopoly purchaser, and this organization had oblast-level organizations.[122]

Because of imperfect competition, which affected grain trade, in late December 1993 Yel'tsin then signed a decree entitled "On the Liberalization of the Grain Market in Russia." This decree was intended to achieve freer domestic food trade. Toward this end, the decree stipulated that "all enterprises for the purchasing, primary processing, storage of grain and production of bread products" that were state property were to be auctioned before April 1, 1994. As a result, Roskhleboprodukt was privatized—as were its oblast-level subunits. Oblast-level organizations continued to work through Roskhleboprodukt agents in order to provide grain to state reserves. Furthermore, according to the "Liberalization of the Grain Market in Russia," any local decision by an organ

Table 17. Grain Harvests and State Purchases in Russia, 1986–1996 (million tons)

	Grain Harvest (clean weight)	Volume Purchased by Regional and Federal Food Funds	Volume Purchased by Federal Food Fund Only	Federal Purchases as Percentage of the Harvest
1986–1990 average	104.3	—	—	33.0
1991	89.1	—	—	25.0
1992	106.8	—	—	24.0
1993	99.0	28.2	11.5	12.0
1994	81.3	12.1	2.2	3.0
1995	65.4	9.5	0.93	1.5
1996	69.2	8.6	0.41	0.59

SOURCES: *Narodnoye khozyaystvo RSFSR v 1990 g.: statisticheskiy ezhegodnik* (Moscow: Goskomstat RSFSR, 1991), p. 448; "Sel'skoye khozyaystvo Rossii v 1993 godu," *APK: ekonomika, upravleniye*, no. 4 (April 1994): 20; *Rossiyskaya gazeta*, January 21, 1995, p. 4; *Sotsial'no-ekonomicheskoye polozheniye Rossii*, no. 12 (Moscow: Goskomstat, 1995), p. 82; ibid., no. 12 (1995), p. 82; ibid., no. 12 (1996), p. 55; and "Sel-skoye khozyaystvo Rossii v 1996 godu," *APK: ekonomika, upravleniye*, no. 3 (March 1997), p. 9.

of executive or representative power that in any way limited or forbade the free trade of grain was declared invalid. The penalties for violating free trade were "the stopping of all forms of federal support to agricultural enterprises located in the territories of the corresponding subjects," and "ceasing to send food and agricultural products from the federal fund."[123]

The lessening of state intrusion into the grain market was evidenced by the purchase targets defined for the federal food fund. In 1993 the goal was to buy almost 12 million tons of domestic grain for the federal food fund, down from the 26 million tons the state had procured in 1992. A brief review of grain harvests and state procurements is shown in table 17.

The table indicates that both the volume and percentage of grain purchased by the federal food fund decreased significantly after 1993, and similar patterns were true for other food products. The table thus demonstrates reduced state control over food trade.

Despite less state regulation of the domestic food market, the state offered several selective incentives to make it advantageous for food producers to sell to state purchasing agents.[124] One lever was the use of state purchase prices. Beginning with the July 1993 price increase for grain and the decision to index purchase prices to inflation, state purchase prices increased significantly during the next few years and were in fact higher than market prices available through commodity exchanges.[125] This was important because even under "market" conditions, it was commonly acknowledged that food producers did not set the price of their produce.[126] The major constraint, though, was that Roskhleboprodukt seldom had money to pay farms for their produce, as state indebtedness to food producers climbed into the billions of dollars. Roskhleboprodukt itself became the subject of corruption and misuse of funds. So any apparent advantage to producers from higher state purchase prices was not actualized.

A second lever was the use of production subsidies, compensations for production costs, and price supplements for grain and other food products sold to the federal fund. According to the Law on Grain, grain producers who concluded a contract with a state purchaser could receive an advance of not less than 40 percent of the value of the sale, based on the guaranteed purchase price. Much of the Law on Grain was gutted by the December 1993 decree on Liberalization of the Grain Market, but the subsequent Law on Purchases and Deliveries adopted in late 1994 allowed for even higher advances—not less than 50 percent of the value of the delivery, 25 percent of which would be paid after the conclusion of the contract and another 25 percent after sowing was completed.[127] Moreover, beginning in 1993, food producers who sold to the federal fund became eligible to receive a range of price supplements per unit of production. Legislation adopted after 1993 offered producer subsidies and compensations for production and operational costs such as machinery, fuels, and mineral fertilizers.[128] Despite these incentives, food producers turned away from state channels of trade.

"Market" Responses from Food Producers

The fact that food producers began to move away from state channels of trade—primarily for nonperishable products—was an indicator of dissatisfaction with the trading conditions offered by Roskhleboprodukt. The primary use of the market was to refuse to sign contracts with Roskhleboprodukt or to find other outlets for food trade, whether or not contracts had been signed with state food funds. Through the first half of 1994, Roskhleboprodukt handled less than 11 percent of winter grains, while market sales accounted for an estimated 45 percent of all grain trade and barter an additional 16 percent.

As a consequence, purchases for the federal food fund during 1994 declined significantly. On January 1, 1995, national reserves were less than one-half their volume as of January 1, 1994.[129] As grain growers turned to nonstate channels, less grain was sold to Roskhleboprodukt. During 1994, for example, Roskhleboprodukt was able to purchase only about 20 percent of its grain target for the federal food fund (2.2 million tons from a planned 10.7 million tons). In 1995, the federal target was 8.6 million tons of grain (plus targets for other products) but less than 1 million tons of grains were purchased by the federal fund, representing less than 13 percent of planned purchases.[130] During

1996, the federal food fund was able to purchase less than one-half the volume it had in 1995 (see table 17).

Farms' withdrawal from state channels of trade and their search for alternative channels are indicated in table 18. Comparing the data in table 18 with patterns of food sales at the end of 1996 reveals significant change. By the end of 1996, state purchases (defined as federal and regional food funds) from all types of producers comprised 12 percent of grain production, 8 percent of potatoes, about one-third of vegetables, just over one-half of cattle and poultry production, and more than 70 percent of milk.[131] A final way that farms used the market against the state was to change production mix to more advantageous products. In this respect, farms in all regions of Russia reduced the area of sown land devoted to grain. For example, the area devoted to grain production declined 63 million hectares in 1990 to less than 52 million hectares in 1997.

Financial Levers and Their Political Impact

Just as state financial levers had an economic impact, so too did they have a political impact. Political organizations were formed in order to undertake collective action against the Yel'tsin administration and his reform policies. The reason for rural political organization was that rural interests were not well represented among urban interest groups or parties. Nor was this situation unique to Russia, as Western scholars have argued that rural interests often make poor coalitional partners, particularly for urban interests, because a rural "win" (increased purchase prices, trade protectionism) means a "loss" (higher food costs) for other members of the coalition.[132] The creation of agrarian interest groups was necessary, therefore, because it would be difficult for agrarian interests, with their differing goals, to be included in urban political coalitions.

The antecedents to rural political organization in Russia occurred in the late Gorbachev period. During 1990, many different rural interest groups, spanning the entire political spectrum, arose because they were not being represented adequately in existing political structures. At the liberal end of the political spectrum one group was the Association of Peasant Farms and Agricultural Cooperatives of Russia (AKKOR), which held its founding congress in January 1990. This voluntary organization was intended to protect and promote the interests of newly created private farmers and leaseholders.[133] Another liberal organization was the USSR Union of Leaseholders and Entrepreneurs, which held its first congress in June 1990.[134]

At the conservative end of the spectrum, the trade union workers in the agroindustrial complex split from the larger trade union movement by forming their own trade union organization. A plenary meeting was held in March 1990, at which it was noted that industrial trade unions often "weakly defend-

Table 18. Structure of Food Trade, 1991–1994 (in percentages)

	Sold to Procurement Organization	Sold to Centrosoyuz	Sold on the Market, in Private Stores and Stalls	Distributed or Sold to the Farm Population	Barter Trade
Grains					
1991	62.3	0.5	13.8	22.0	1.4
1992	63.9	0.3	13.1	20.5	2.2
1993	63.1	0.2	12.4	21.0	3.3
1994	10.8	0.2	45.7	27.7	15.6
Potatoes					
1991	52.7	16.0	18.6	11.7	1.0
1992	50.0	10.5	24.3	13.8	1.4
1993	45.0	7.0	28.8	17.3	1.9
1994	12.9	0.4	51.0	28.8	6.9
Vegetables					
1991	71.8	12.1	10.7	5.1	0.3
1992	65.5	8.0	18.3	7.8	0.4
1993	65.6	4.9	19.3	9.6	0.6
1994	52.0	2.7	36.2	8.5	0.6
Cattle					
1991	80.7	1.1	7.0	10.5	0.7
1992	78.0	1.2	8.2	11.8	0.8
1993	77.2	1.2	8.1	12.4	1.1
1994	66.8	1.2	11.9	17.8	2.3
Pigs					
1991	84.4	0.9	6.6	7.6	0.5
1992	79.6	1.0	9.3	9.4	0.7
1993	79.2	0.8	8.4	11.0	0.6
1994	73.1	0.7	12.2	12.8	1.2
Poultry					
1991	88.1	0.4	6.1	3.7	1.7
1992	85.0	0.4	9.2	4.3	1.1
1993	91.2	0.3	5.0	2.9	0.6
1994	88.4	0.2	7.3	2.9	1.2
Milk					
1991	97.7	0.3	0.7	1.2	0.1
1992	95.9	0.2	1.9	1.8	0.2
1993	96.4	0.1	1.6	1.7	0.2
1994	94.6	0.1	3.1	2.0	0.2
Eggs					
1991	92.2	1.6	3.9	1.9	0.4
1992	83.2	1.7	11.2	2.9	1.0
1993	91.2	1.0	5.2	2.0	0.6
1994	88.8	0.7	7.6	2.1	0.8

SOURCE: Data from Goskomstat.

ed the interests of the Russian peasant and workers in agricultural enterprises." At the meeting, a chairman for the rural trade union was elected, Aleksandr Davydov, and a commission was formed to draft new union statutes and to formulate concrete measures to improve facilities for rural workers.[135]

Agrarian deputies who attended the Third Congress of People's Deputies at the end of March–early April 1990 met separately and expressed the need to unify peasant and rural interests. One agrarian deputy—A. F. Veprev, former chairman of the USSR Supreme Soviet for Agrarian Questions and Food—explained both the need to unify and the political weakness of rural interests when he commented that "there is a danger the peasantry, deprived of their own political association, will be separated into other parties; . . . a peasant party will be joined by those to whom the idea of the peasantry's resurrection as a class are close; . . . unless we have political power we will be unable to really struggle for the peasantry's interests."[136]

Out of that meeting the basis for the Russian Agrarian Union was laid. The Russian Agrarian Union held its founding congress in Moscow on April 26–27, 1990, thus becoming a formal parliamentary faction although it never did become a formal political party. The stated purpose of the Russian Agrarian Union was to unite collective and state farmworkers, leaseholders, and peasant farmers, as well as food-processing and agricultural workers, in the trade union of agroindustrial workers. At its founding congress, some 736 delegates attended, as well as Politburo members N. Ryzhkov and V. Vorotnikov. The Russian Agrarian Union claimed it represented some 14 million workers throughout the agroindustrial complex, and although not a political party, it reserved the right to introduce legislation and run candidates in order to defend rural interests. The Russian Agrarian Union—though initially an organization uniting private farm, state farm, collective farm, and agroindustrial workers' interests—was more interested in defending the interests of and securing financial support for existing agricultural enterprises. All but about 15 of the 205 delegates who were elected at the founding congress were from a local party apparatus or were farm managers (the precise occupational status was never published).[137] For this reason AKKOR soon resigned from the Russian Agrarian Union and aligned with state policies to promote private farming.

In general, the Russian Agrarian Union was quite conservative and the vast majority of delegates represented the local party apparatus and farm managers. The Russian Agrarian Union opposed land reforms, arguing that reform from above violated farm sovereignty and stating that the "Union will decisively oppose any unlawful interference in the definition in the form of farming, organization of production and management, and will steadfastly observe the Leninist principle: 'Do not dare command the peasant.'"[138] The common denominator for groups in the Russian Agrarian Union was twofold: first, they wanted to defend the collective and state farm system.[139] Second, they were interested in promoting the socioeconomic interests of rural dwellers and, toward that end, lobbied for higher investments and financial support for the agricultural sector. For example, at its second congress, held in June 1993 (and attended by 700 delegates in Moscow), farm chairmen complained about the deteriorating standard of living in the countryside, and other speakers lobbied for higher domestic grain prices, increased subsidies for meat and milk producers, and continued food subsidies for the population.

The Rise of the Agrarian Party

The political party that grew out of the peasant and agrarian unions in the postcommunist period is the Agrarian Party of Russia (APR).[140] The APR is a formal political party that ran independent candidates in the December 1993 and December 1995 elections. The APR held its founding meeting in Nizhniy Novgorod in December 1992, and its founding congress was held in February 1993, attended by some 250 delegates from 48 regions. The delegates to the congress included state and collective farmworkers, farm managers, private farmers, workers in food-processing branches, rural intelligentsia, scholars, and other citizens of Russia, indicating that this party was more inclusionary than AKKOR. In late 1994 the party held its third congress, at which its party program was adopted.[141] The party also has its own newspaper, *Zemlya i trud*, which is published weekly.

The rise of the APR was important because it constituted the primary rural organization that opposed the state's agrarian reform and land reform policies. A representative example of APR differences with reform policy is indicated in table 19.

The APR led protest demonstrations and meetings in Moscow, submitted open letters to Yel'tsin and Prime Minister Viktor Chernomyrdin, and attempted to influence national legislation.[142] In addition to favoring higher levels of resource allocation to the countryside, the party called for parity of prices between agricultural and nonagricultural goods; a reduction of taxes on rural food producers; a reduction of the value-added tax on agricultural products; an end to speculative reselling of food products; a reduction in food imports and the ending of all subsidies for foreign food imports; compensations and credits for fuels, sowing, and harvests; and guaranteed purchases of production that cannot be sold on the wholesale market.[143] The fact that rural life had deteriorated significantly during the years of reform led to notable strength in the post–December 1993 State Duma.

Table 19. Policy Positions on Agrarian Reform

	Agrarian Party	Government Policy
Private ownership of rural land, without size restriction	Against	For
Unregulated land market	Against	For
Unregulated land use changes	Against	For
Close state and collective farms	Against	For
Subsidized credits to food producers	For	Against
Increase state support to agriculture	For	Against
Price parity between industrial and agricultural products	For	Against
Retail food price regulation	For	Against
Agricultural purchase price guarantees[a]	For	For
Protectionism for domestic producers	For	Against
Increase federal funding for rural infrastructure	For	Against

SOURCE: Author.
  a. State policy favoring purchase price guarantees would apply only to production sold to state food funds, not to the food market in general.

Party Strength

In 1995, the APR was organized in seventy-nine oblasts and krays and had more than seven hundred raion and more than three thousand primary party organizations. Party membership increased from around a hundred thousand to an estimated three hundred thousand members between the end of 1993 and mid-1995.[144] In several oblasts in 1994 the party was organized in almost all raions, including Altai kray, Saratov Oblast, Orel Oblast, and Kursk Oblast.[145] APR leaders characterized the party as especially strong in Moscow, Ryazan, Kaluga, Kostroma, Kursk, and Rostov Oblasts, in Altai and Krasnodar krays, and in the Republic of Bashkir.

The December 1993 elections reflected support for the APR, which received more than 4 million votes nationwide, or about 8 percent of the total number of votes cast. As a result, about thirty-five APR members were elected to the State Duma. The deputy head of the APR, A. Mikhailov, described the "agrarian faction" in the State Duma as having a base of fifty-five persons.[146] In addition to building party organizations from the ground up, the APR also had a number of its members holding important positions in the government from December 1993 to December 1995:

1. Aleksandr Zaveryukha, deputy prime minister in charge of agriculture and, between January and May 1996, the minister of agriculture.

2. Aleksandr Nazarchuk, the minister of agriculture from October 1994 to mid-January 1996.

3. V. Zvolinsky, the chairman of the Committee on Agrarian Policy in the Federation Council.

4. Gennadii Kulik, the deputy head of the Committee on the Budget, Taxes, Banks, and Finances within the State Duma.

5. A. Chernyshev, head of the State Duma Committee on Agricultural Questions.

6. Agrarian representatives are deputy heads of ten other Duma committees.

7. Ivan Rybkin, chairman of the State Duma.

Due to a number of strategic mistakes, the APR fared quite poorly in the December 1995 election. The party failed to clear the 5 percent threshold in order to be assigned seats from the party lists and managed to win seats only from single mandate districts. As a result of the electoral losses, the head of the APR was no longer in the State Duma; the head of the Agrarian Union lost his seat; and the minister of agriculture, who was an APR member, was fired in January 1996.

The APR did win 20 seats in the State Duma in the December 1995 election (out of 450), a sum that represents a significant decline from the 55 seats it had controlled after the December 1993 election. Despite its electoral defeat, in mid-January 1996 the APR formally announced an agrarian faction in the Duma. According to parliamentary rules, a faction must have a minimum of 35 members, so in order to form this faction a number of deputies from other parties and independents joined the Agrarians. Following the December 1995 parliamentary election, an APR member was once again named to head the Committee on Agricultural Questions in the State Duma, from which a significant percentage of agricultural bills originate.[147]

Thus, in response to the state's reform program a rural opposition arose. The rise of a rural opposition was an inevitable result of ideological differences. The fact that the APR existed is not evidence in itself of "failed" reform policy. However, the manner in which state financial levers were used and the conditions they created made it easier for the APR to evolve into the dominant rural political organization.

Rural Interests and State Policies

Rural interests—both AKKOR and the APR—undertook different forms of collective action in response to state financial policies. Although the two groups favored a different allocation of resources within the rural sector, on certain issues they shared common interests. The intent was to obtain policy goals that would benefit rural interests. When AKKOR and the APR negotiated jointly, there were selected instances of limited success. Even when limited success was obtained, however, it was either short-lived or offset by other actions as the following examples will show.

"Successful" Collective Action

The first instance of limited success was in July 1993 when AKKOR and the APR were instrumental in negotiating a large increase in state grain purchase prices and the indexation of these prices to inflation. In real terms, AKKOR and the APR were able to achieve almost a doubling in the price of hard wheat (class 3), from $32 a ton in May 1993 to $59 a ton in July 1993 and $71 a ton in August 1993. Other purchase prices increased significantly as well. Class 1 wheat, for example, increased from $55 a ton in May 1993 to $94 a ton in July 1993 and $112 a ton in August 1993.[148] State concessions on prices and indexing lasted only three months, however, and were ended effective October 1993.

A second example of collective action occurred when AKKOR and the APR combined lobbying efforts in order to persuade the government to protect domestic producers by imposing higher tariffs on food imports or even to restrict imports. The goal was to protect domestic producers from foreign competition, whose prices were often lower (in part due to state subsidies) than those for Russian farms. By late 1993 explicit statements by high-ranking government officials indicated that a change in Russian food import policy would be introduced in 1994. In mid-November 1993, during a visit to Stavropol kray, former Deputy Prime Minister Aleksandr Zaveryukha announced that in 1994 the Russian government would not import foreign grains.[149] One should note that due to decreases in domestic livestock herds, demand for grain had fallen. However, Zaveryukha explained that the purpose for the policy was to strengthen state and collective farms, agroindustrial enterprises, and peasant farmers; to stabilize the domestic grain market; and to stimulate higher domestic grain production. He also added that wool and flax imports would be ended, and that the government intended to reduce imports of milk, meat, and other products. As a result of this more protectionist policy, Russia imported one-half as much meat, coffee, grain, and agricultural machinery in 1993 when compared to 1992, and overall Russian food imports declined 30 percent in 1993.[150]

In addition to the government decision to end grain imports and to curtail other imports of food products, in March 1994 an import duty was announced for food imports for the purpose of protecting domestic food producers.[151] Import duties averaged 15 percent of the custom's value of meat products, fish, butter, milk products, fresh fruits and vegetables, white sugar, grain and grain products. These import duties went into effect on July 1, 1994.[152] However, the "victory" of trade protectionism desired by rural producers was not long-lived.

Evidence suggests that the government was much more concerned with urban interests and securing adequate food supplies to meet urban demand. In September 1994 food import duties were rolled back an average of 6 percent, perhaps in response to urban complaints.[153] As domestic production continued to decline, foreign imports rose, particularly of meat and meat products, setting off consternation among government officials and alarm among rural conservatives.[154] As a result, in July 1995, food import duties once again were increased. The import duty on meat rose from 8 to 15 percent of its value, for butter from 15 to 20 percent, for poultry from 20 to 25 percent, and for sugar from 20 to 25 percent.[155]

Even with increased duties, food imports increased after 1993, particularly for expensive high preference items such as meat and dairy. The United States alone accounted for nearly one-third of Russia's meat and poultry imports during 1994. Meat imports totaled 288,100 tons in 1992, dropped to 85,100 tons in 1993, but then increased to 358,400 tons in 1994 and 504,000 tons in 1995. Poultry imports increased without interruption: from 45,700 to 495,900 tons from 1992 to 1994, to 824,000 tons in 1995.[156] Food imports as a percentage of the total volume of all imports increased from 22 to 28 percent from 1994 to 1995. During 1994 and 1995 the nation spent more on foreign food purchases than was paid to domestic food producers.[157]

Rural interests were unable to achieve protection through lobbying, and thus, at the end of 1995, a draft of a law on food security was circulated, which would have curtailed imports and led to greater domestic protection. This law was never adopted, however. Within the general anti-import and anti-Western mood developing in the spring of 1996, trade protectionists raised fears over the quality and health standards of U.S. chicken imports. Although the poultry industry had been hard hit by reform (as had all processing enterprises), this issue was purely political. Seen as a "last stand" by trade protectionists, this effort—if it had been successful—would have meant that other food import groups were to be targeted in the future. However, after several weeks of high-level negotiations, the issue was settled and Russia agreed to continue poultry imports from the United States. Agrarian interests lost out in three ways: food imports continued to increase as a percentage of all imports; legislation to protect domestic producers was not adopted; and industry-specific attempts at trade protection also failed.

On the whole, despite rural collective action, rural interests were not able to offset the effects of ending the social contract or the direction of state rural social policy. Despite representation within state political institutions, working through the legislative system was not effective in ameliorating the effects of state financial levers. Despite the alleged strength of the agrarian lobby, in fact, rural interests did not win on a single issue of disagreement with government policy.[158]

Rural interests were politically weak compared to state capacities and instruments of policy. For example, from 1988 to 1991 subsidies to agriculture constituted almost 16 percent of national GDP, but since 1992 they have totaled only 5–6 percent. In addition, capital investments to agriculture declined by a factor of 20 in comparison to 1991. Furthermore, the APR was unable to achieve compliance with financial resource allocations that were approved in federal budgets. The political weakness of the APR to prevent the effects of state financial levers was evidenced by the admission of A. Burykov, the deputy head of the APR and a former member of the State Duma, who revealed that although agrarian deputies were unsatisfied with the 1995 budget, they voted for it anyway because they feared even worse consequences for agriculture if they did not.[159] The real political story was one of rural weakness and inability to offset the state's use of financial policies that adversely affected rural interests.

Conclusion

In the Soviet period, the state determined the economic environment in which farms operated through state and party planning structures. The economic environment in turn affected incentives at both the farm and the individual level. Dysfunctional incentives more than anything else influenced farm operations, productivity, and efficiency. In the post-Soviet period, state financial levers were also important, as planning (administrative management) was replaced by "economic levers."

In designing economic policy, it is often the case that economic strategies and their levers collide with social policies, and both are affected by politics. The ending of the social contract—which was a social policy designed to serve political ends during the Soviet period with minimal attention to economic costs—ushered in an economic strategy that reduced financial flows into the countryside. On one hand, the ending of the social contract could be justified on economic grounds. On the other hand, the combination of economic policy and politics had unintended as well as intended consequences for the agrarian sector. Although the Soviet social contract had contributed to some of the economic deficiencies in the agricultural sector, the ending of the social policies implied in the contract went beyond economic imperatives. The contract ended before equalizing urban-rural living standards and constructing an infrastructure capable of supporting a modern agricultural system.

The ending of the social contract signified a clash between macroeconomic policy and social policy. During the contemporary reform period, reduced financial flows were intended to force farm transformation but instead contributed to the "crisis" atmosphere in Russian agriculture. Farms cut down their livestock holdings and curtailed acreage planted, and output declined significantly. As a result, meat production and consumption declined (which some Western economists have noted was simply a correction in excess demand), but consumer demand was subsequently satisfied through meat imports. As agriculture in general became increasingly unprofitable and unable to meet production costs, farm debt increased. Farm debt was compounded by high interest rates and reduced consumer demand, which forced farms to adapt by further curtailing production. A vicious circle developed, and ultimately the state had to write off some farm debt. Thus, we see that the dilemmas of farm transformation gave rise to Soviet-type behaviors that in the end undermined reform goals. The problems of farm transformation were more complex than merely reducing budgetary outlays to the collective sector, but instead, a relatively simplistic approach was adopted, which did not have intended effects and which gave rise to numerous undesirable consequences.

The ending of the social contract also had an effect on rural living standards. Urban-rural wage inequality increased to levels not seen for at least fifty years. Inequalities in nonmonetary living standards and housing amenities also increased, with rural life growing more difficult and unattractive. As rural conditions deteriorated, rural organizations arose to protect the broader interests of the countryside. In particular, the Agrarian Party espoused prorural policy positions, which resonated with rural dwellers and which in turn translated into a rather significant—and surprising—electoral showing in December 1993. The APR operated both within and from outside the state political apparatus. The actions of the APR were important because it worked for policies that frustrated attempts to transform the collective sector. The APR was also instrumental in organizing rural responses to the state's financial policies. With few exceptions, however, the rural sector was not able to defeat the power of an urban-government alliance.

It is clear, therefore, that a multitude of consequences arose from the termination of the social contract. But did the effects have to be as grave as we have seen in this chapter? Probably not. Inasmuch as ending the social contract had economic imperatives, there was no overriding economic impetus to turn financial levers so sharply against the rural sector in so short a time. We have seen that the state strategy was to use financial levers to force farm transformation. But the real questions lie at the heart of that strategy: Can farms transform in a hostile economic environment? Or has production simply declined without transformation? In the Russian case, it is clear that the latter occurred.

5

Land Reform and the Development of Private Farming

In addition to enterprise privatization within the collective agricultural sector, a second component of Russian agrarian reform was the development of private farming.[1] State reform measures defined rural social policy in the private sector, just as they affected the collective sector. In the next two chapters we shall analyze the development of private farming in Russia.

During the Soviet period the state controlled the private sector—primarily personal plots—through various limitations imposed from above, which were reinforced by local officials. During Gorbachev's rule, conditions were liberalized for personal plots, but his main reform impetus was to promote leasing within a collective farm structure. Meanwhile, Yel'tsin's Russia moved toward land privatization and the creation of a stratum of private farmers whose agricultural performance, it was hoped, would exceed that of the collective sector. By the late Gorbachev period, Russia had assumed the leading position in developing a rural private sector that moved beyond leasing and liberalized personal plots. Dating from the earliest reform initiatives on land privatization in 1990, during the next four years state legislation facilitated the development of a private agrarian sector comprised of private peasant farms, privatized food-processing plants, and liberalized conditions for agricultural trade.

In this chapter we shall analyze how the post-Soviet state exerted an influence on the newly emergent private farming sector. An understanding of land reform and privatization in the post-Soviet period must begin with the role of the state and the nature of the reform environment created by state-sponsored reform legislation. The manner in which the state defined reform legislation had direct relevance for creating incentives to undertake private farming and for defining the environment in which those farms operated. The questions I intend to address in this chapter include:

1. How did state-sponsored incentives influence the development of the rural private sector?
2. How did reform legislation influence the operation of the private sector?
3. To what extent did state actions affect the popularity of private farming?
4. How successfully did reform legislation address issues of egalitarianism within the rural sector?

Private farming was expected to yield economic and political benefits to post-Soviet society. In the economic sphere, private farming was expected to utilize labor and productive inputs more effectively than collective agriculture, which was notorious for its high cost and wastage. In the political sphere, private farmers and their interest group representatives were expected to provide a counter to the political influence of collective farm managers and the entrenched agricultural bureaucracy. Private farming was also consistent with the diffusion of political power away from Moscow and the development of a pluralistic democratic society with an inclusionary political system. Barrington Moore, in his classic book, argued that there is an inherent link between the type of agrarian system and the nature of the political regime.[2] Thus, private farming would complement other reform institutions in post-Soviet Russia and have importance for the overall transformation and democratization of society.

Private Farming and the "Weak State"

One of the most prevalent explanations for the difficulties encountered by private farmers asserted that Russia was a "weak state." This view blamed obstructionist behaviors by state and collective farm managers and others opposed to the transformation of the countryside. The weak state view argued that the state was weak because it could not break conservative opposition to rural transformation.

That the state occupied an advantageous position in society was overlooked in the weak state approach. First, agrarian reform legislation was defined from above, not below, and this had an enormous impact on how reform unfolded. State-defined legislation defined the very nature of reform. The nature of state legislation defined the incentives that rural actors faced and framed the environment in which rural actors operated. Thus, the state's first advantage was to define the nature, course, environment, and pace of reform.

The second advantage was that the state used various financial levers. The state's influence over rural social policy affected production incentives, production potential, and the cost of production—in other words, the core issues in food production. Thus, rural reform from above defined the environment and conditions in which private food producers operated.

To be sure, early in the reform process (that is, during 1990–1991 and into 1992), the weak state approach did help explain some of the early results of land privatization. There is strong evidence that farm managers and local officials tried to obstruct the creation of private farms. During this period we should hasten to point out that there were other factors hindering land reform as well.[3] Nonetheless, during the early period there is no question that farm managers and local officials attempted to block or distort the aims of privatization and rural marketization.

Early reform legislation in the RSFSR made it relatively easy for farm managers to obstruct departure from a farm or the acquisition of land for private farming. Originally, a member of a collective farm (kolkhoz) or a worker in a state farm (sovkhoz) who wished to leave the farm and create an independent farm, submitted an application to the chairman of the kolkhoz or the director of the sovkhoz. A general meeting of the kolkhoz considered the member's request to leave the farm and made a decision within a month.[4] The farm's decision was transmitted to the rural Soviet of People's Deputies. If the rural soviet concurred with the decision, it was sent to the raion Soviet of People's Deputies, who then withdrew the land plot from the parent farm and transferred it to individual use. The land plot, to be allocated as a single area, was to have a value equal to the farm's overall average land value; if not, then tax and other financial concessions were to be granted.[5]

Private farming was in general constrained, however, by the difficulty of leaving the farm and the shortage of land to be used for such purposes. Oppositionist behaviors during the early period of reform were chronicled by reports in the press, by personal accounts in agricultural newspapers, by actual interviews with farmers, and by statistical data on land code violations compiled by the Russian Committee on Land Reform and Land Resources (Roskomzem). Interviews with peasant farmers in both Rostov and Kostroma Oblasts almost always yielded the answer that land was "very hard" to obtain, sometimes taking a year or more for the process to be completed. One peasant farm family whom this author interviewed in Kostromskoy raion battled with a collective farm for eighteen months, finally having to go to court to receive land. Another peasant farmer from the same raion had struggled for eleven months to get land. After this "victory," he was continually harassed by the nearby collective farm from which he was leasing land. He was still linked to the parent farm for water, electricity, gas, and so on, which made him vulnerable, similar to many other private farmers. One time he went to Moscow for a meeting of peasant farmers and when he returned he found the collective farm had turned off his electricity and caused the death of twenty-eight young pigs. On another occasion the collective farm forbade him to gather potatoes that were rotting and that would not be harvested by the farm. He needed to use the potatoes for livestock feed, but the collective farm manager preferred to let the crop rot in the ground rather than help this particular farmer.[6] This case may have been extreme, but it was representative of the types of problems faced by private farmers early in the privatization process.

Statistical data also chronicled farm manager obstructionism to land, but this type of opposition declined over time. In 1992, for example, many farms refused to reorganize, even in regions with close proximity to Moscow where enforcement was stricter.[7] During 1992 throughout Russia there were 424 fines levied for "procrastination" in allotting land, and 7,575 fines for actual violations of the land code. Of fines for procrastination, 47 percent were in regions with good soil (Central black earth, Volga, Northern Caucasus, and Urals regions), and 46 percent of land code violations occurred in those same regions.[8] That farm managers and rural agricultural officials obstructed reform should be of little surprise, since elites seldom willingly or easily yield their power and privilege.

By the time land reform began to enter its second, more mature stage, the weak state approach no longer was an adequate tool for understanding land reform realities. Changes in reform legislation made it more difficult for farm managers to block departure from a farm. Surveys of private farmers by the World Bank during 1992 in five oblasts show that the price of inputs—not access to land—was the primary problem confronting farm operations.[9]

While it is certainly true that early in the land reform process farm managerial staff and agricultural officials at the raion and oblast level obstructed land reform and private farming, their resistance decreased over time and by 1994 was no longer the main constraint. A second survey in 1994, sponsored by the World Bank with cooperation from the Agrarian Institute in Moscow, polled farm managers, farm workers, and private farmers. According to the results from that five oblast survey, in 1994 85 percent of farm managers supported the allocation of land for private farming.[10]

Among the questions asked was "why not become a private farmer?" Only 6 percent answered due to restrictions on buying and selling land, and 15 percent cited inadequate land. Instead, economic and psychological reasons dominated: 74 percent cited insufficient capital, 60 percent mentioned difficulty obtaining farm inputs, and 56 responded that they were afraid of risk.

My own fieldwork in Kostroma Oblast confirms a decrease in oppositionist behaviors over time, reflected in a sharp decline in fines as time progressed. By July 1, 1993, there had been only one fine for "procrastination" in land distribution, and forty-nine violations overall of the land code. A survey of 255 private farms in Kostroma Oblast during 1993 showed that only 7 percent "had difficulty" obtaining land.[11] By July 1, 1994, there had been no fines for procrastination and only thirty-six violations of the land code in the oblast, representing a decrease of 38 percent from 1992.[12] We should note that land code "violations" most often occurred when individuals illegally took land without permission and began to use it. During 1993, for example, some 41 percent of land code violations throughout Russia occurred as a result of the illegal taking of land; another 34 percent was because of nonproductive uses of land, and 16 percent because of unlawful structures on land.[13] Not all "violations" therefore were by state or collective farm managers.

Thus, the weak state approach accounts for only a portion of the difficulties faced by private farmers. Over time, reform failure is largely explained by the effects of state-defined reform. It is now necessary to turn to an analysis of reform legislation. The rest of the chapter is divided into two main parts: in the first part we shall examine four key aspects of reform legislation; in the second part we shall analyze the impact of that legislation.

Reform Legislation and Land Distribution for Private Farms

Because land was the basic economic factor in creating a new stratum of private farmers, several issues surrounding land privatization and distribution were vitally important. State-defined reform legislation was important because of its fundamental effect on who would get land, the quality of land, how much land could be obtained, and what an individual could do with his land. These aspects would play a key role in determining the economic viability—and ultimately, the success—of the rural private sector.

In the Russian Republic a series of laws were adopted in 1990 that laid the juridical basis for private landownership and private peasant farming.[14] A draft of peasant farm legislation was first published in the summer of 1990 entitled the Law on Peasant Farms. Ownership of private property for farming and the hiring of labor were legalized in the final version of the law (after changes and additions) when it was adopted in December 1990.[15] Although RSFSR legislation permitted private ownership of land, early legislation restricted land sales through various moratoria.

Also during November and December 1990 a number of other laws were adopted by the RSFSR parliament that legalized private landownership. In November 1990 the RSFSR Law on Land Reform ended the state monopoly on land and allowed for the transfer of land to individuals. This law indicated that the state would promote the development of all forms of farming and permitted the state to define the size of land plots to be transferred to the population.[16] There was also the possibility of long-term leasing (for between five and fifty years) with inheritable rights, but this was considered leased land, not privately owned. Private ownership and different forms of land tenure were codified in the Law on Ownership in the RSFSR, which was also adopted at the end of December 1990.[17] From these legal acts in late 1990, different forms of landownership were codified and different types of land tenure were established. The 1990 legislation established three main types of landownership (state, collective, and individual), each with an equal right to exist.

Land tenure consisted of lifetime use with the right of inheritance *(vladeniye)* and actual ownership *(sobstvennost')*. At first, lifetime use with rights of inheritance was the most popular form of land tenure, as about 58 percent of all private peasant farms in the RSFSR during 1991 were based on vladeniye. Another 24 percent were held in ownership, and the remainder used leased land.[18] Land tenure patterns changed significantly during 1992, however, with a significant increase in leased land for farms (32 percent) and land held in sobstvennost' (42 percent) accompanied by a large decline in vladeniye.[19] A decree signed by Yel'tsin in October 1993 allowed all those with land held in vladeniye to convert it to sobstvennost' up to the norms established for that raion, and therefore vladeniye declined as a main type of land tenure.

The Land Code of the RSFSR, adopted in early 1991 and entered into force in April of that year, unified many of the legal measures that had been adopted in previously adopted laws. The Land Code covered all types of land, the process of obtaining land, land use rights and responsibilities, financial aspects such as rent payments and land taxes, and it established principles for the operation of peasant farms and other private farming activities.[20] Finally, the new Civil Code—adopted by the State Duma and Federation Council in November 1994—codified the right to land and land uses in conformity with the December 1993 constitution.[21]

Who Could Obtain Land

Russian land reform started from scratch, which is to say that there was no "private" rural sector except for small-scale personal plots prior to the 1990 legislation. One of the first tasks was to decide how to transfer land from the state and collective sector to persons wishing to undertake private farming. In general, there were three main methods for prospective private peasant farmers to obtain land: to lease land from a parent farm; to receive land from the state land fund; and to acquire land from state and collective farms upon departure from the farm. (There was scattered evidence of a fourth source, the expansion of private plots, but in general this method was superseded by the other three.)

It is significant to note that just as land shares were distributed equally among collective and state farm members, the distribution of land for private peasant farming was also egalitarian and, further, was not rigorous in screening out unqualified applicants. The legal bases for land acquisition were established by the Law on Peasant Farms adopted in Russia on December 27, 1990, which stipulated the right of "every able-bodied citizen" who possessed "specialized agricultural knowledge or past specialized training" to organize a peasant farm.[22] In the case of many land claims, preference was given to citizens who lived in that locality.[23] Chapter 10, article 58 of the 1991 RSFSR Land Code added that any citizen could receive land who was at least eighteen years old and who had experience in agriculture and corresponding skills, or who had past specialized training. The requirement for agricultural knowledge was often ignored during 1991–1992, and most private farms were started by former urban residents. For example, in September 1991 a published account based on government surveys indicated that on average throughout the country less than 28 percent of private farms had been created by former state and collective farmworkers (there was, of course, regional variation).[24]

Aware that the "wrong" people were obtaining land, at the end of 1992, Minister of Agriculture V. Khlystun indicated that land distribution requirements would be stiffened.[25] Nevertheless, at the national level there appeared to be a basic tension between efficiency and equity, suggested by a Yel'tsin decree in April 1993 that promised land for small-scale agricultural operations to "every citizen." First on the list for this new distribution were to be former military personnel who by that time had been demobilized in large numbers, and stories abounded of military families living in tents due to a lack of adequate housing.[26]

At the regional level, in areas where land was in short supply but demand was high, there is evidence that distribution requirements were in fact tightened. In Moscow Oblast, for example, in July 1993 a resolution was adopted making it more difficult to obtain land for private farming.[27] Part of the July resolution (No. 6/54) simply conformed with statutes of the 1991 Land Code and 1990 Law on Peasant Farming. Although these provisions were already on the books, widespread evidence suggested that they had been widely ignored. In addition to meeting age requirements (ages eighteen to sixty-five) and the requirement of being physically able to perform agricultural work, the new requirements for receiving land specified that a person had to possess a permanent residence permit *(propiska)* for the oblast, which in effect precluded persons from other oblasts receiving land.

Furthermore, the educational and practical work requirements were stiffened. A person had to fulfill one of three conditions in order to establish eligibility for land. First, the person was to have either a higher or a specialist education in one of the leading branches of agriculture, as well as practical experience in agricultural production for not less than two of the last ten years. Second, a person could obtain specialist instruction and preparation in agricultural production and complete two years of practical work. A third option was intended for those without much education and allowed the person to receive land if he had worked in a main branch of agriculture for at least five years. In addition it was necessary to complete a course of special preparation for private farming.[28]

Quality of Land

The quality of land that was obtained by prospective private farmers was directly linked to the source of that land. The primary sources of land for the creation of private farms were leased land from the parent farm, land from the state land fund, or land from state and collective farms.[29] The latter two sources provided an opportunity to obtain a land plot free of charge for the operation of a private farm (and will be discussed below).

In general, we should note that laws regulating the source of free land plots for peasant farming underwent considerable change during 1990–1991. During 1990, only unwanted or unused land was available to prospective farmers, and these lands were to be used to create a state land fund with subdivisions down to the raion. In January 1991 Mikhail Gorbachev issued a presidential decree that ordered an inventory of "irrationally used land in kolkhozy, sovkhozy, forestries, and other land users, including land given to ministries and departments of the USSR."[30] The purpose of this decree was to increase the availability of land for distribution; but the quality of this land was suspect and comprised a very small percentage of agricultural land, about 2 percent of all agricultural land. Thus, an early approach was to give peasant farmers unwanted or unused farmland that obviously was of inferior quality.

The Russian Republic adopted a more radical approach. Two months after Gorbachev's decree, in March 1991, a resolution—passed by the Presidium of the RSFSR Supreme Soviet and Council of Ministers and signed by Boris Yel'tsin—gave executive committees and the Presidium of local Soviets of People's Deputies the right to withdraw up to 10 percent of a farm's land and allocate it to citizens for cooperative farming, peasant farming, or for the operation of private plots and collective gardens. Approval from the local Soviet of People's Deputies would come after the fact of withdrawal.[31] This was a method to make better land available and to prevent conservative rural soviets from blocking this transfer of land. This March resolution represented the first attempt to open up state land for privatization without farm approval, although its impact was still blunted by the obstacles to leaving a farm.

Land Funds. During the earliest stages of land reform a primary source of free land was the state land fund. Land funds existed at the raion level. To create a land fund, the raion (or city) Soviet of People's Deputies withdrew land from kolkhozy, sovkhozy, and other land-using enterprises. The land in this fund included agricultural land that had been abandoned or converted into less valuable land; land of all categories that was not used for a special purpose; agricultural land that had not been used for one year, or land that had a non-agricultural purpose and had not been used for two years; land of agricultural enterprises whose productivity for the last five years was more than 20 percent below the norms of similar land; and land that had not been purchased and was still in the possession of the bank.[32] With these stipulations it is easy to see that often the best land was not included in a land fund and not available for private farming, although we should note that land on which it was not possible to grow agricultural products was not included in the land fund.[33] For example, in Kostroma Oblast raions with the most advantageous location and proximity—key considerations when ascertaining transportation and production costs—had the least amount of land available in their raion land funds.[34]

Land funds were intended to provide land for citizens who were not members of agricultural enterprises but who wanted to become private farmers.[35] In most cases non-farm personnel did not have access to farmland and received land from their raion land fund. The one exception was if a state or collective farm disbanded. If land remained after distributing land to farm members, up to the established norm, excess land then was transferred to the raion land fund.[36] Recipients of land could be either local citizens or persons who relocated to other regions in order to receive land plots. To receive land from this fund an application was submitted to the local or city Soviet of People's Deputies (and after October 1993 to the rural or city administration).

Land funds in general were rather small. In Moscow Oblast by mid-1994 the land fund consisted of 195,000 hectares, or about 11 percent of all agricultural land, of which 144,200 hectares were considered arable agricultural land.[37] In Kostroma Oblast as of July 1, 1995, the total amount of land available in the oblast land fund was only 63,200 hectares, of which 43,800 were considered arable.[38] These figures represented 3.6 percent of all agricultural land and 5.8 percent of arable land in the oblast.[39] We should also point out that agricultural users of all types competed for land from land funds. In other words, not only private farmers received land from this fund. In fact, private farmers obtained rather small percentages of land from land funds.

Land from State and Collective Farms. The second source of free land was from state and collective farms. Whereas land funds were intended to provide land to non-farm members, land withdrawn from state and collective farms was intended to be distributed to farm members who wished to engage in agricultural production outside the parent farm. The Russian Republic adopted several measures that put it in the vanguard for making state and collective farmland available for private farming while the Soviet Union still existed. In May 1991 recommendations were approved for the breakup of unprofitable state and collective farms for the purpose of making their land available for private farms. These recommendations then were incorporated into the December 1991 legislation on the reform of state and collective farms. Also in 1991, a land tax was adopted that imposed a tax per hectare of land held by state and collective farms. The idea was to penalize state and collective farms for holding onto land that might be cultivated by other agricultural users.

In effect, the land tax acted as a differential rent to extract a higher charge for areas with better land. The purpose was to make fertile lands available to private farming by making it very expensive to retain this land if the farm was not using it; and this tax was introduced at a time when the early results of land reform indicated that nearly two-thirds of peasant farms were being started in non–black earth regions with relatively poor soil.[40] The land tax was differentiated by land quality, with southern areas having a tax much higher than in non–black earth areas where the tax per hectare was only ten rubles per hectare (a little more than sixteen cents per hectare at prevailing exchange rates at the time).[41]

In general, however, it was not clear whether the land tax was successful in freeing up land for private farming. One land official explained to me that the land tax was not a factor at all because it was so small. Farms did not incur a significant financial penalty for holding onto land. As inflation raged during 1992–1993 the land tax became even less important in affecting economic behavior. In late 1993 the land tax per hectare in the non–black earth zones of Russia was many times less than the cost of a newspaper or of a metro ride in Moscow. In November 1994 a base land tax was adopted, establishing average land tax levels for oblasts based on land quality and other factors (and within oblasts raions had different land taxes, but the mean of all raion land taxes had to conform to the federally mandated oblast level land tax). In addition to a base land tax rate there is also a coefficient by which the base tax is multiplied. The purpose of the coefficient was to index the tax rate to inflation. In order to keep up with inflation, starting in 1994 and continuing thereafter, a new coefficient was introduced by which the land tax would be multiplied. For example, a multiplier of 1.5 times the land tax was introduced in 1995 for rural land,

and again in 1996, making the land tax for rural land equal to the following equation: Raion land tax times 1.5 times 1.5 times number of hectares.[42]

The origins of land availability from state and collective farms can be traced to the December 1990 Law on Peasant Farming, which took effect on January 1, 1991. This law permitted farm members to receive property shares and land shares, which, if they decided to leave the farm, would be distributed to them free of charge for private farming operations. At first, access to good quality farmland was restricted because of the complicated process for leaving the farm, which permitted abuses and obstructionism by local officials and farm chairmen who did not want to lose farmland or workers.[43] There were many stories in the press of farm members who were not allowed to leave the farm, or who, upon leaving, were assigned extremely poor quality land in remote locations.

Over time the process for leaving the parent farm with land and property became simplified. Yel'tsin's December 1991 decree "Urgent Measures for the Realization of Land Reform" was of central importance because it streamlined the process for obtaining farmland. This decree stipulated that "leaders of farms will assign land to workers and to members of his family within one month of the time of receiving the announcement [of his desire] to establish a peasant farm."[44]

After December 1991, two possibilities to receive state or collective farmland existed. In the first method, a reorganized farm distributed land shares to all those who were eligible to receive them. However, the way the reorganization system worked, land shares were conditional and only those persons who actually left the farm received their land in-kind and were assigned specific plots of land. In the case of land distribution in-kind, the oblast Committee on Land Reform (now called Committee on Land Resources and Land Surveying) designated specific land plots to each applicant using a standardized procedure that numbered land plots and then assigned them to persons in the same order as the applications were made.[45] The land allocated to persons leaving a farm was to be at least of equal quality to the rest of the farm (this provision was often ignored as departing farmers were given marshy or rocky soil, or forest). A second method for receiving land from a state or collective farm was if farm members elected to disband the farm. In this case, all members of the farm received equal land shares, irrespective of age, ability, occupational status, or how long they had worked on the farm.

Yel'tsin's December 1991 decree was quite successful, as the number of private farms increased dramatically during 1992. Nonetheless, it was felt that landownership by farm members had proceeded quite slowly, so further impetus to opening up farmland was given by Yel'tsin's decree on the "Regulation of

Land Relations and the Development of Agrarian Reform in Russia," which was signed on October 27, 1993, and entered into force upon publication.[46] Although in many respects the October 1993 decree duplicated much of what had already been legislated, it also went beyond previous legislation in significant ways.[47]

One of the most important aspects of the decree was to create a foundation for a land market based on the freedom to buy and sell agricultural land. In order for a land market to operate, landownership had to be institutionalized, and thus members of state and collective farms were to receive a certificate of landownership for a share of the farm's land.[48] This certificate was a legal document giving landownership to the holder, as well as serving as the basis for purchases, sales, leasing, or mortgaging land. The right to purchase land was significant because it meant that a person outside that farm, a person who had relocated to that area, or even an urban person would have the ability to obtain land other than that in a land fund. Granted, the decree stipulated that other members of the collective had first priority in obtaining land or land shares for sale, but in principle this decree established a rudimentary land market.[49] The distribution of certificates proceeded quite successfully, as by mid-1995 in Kostroma Oblast 85 percent of private plot operators, 98 percent of private farmers, and 100 percent of land shareholders in agricultural enterprises had been issued landownership certificates.[50]

Statistics indicated that similar percentages had been attained nationwide. On July 1, 1996, the following percentages of landholders had received land certificates nationwide: in agricultural enterprises, 96.2 percent; on private farms, 79.2 percent; private plot operators, 81.4 percent; collective garden plot operators, 65 percent; and holders of land plots for individual dwelling construction, dacha plots, and garage construction, 45.2 percent.[51]

Because Yel'tsin's October 1993 decree established the basis for a land market, in December 1993 another decree abolished free land from land funds for urban dwellers. Starting in late December 1993, urban dwellers no longer were given land free for agriculture but had to purchase it, except for private farming. In addition, oblast-level land norms that had capped how much could be bought were also abolished, and henceforth land purchases were regulated primarily by what the purchaser could afford.[52] Rural dwellers and farm members continued to obtain land and property from state and collective farms free of charge, up to established norms, until the end of the farm reorganization process. After farm reorganization was completed in early 1994, a farm member no longer could withdraw from the farm with land, and the rural land market was expected to fill the demand for land.

How Much Land Could Be Obtained

Both urban and rural dwellers were limited in the amount of land they could obtain free by land norms that were established by raion soviets. These norms regulated how much land each person could obtain free of charge. Raion land norms were to reflect the number of eligible recipients and the quality of land.[53] In the rural sphere, the amount of land actually distributed free of charge to a farm family could not exceed the norm per person but might be less than the norm, and there were reports of land allocations less than established norms, which put additional downward pressure on the size of private farms. The size of the land plot assigned to a prospective farmer took into account the farmer's specialization, the number of farm members, and other factors such as the type of farming operation (grain growing, animal husbandry, fish, reindeer raising, and so on). The quality and location of the land were also considered when assigning land quantities.

Sizes of land plots for other forms of private agriculture (private plots, collective fruit and vegetable gardening, dacha plots, and so on) were also established at the raion or city level, and these plots likewise were transferred to users free of charge. Different sizes were established for rural and urban owners, with urban-owned plots usually capped at 0.15 hectares whereas a rural dweller could obtain several hectares.[54] These ranges also varied by location of the land. In Kostroma raion, up to 0.15 hectare of rural land could be obtained free, whereas inside city limits only 0.09 hectares could be obtained free. In Moscow Oblast the limits were smaller because of the larger population base competing for available land.[55]

Thus, land plots that were transferred free of charge and land plots that were privately owned for small-scale agriculture continued to be regulated in size. However, additional land above free land norms could be purchased at prices established by the Russian Ministry of Agriculture and the Ministry of Economics and Finances until Yel'tsin's October 1993 decree, and then after October 27, 1993, at free market prices. Conversely, if a person received free land for private farming and then wanted to convert it to a private plot, for instance, then local land norms for plots would apply and any excess land would either have to be purchased or would be withdrawn from the owner and would revert to the raion land fund.[56]

Freedom of Sale and Purchase of Land

Under the Soviet system, the state owned all land and individuals were precluded from either buying or selling land. Even as Gorbachev attempted to move the Soviet Union away from Stalinist-era institutions, he was unwilling to embrace private landownership, although during his rule legislation was adopted that permitted long-term land leasing.⁵⁷ As we noted previously, the USSR Law on Land, adopted at the end of February 1990, provided for the leasing of land in order to create independent individual farms but did not embrace landownership, so still there was no land market.⁵⁸ Later in the Gorbachev period, the Russian Republic government further advanced land reform, both in terms of private ownership in general and landownership in particular. In December 1990, for example, the RSFSR adopted a Law on Property that was different from the USSR Law on Property in that it distinguished between private *(chastnaya)* and state property. In December 1990, the first legislation allowing persons to buy and sell land since the 1922 Land Code was adopted at the Second Congress of People's Deputies in the RSFSR. This legislation, however, placed a strict (ten-year) moratorium on land sales, and only the state was permitted to buy land. No genuine land market could develop as a consequence of these restrictions.

The 1990 legislation was in effect for about a year until Yel'tsin's Urgent Measures Decree of December 1991. This decree allowed workers who left a state or collective farm to exchange their shares for land and property, or to lease those shares. The decree also allowed farmers to sell their land plot "to other citizens" if they were retiring from farming, had inherited the plot, intended to move in order to farm elsewhere, or would invest the proceeds of sale in the countryside. In short, owners of land plots or land shares were allowed to sell their land under certain conditions, and workers in agricultural enterprises who had land shares could sell them to other workers of the farm or to the farm at free prices.⁵⁹ Even with this limited right, the local soviet was empowered with control over the process, including preparation of the documents and veto power over the sale. Thus, Yel'tsin's Urgent Measures Decree of December 1991 went further than the 1990 laws but ultimately failed to introduce free land sales.

In April 1992, the Russian Sixth Congress of People's Deputies rejected a constitutional amendment that would have permitted land sales. Ignoring the rejection, in October 1992, Yel'tsin signed a decree that allowed the sale—on an experimental basis—of private garden and dacha plots in Ramenskiy raion in Moscow Oblast (not far from the city of Moscow).⁶⁰ According to the decree, the money derived from land sales was to be divided between the local soviet where the land sale had occurred, the oblast government, and the Russian government. Later that month the first land sales were conducted for the construction of individual dwellings.⁶¹

Following this experiment, in December 1992 the Russian Supreme Soviet finalized a law that legalized land sales, but with conditions attached.⁶² These restrictions concerned land use, the price of land, and the size of the land plot.

First, the land could only be sold for purposes of subsidiary agriculture *(lichnoye podsobnoye khozyaystvo)*, as plots around dachas, for gardening, and as plots around individual housing. If the land being sold was used for these purposes, then it could be sold without a moratorium, provided that the purpose for the land use was not to change. In other cases, if the land was sold for purposes other than those stated above and had been received free in ownership, then the land had to be held for ten years before it could be sold. If the land had been purchased, then a five-year wait was required.

Second, the price of land was to be at negotiated market prices. However, market prices applied only to the sale or purchase of land up to the norms regulating plot sizes established by local soviets. If the amount of land being sold was above the established raion norm, then the owner could retain the rest in lifetime use with the right of inheritance or, if he wished to sell, could do so through the local soviet at a negotiated price. Further restrictions stipulated that the size of the land plot bought or sold could not exceed the norms established by local soviets when land was redistributed. This represented progress but still, as a land market, was extremely limited.

In July 1993, however, after long debate, the Russian Supreme Soviet adopted new "Bases" for the land code. These new land bases went further toward laying the foundation for a land market than any previous legislation coming out of the legislative branch (an unrestricted land market was still not permitted). According to these bases, owners of land were permitted to mortgage their land, to exchange land, to sell it or give it away, and to purchase additional land plots. Owners were protected from having their land taken by the bank in the event the debt was not repaid. The nonagricultural uses for which urban dwellers could obtain free land plots were expanded to include, for example, dacha construction, construction of individual dwellings, and entrepreneurial activities not prohibited by law. Free land sales without moratoria were permitted with the proviso that land use would not change, and the previous procedure for land sales was simplified.[63] Private farmers were still limited in selling their land, however, having to wait ten years if they had received the land free or five years if they had purchased the land.[64]

The concessions made by the Supreme Soviet were not in operation for long before they were invalidated by the dissolution of the Supreme Soviet, followed by Yel'tsin's decree of October 27, 1993,[65] and then by the new constitution that was ratified in December 1993. Yel'tsin's October 1993 decree was important to the creation of a land market in that it annulled any remaining moratoria on land sales, and in 1994 a rudimentary land market arose.

The October 1993 decree also contained important provisions for persons who owned land shares or plots of land. The decree allowed for the free sale, exchange, or barter of land shares. The October decree also gave owners of land shares the right to mortgage, lease, or bequeath their shares. Exchanging land plots would conceivably make it easier to consolidate land plots that by law were to have been issued as a "single mass" but often were scattered. The mortgaging of land would allow a farmer or group of farmers who created a multifamily farm to raise capital in order to purchase machinery or to cover the cost of inputs through the growing season. The ability to mortgage land was especially important given the difficulty of obtaining credit after September 1993. The ability to lease land would, in theory, permit older share recipients to allow younger, potentially more productive farmers to cultivate agricultural land. Land share leasing served as an important source of passive income for pension-age individuals. Both of these provisions were important because of the rural demographic structure (examined earlier).

It is important to note that the October 1993 decree itself did not create a completely free land market. Although it granted the right to sell land or land shares, the actual mechanism regarding the sale of land shares contained some ambiguity as to whom the land share owner had the freedom to sell his land shares. The October 1993 decree indicated that an owner had the right to sell his land shares to "other members of the collective, or to other citizens and juridical persons" so long as the land would be used for agricultural production (point 5). Many people, including land shareholders, interpreted this to mean that unrestricted land sales were permitted. What was often ignored, however, was the very next phrase of the decree: "At the same time, members of the collective have preference in obtaining land shares over other purchasers."

Nikolay Kalinin, head of the legal department at Roskomzem, addressed this ambiguity when he was asked by a land shareholder in Kaluga Oblast (whose farm had converted to a joint stock farm) how to complete a land transaction with two Muscovites who had agreed to buy his land shares of four hectares. Kalinin answered that land shareholders had the right to sell their land shares to other members of the agricultural enterprise, as well as to other persons and juridical entities (organizations) as long as the land would be used for agricultural purposes. However, "members of the collective have the primary right to acquire land shares before other buyers." If other farm members "refuse to exercise their preferential rights to purchase those land shares, the seller has the right to sell them to any other person."[66]

The second limitation on the emerging land market concerned land use. When land changed, owners' land use was not to change; that is, agricultural land was to remain agricultural land. However, the decree allowed land use to change in "special circumstances," that is, to convert agricultural land to urban purposes. In this case an application was submitted to the raion administration who decided whether to allow land use to change. The raion administration's decision was then subject to approval at the oblast level.[67] In reality, land shares were sold to persons other than members of the agricultural enterprise. In Kostroma and elsewhere land shares that changed hands sometimes led to land use conversion, as witnessed by the construction of large private homes in rural areas and suburbs of cities. The land use change process is administered by the local oblast administration, so the degree to which land use change will occur varies by locale and location of the land in question.

Various press accounts indicated that Yel'tsin bowed to political pressure from conservatives by including the restriction on land use, with the intent to protect agricultural land from being converted into nonagricultural land. Nonetheless, the ability to change land use incensed rural conservatives, who charged that the decree was an announcement of "civil war" between urban and rural interests. Rural conservatives complained that allowing land use to change would lead to land speculation and discriminated against rural dwellers who had no money; and "the worst danger" was that the decree would lead to land being taken out of production, thus "dooming" agricultural production and all food producers.[68]

The liberal response, coming from a high-level government reformer, suggested that this fear was overblown. Evgeniya Serova noted that, in principle, suburban land surrounding large cities should be used for family dwellings, parks, and recreation. Nonetheless, the fact that large cities were ringed by state and collective farms, dachas, and land for private plots and other small-scale agriculture made it very difficult to change land use in the near future, for two reasons: first, because cities such as Moscow received a large portion of their food from food producers within close proximity to the city; and, second, a city wanting to expand would have to purchase land from the owners, since farms and agricultural land had been privatized, and many cities such as Moscow simply had no money to do so.[69] Thus, to maintain the flow of food from surrounding areas into large cities, restrictions were accepted on land use changes, even if liberals disagreed in principle with those limitations.

The right to change land use and the nature of the land market remained two of the most contentious issues dividing liberals and conservatives. During 1995 there were acrimonious political debates as to whether land use changes should be freely permitted and about the nature of the land market. In March 1995 the State Duma finally voted on a version of the land code that had removed all restrictions on land sales and land use. This version was defeated by the bloc of Communists and representatives of the Agrarian Party.[70] After changes in the code, in mid-June 1995 the Duma again rejected a draft land code, this time led by Gaydar's Russia's Choice Party and Zhirinovsky's Liberal

Democratic Party. After a day of further discussion and corrections, another vote took place in which the number of "for" votes increased, but the draft was still defeated, falling short by sixteen votes. In July 1995 the State Duma finally did approve a draft land code by a vote of 230 to 43,[71] although the liberal press vehemently condemned the version for the restrictions it contained on landownership and right of sale.

To break the standoff between the government and the Duma, Prime Minister Chernomyrdin threatened to hold a referendum on private ownership of land in the fall of 1995, which conservatives feared because "there are many more city than rural dwellers and the urban population does not have a very good understanding of agricultural problems."[72] However, the referendum never occurred. Following the parliamentary victory by the Communists in December 1995, liberals and conservatives appeared hopelessly deadlocked.

By late 1997 the battle over the land code had already extended more than two years, with the core issues unresolved. Conservatives continued to press for restrictions on free land sales and land use changes, while proreformers were just as determined to introduce a liberal land market. As a result of this standoff, a land code proved impossible to get adopted. The version of the draft land code that was adopted by the State Duma in 1995 was subsequently shelved under threat of a presidential veto. In May 1996 another version of the land code was passed by the State Duma by a vote of 288 to 16, but it was even more restrictive than the first, "practically forbidding free land turnover." This version of the land code did not even allow the free sale of purchase of land or mortgaging of land. Land shares could be exchanged, but only among members of a farm.[73] This version did not even have the support of the new minister of agriculture, Viktor Khlystun, who was reappointed to the post in May 1996. As a result, this version was rejected by the Federation Council in July 1996. Members of the State Duma attempted to gather support in order to overturn the decision of the Federation Council, but were only able to muster 269 votes instead of the 300 that were required.[74]

In the fall of 1996, a compromise commission was formed, including fourteen members from the State Duma, four from the Federation Council, and two representatives from the government. The commission reportedly worked on a version that would allow free land sales, but only under very restrictive conditions.[75]

The commission reportedly reintroduced the concept of a moratorium prior to selling. According to published reports, free land sales would be allowed only after the land had been worked by the owner for a period of ten years.[76] After months of work by a mediation committee, in June 1997 the Duma passed another version of the land code. On June 11, 1997, the State Duma adopted a controversial land code outlawing the sale of arable land. This newest version of the land code passed with 285 votes for, 10 against and six abstentions. Supporters of this version in the Duma stated that they have nothing against the constitutionally sanctioned right to private ownership of land, but they continued to insist that the free sale of farmland land must not be allowed. "In accordance with the code passed, the sale and purchase of arable land shall not be permitted," said chairman of the State Duma's Committee on Agricultural Questions Alexei Chernyshev, one of the authors of the document.[77] Within a few weeks Yel'tsin indicated that he would not sign this version of the land code, thus effectively indicating his veto. By late 1997 Russia remained without a land code.

The conservative position was reflected in an interview with Yuriy Lebedev, the head of the APR in Kostroma Oblast in July 1995. Rural conservatives argued that land sales and land use were regulated in many nations in the world and that Russia should be no different. Further, rural conservatives discerned a basic urban-rural divide, and they feared that an unrestricted land market would deprive rural dwellers of the ability to purchase land because the countryside in general was impoverished. In the interview Mr. Lebedev indicated that, unless restrictions on the land market were included to protect rural dwellers, "there will be no new land code."[78]

The Impact of Reform Legislation on Private Farming

The nature of state interventions in the rural sector changed during the transition from a communist to a postcommunist nation, but the impact of state-defined measures remained significant. Reform legislation was critically important to how land privatization and private farming developed. State-defined legislation influenced who got land, how much land, and what a person could do with his land—in short, aspects that were central to the economic environment and the productive potential of the private farming sector as well as to the operation of rural land market. In order to understand the centrality of state-defined legislation on rural social policy, we shall now explore the impact of state-sponsored reform on private farming.

Who Became Private Farmers

The composition of private farmers underwent a significant change during the course of the first five years of land privatization, even as the average number of persons per private farm remained constant at about three to four persons. We saw above how reform institutions permitted both urban and rural persons to obtain land for private farming (each had different land sources). Although a key criterion for non-farm personnel to obtain land was to be agricultural training or experience, in point of fact this was seldom enforced. In practice, a common way around the stipulation requiring "experience in agriculture" was to enroll in a course intended for new peasant farmers. Even those individuals who did have prior experience in agriculture often were unprepared to operate a private farm. Soviet farms had strict divisions of labor so that a person might know one aspect of farming very well, but he or she might be completely ignorant in other aspects. For this reason, extra training was especially useful for those who had worked in agriculture but who needed additional education in farm management, bookkeeping, and many other aspects of running an independent farm.

The problem of preparing individuals for private farming was addressed in various ways. Localities were at the forefront in training prospective private farmers. In Kostroma, a course of study for new private farmers was begun in early 1992 by Germans in association with the local agricultural institute. The course of study at the agricultural institute in Kostroma was offered twice a year and ran about a month; the first class had nine persons. Technical and instructional help was also supposedly available from peasant farmer associations in Kostroma (AKKOR, the largest, and its main competitor, Assistance), but interviews with peasant farmers indicated that often little help was forthcoming.[79]

In early 1993 a national program for training private farmers was adopted. A resolution signed by Prime Minister Chernomyrdin in February 1993 established a federally funded program that was to be created by the Ministries of Agriculture, Science, and Education. It was hoped that, annually, at least five thousand persons at agricultural institutes and universities and two thousand persons at colleges would be trained to become private farmers. The program was to offer courses on agricultural production, management, and marketing, as well as other basic skills such as accounting. In addition, the program established plans to hold seminars by foreign agricultural experts and farmers and created regional consulting centers in order to provide services to private farmers.[80]

Prospective farmers, existing private farmers, and agricultural personnel in general could also receive training through local or regional academies and institutes of higher agricultural education. Without going into exhaustive detail about the educational structure, Kostroma offers a useful example how this training program operated.[81] In 1993 the Academy of Management and Agrobusiness was created in Kostroma. The professors who created this academy had previously taught at the Kostroma Agricultural Institute, which became an "agricultural university" in 1994, bringing more prestige and higher wages, but changing little else.[82] The academy in Kostroma was one of six in the country as of mid-1994 and was part of the Main Administration of Higher Education Institutions under the Ministry of Agriculture, which funded the training programs.[83] The academy employed only three or four full-time faculty but hired lecturers from the agricultural university, other research institutes, the oblast agricultural administration, and farming enterprises in order to conduct training and retraining classes.[84] The Ministry of Agriculture defined the curriculum for the Academy of Management and Agrobusiness, and the courses were taken primarily by collective farm and joint stock farm management personnel and specialists, accountants, economists, agronomists, and veterinarians on those farms, and by various specialists in the oblast agricultural administration. Courses vary in duration, usually lasting from one month to one semester, according to the subject being offered. Upon completion of the course the student is presented with a certificate *(udostovereniye)*, which indicates when, where, and what subjects were studied.

In addition to courses for persons employed on farms and in the agricultural bureaucracy, there were also courses for those desiring to become private farmers. Among the courses at the Kostroma Academy was an offering on "The Preparation of Peasant Farmers." In 1993, 135 individuals enrolled in this course (out of a total of 1,862 students). This course included 218 hours of instruction over a four-week period. The structure of the course was as follows:

1. The Economic Bases of Organizing Private Farms (twelve hours): This section covered such things as the distribution of land plots, and the nature of farm property, property insurance, farm income and expenses, and tax obligations.

2. The Legal Bases Regulating Farm Activities (ten hours): This section covered contracts, laws governing private farms, legal responsibilities, and relations between farms and service organizations.

3. Productive, Financial, Commercial, and Other Types of Farm Activities (fifty-two hours): This section covered such things as product quality, health standards, veterinary standards, types of marketing, how to market production, relations with food processors and suppliers of material inputs, credit and financing farm operations, accounting, how to apply for credit at a bank, and farm construction.

4. Production of Plant Products (fifty-two hours)

5. Production of Animal Husbandry Products (fifty-two hours)

6. Preparation and Storage of Production (seventy-two hours).[85]

In addition to various training opportunities, there were numerous publications, handbooks, and other literature for new farmers. In Kostroma, for example, prior to the creation of the Academy of Marketing and Agrobusiness and the establishment of a federal training program for farmers, the peasant farmer association Assistance during 1992 published a trilogy of pamphlets to help new farmers. One was entitled "I Would Like to Enter Farming" and covered such topics as how to become a landowner, how to receive credit, purchase prices for products, taxes, and prices for farm machinery. The second pamphlet was called "How to Become a Farmer" and covered the Assistance regulations and program, a short section on governmental resolutions on private (individual) farming, a membership form for the association, and accounting information for income and credits. The third pamphlet was entitled "Laws for Farmers" and included the most important laws, decrees, and resolutions relating to individual farming and the social development of the countryside. There were also a number of different books available from Moscow publishers that offered comprehensive advice and background knowledge for prospective farmers. These publications, with a press run of between two and ten thousand, offered both technical and general information.[86]

In addition to the fact that agricultural experience often was lacking, when land was distributed during 1990–1991 it was relatively difficult for farm members to leave the farm and obtain land. These two factors affected the composition of private farmers. Early in the land distribution process, an overwhelming percentage of private farmers were former urban dwellers, and this was reflected in the skill level of private farmers.[87] Some sources estimated that between two-thirds and three-quarters of early private farmers were not former farmworkers.[88] In the fall of 1991 the vice president of AKKOR, the main organization for peasant farm interests, revealed that about 25 percent of new peasant farmers were former specialists and management personnel of state and collective farms, another 25 percent were workers from state and collective farms, and the rest were urban dwellers of various professions who "did not have the practical skills" to conduct agricultural operations; such people were said to number 70–80 percent of new peasant farmers in non–black earth oblasts.[89] As a result, early private farmers did not have the agricultural background to be effective producers. In a survey of 234 private farms in the RSFSR during early 1991, only one-third had received specialized agricultural training and more than 60 percent had no agricultural education at all.[90] Urbanites-turned-farmers had an impact on farm production: in July 1992, former vice president A. Rutskoy noted that of the 150,000 private farms in existence, only 3,000 were commercially viable.[91]

Beginning in 1992 it was clear that changes in reform legislation had an impact on who was becoming a private farmer. Following Yel'tsin's decree in December 1991, which streamlined the process to leave a parent farm with land, an increasing percentage of former farmworkers had taken up private farming.

Evidence from surveys in selected oblasts confirmed the changing nature of private farmers. For example, a survey of 190 peasant farms in Kostroma Oblast during 1992 revealed that 40 percent of private farms had been created by former workers on state farms and another 27 percent were former kolkhozniki.[92] In Kaluga Oblast (also in the Central region), former agricultural workers started 826 of the oblast's 1,173 private farms (70 percent) as of January 1, 1994. Another 114 farms were begun by rural dwellers who were, however, not former farmworkers.[93]

Trends in Kostroma and Kaluga Oblasts were supported by national data. In a survey of sixty thousand private farms throughout Russia during 1993, Goskomstat discovered that 84 percent of all private farms were organized by rural dwellers, and of that number 86 percent were run by persons who had previously worked in agricultural production.[94] Western analyses based on nationwide data supported these conclusions.[95]

Improvements in the composition of farm personnel, together with state-financed training, improved agricultural performance in the private sector. For example, selected regions reported that, already in 1992, private farm yields had surpassed those on collective farms.[96] Thus, the entrance of more former farmworkers into private farming had a beneficial impact on private farm production.[97] Although regional and national surveys indicated private farmers were struggling with high prices for inputs and construction materials, suffering from shortages of seed, fertilizer, inadequate roads, and other aspects of rural infrastructure, and finding it difficult to obtain loans and credits, private farm output did rise, and part of the reason was because more experienced people were becoming private farmers.[98] Production and sales trends from 1992 through 1996 by Russian private farmers are indicated in table 20.

In considering table 20, we should note that production percentages are difficult to determine exactly since estimates from Goskomstat and peasant farm groups such as AKKOR differ significantly, with AKKOR claiming much more production than Goskomstat. For example, whereas AKKOR maintained that private farmers contributed about 10 percent of gross agricultural production, Goskomstat calculated that private farms contributed only about 2 percent in 1995 and 1996.[99] In addition, private farm output expressed as a percentage of national output is somewhat misleading simply because production from the non–private sector fell so dramatically after 1991. Despite disagreements over total food contribution by private farmers, the main point is that changes in reform legislation had a beneficial impact on land privatization by creating opportunities for the "right" people to become private farmers who could use their knowledge to increase production despite a large number of economic obstacles.

Table 20. Percentage of Food Production and Food Sales from Private Farmers in Russia, 1992–1996 (for selected commodities)

	1992	1993	1994	1995	1996
Grain production	2.1	5.2	5.1	5.0	4.9
Sugar beet production	2.0	3.9	3.5	3.7	3.5
Sunflower seed production	5.8	9.9	10.2	14.4	13.1
Potato production	0.8	1.0	0.9	0.92	0.93
Vegetable production	0.8	1.0	1.1	1.3	1.1
Meat and poultry production (dead weight)	0.7	1.1	1.6	1.6	1.7
Milk production	0.5	1.1	1.5	1.5	1.6
Egg production	0.1	0.2	0.3	0.4	0.4
Meat and poultry sales	0.4	0.8	0.7	1.5	1.6
Milk sales	0.3	0.5	0.5	1.5	1.5
Egg sales	n.a.	n.a.	n.a.	0.4	0.4

SOURCES: Goskomstat Rossii, *Itogi khozyaystvennoy deyatel'nosti krest'yanskikh (fermerskikh) khozyaystv Rossiyskoy Federatsii v 1994 godu* (Moscow: Goskomstat, 1995), pp. 9–11; Goskomstat Rossii, *Proizvodstvo produktov zhivotnovodstva v Rossiyskoi Federatsii za 1996 god* (Moscow: Goskomstat, 1997), various pages; and Goskomstat Rossii, *Rossiya v tsifrakh 1997* (Moscow: Goskomstat, 1997), pp. 297, 299–300.

Quality of Land

The evolution of reform legislation was also important for opening up better quality land for private farming. Early in the land distribution process, there was definite evidence that private farms were being assigned poor quality land and land located in remote areas. Obviously, not all remote land was of poor quality, but remote land often lacked basic services and infrastructure, and these considerations affected farm marketing opportunities and ultimately farm income. A survey of more than forty thousand private peasant farms in 1992 showed that about 57 percent of all private farms in Russia were over twenty-one kilometers from a raion center.[100] Distance in and of itself is not a problem if there is a decent road system, but in rural areas hard-paved roads often do not exist. To reach one peasant farmer in Rostov Oblast whose farm I visited, we drove for almost two hours along a dirt "road" through endless farm fields. I was amazed when we arrived—there were no mail boxes, no addresses, no land marks of any kind to identify a private farm, just a small house and a dilapidated building for livestock basically in the middle of nowhere. The farmer himself had constructed the road leading from the "main" road to his house. I could not imagine how those roads were passable when it rained.

In addition, other basic services were lacking for private farmers. For exam-

Table 21. Creation Rate of Private Peasant Farms, Russian Federation, January 1991–July 1994

	Net Number of Farms Created (registered)
January 1, 1991–June 30, 1991	20,727
July 1, 1991–December 31, 1991	24,611
January 1, 1992–June 30, 1992	78,086
July 1, 1992–December 31, 1992	55,807
January 1, 1993–June 30, 1993	74,456
July 1, 1993–December 31, 1993	11,853
January 1, 1994–June 30, 1994	15,845

SOURCE: Author's calculations from Goskomstat data on private farms.
NOTE: Farm creation corresponded to the time of the year, with the first part of the year more popular than the latter part. For that reason it is best to compare corresponding time periods, for example, all January–July periods and all July–January periods.

ple, in Kostroma Oblast in mid-1991 it was reported that only 15 percent of private farms were located less than one kilometer from a telephone, 47 percent were located between two and six kilometers from a phone, and 29 percent of private farms were located at least seven kilometers from a phone (and this statistic did not even address the quality of transmission or how often the phones were operable).[101] By January 1, 1995, most private farms in the oblast remained poorly equipped. Over one-half of farms had no running water, 20 percent had no electricity, 40 percent had no normal access roads, and only 10 percent had animal sheds.[102]

The quality of land available affected the regional distribution of private farms. As a result of limited access to "good land," during 1990 and most of 1991 a majority of peasant farms developed in non–black earth regions.[103] It was only after Yel'tsin's December 1991 decree that state and collective farm property became easily available to persons wishing to leave a parent farm and begin private farming. Once farm managers had to allow workers to leave unhindered and to receive land, a fundamental change in the pattern of farm development became evident. Beginning in 1992 a rapid growth in the number of registered private farms occurred, as well as a change in the location of new farms. Thus, the number and location of private farms changed rapidly as reform legislation changed. The growth rate in the number of private farms to mid-1994 is indicated in table 21.

The period of rapid expansion in the number of private farms occurred during 1992 into 1994 before beginning to stagnate thereafter. The table depicts the explosive growth that occured during 1992 and 1993, and then reflects the slowdown that started in 1994, which continued through at least 1997. What table 21 suggests is that, during the early period of reform, the combination of favorable economic conditions (see chapter 6) and liberalized legislation which made it easier to depart a farm contributed to the rapid expansion in the quantity of newly created private farms. During January–July 1992 and January–July 1993, over ten thousand farms a month were being created, a fact that led minister of agriculture V. Khlystun at the end of 1992 to predict that soon Russia would have five hundred thousand private farms.[104] Besides the tremendous growth in the number of private farms, private farming became much more prevalent in the black earth areas of Russia. Since January 1992, the most notable trend was the shift in the numbers of private farms away from northern areas and to southern areas. Specifically, a significant increase occurred in the number of private farms in the Northern Caucasus region—one of the richest farming areas in Russia where yields have been consistently higher and costs of production consistently lower. Table 22 presents data on the distribution of private farms and hectares by region from the end of 1990 to the end of 1994.

The table depicts a desirable trend—the growing importance of private farms in "good" areas and a decrease in areas where land quality and climate are less favorable. Thus, we may conclude that the manner in which reform legislation was defined regarding the quality of land also had a positive impact on land privatization and private farming.

Table 22. Development of Private Farms by Region in European Russia, 1990–1994 (December of each year)

	1990		1991		1992		1993		1994	
	A	B	A	B	A	B	A	B	A	B
Northern	3.2	2.4	2.6	1.5	2.0	1.1	1.4	0.93	1.6	0.83
Northwest	9.0	5.5	4.7	2.2	4.0	1.6	3.8	1.4	4.1	1.3
Central	9.7	13.7	11.8	10.3	11.6	8.7	10.7	7.4	10.9	6.9
Volga-Vyatka	2.2	2.4	3.6	2.4	3.1	1.8	3.1	1.7	3.1	1.8
Central black earth	0.47	0.73	5.6	5.2	5.8	5.8	5.3	5.0	4.6	4.5
Volga	7.7	21.8	12.5	26.1	15.0	26.4	13.9	26.8	13.1	26.5
Northern Caucasus	37.2	12.3	18.4	8.9	22.3	11.1	24.8	11.7	26.3	12.1
Urals	8.2	11.0	13.0	13.5	12.6	13.8	12.7	14.2	12.2	13.8

SOURCE: Author's calculations from Goskomstat data on private farms.
KEY: Column A: Percentage of all private farms in Russia that are located in that region.
Column B: Percentage of total private farmland in Russia that is located in that region.

The Quantity of Free Land

In contrast to the Baltics and most East European nations, the "Russian model" of land distribution was not only notably different but also much more egalitarian.[105] Russian land distribution was governed by a "norm" for land plots to be distributed for the creation of peasant farms. During the Russian land distribution process there was a base amount of land that each prospective farmer received free, an amount that was established by each raion soviet.[106]

Furthermore, the size of land norms was related to the quality of the land as well as the number of claimants, reflecting the production potential of the land and its desirability. It was often the case that the better the land quality, the smaller the land share. Land plots were by law to be issued "as a single mass" but in reality widespread evidence suggests that farmers often found their plots scattered. A land reform official in Kostroma Oblast in an interview said that private farmers received "some good and some bad land" in different locales.[107] Newspaper articles and letters to editors indicate that a dispersed pattern of land distribution was a common practice.[108] A survey by the World Bank of five oblasts in 1992 found that the average land share consisted of 2.3 parcels of land.[109]

In general, land plots that were distributed free of charge to private farmers were quite small. Considerable evidence supports the conclusion that land shares were quite small throughout Russia. A World Bank–commissioned survey found that in 1992 land shares averaged seven hectares in Pskov Oblast, about eight hectares in Orel Oblast, just over eleven hectares in Rostov Oblast, under thirteen hectares in Saratov Oblast, and twenty hectares in Novosibirsk Oblast.[110] In Kostromskoy raion (Kostroma Oblast), which has the best proximity to the central market in the city of Kostroma, the norm for free land to start a peasant farm was established at 3.2 hectares per person in 1992 and changed only slightly thereafter. In outlying regions of Kostroma Oblast, where land quality, roads, and access to market were worse, norms for free land were larger.[111] In 1993 conversations with peasant farmers and land reform officials in Salsk raion (Rostov Oblast), one of the best agricultural areas in all of Russia, indicated that the land norm was ten hectares per farm member.[112]

Farm size was an important consideration because small land plots affected production potential. Although many experts believed that the optimal farm size was between fifty and one hundred hectares, in point of fact at the beginning of 1996 more than 50 percent of private farms throughout Russia were less than twenty hectares in size and another 20 percent were between twenty-one and fifty hectares.[113] Russian agricultural academics, citing experiences in the United States, Germany, and Holland, have argued that large farms enjoy labor productivity that is 1.5–2 times higher than on small farms.[114] Farm size, then, is a critically important variable to the success of private farming. How do we explain relatively small farm sizes and the egalitarian method of land distribution in Russia?

This outcome—small land norms and small private farms—was the result of deliberate state policy. In December 1992, minister of agriculture V. Khlystun stated that he would continue "to insist that it is right to have average quotas for shares of property and land, retaining an egalitarian approach for all peasants."[115] Furthermore, the entire political spectrum of rural political parties, interest groups, and organizations advocated the preclusion of "latifundia" (large land estates).[116] Thus there was a basic egalitarian predisposition to land distribution that had widespread political support.

In fact, when reviewing land legislation, it is important to note that the method of land share distribution for farm members actually became more egalitarian over time. In December 1990 when legislation was originally adopted, land shares for state and collective farm members were based upon length of service to the farm and total labor input.[117] This method of land share distribution meant that older farm members received larger land shares than did younger farm members.

Yel'tsin's decree "On the Procedure for Reorganizing State and Collective Farms" of December 29, 1991, for the first time allowed pensioners and providers of rural social services to receive land shares. This change increased the number of claimants and put downward pressure on land-share sizes within a farm. Available evidence indicates that the shares of these groups were not significantly smaller than those of farm employees.[118] Following Yel'tsin's decree, the process of land distribution to farm members became more egalitarian. A detailed resolution adopted in September 1992 elaborated on Yel'tsin's decree, and the process of land distribution was equalized so that everyone on the farm—including pensioners—received the same size land plot irrespective of length of service or labor input.[119]

Because all farm personnel had equal claim to a farm's land, shares for free land were quite small throughout most of Russia. Furthermore, although land shares could be sold or leased, there was no guarantee that the "best" workers would end up with larger landholdings. The approach to land redistribution in which land shares were distributed irrespective of ability or skill was criticized by many within Russia. One Russian academic for example argued that "practically anyone wishing to do so has the right to receive land and start a farm.... We must clearly and openly recognize that not everyone has the right to land, but only the best of the best, those who are experienced and who are trained professionals."[120] The provision that farm members could exchange or sell their shares to other farm members was hoped to put more land in the "right" hands.

The Russian Land Market

By early 1994 collective farm reorganization was essentially complete, which meant that withdrawals with land would no longer be possible. Henceforth, farm members would receive the cash value of their land share if they wished to depart, and rural dwellers would receive land allotments from a raion land fund. With free land from parent farms no longer available, a land market would be extremely important in order to facilitate the development of private farming in Russia.

Yel'tsin's October 27, 1993, decree created the conditions for a land market, but the land market that emerged, at least in its earliest stage, was quite different from what people had expected because of the restrictions on land use and the land market itself. The land market was restricted in two important ways, both of which were affected by state-defined legislation. The first concerned land use and the nature of the market; the second concerned rural demographics and the land market. The first restriction, concerning land use, stipulated that the purpose of rural land use could not change without permission from the administration having jurisdiction over the land plot.[121] It also stipulated that other members of the farm had the first opportunity to purchase land or land shares. These restrictions meant that, at least according to the law, it was difficult for rural land to be converted into urban land use. Under the provisions of the October decree, urban dwellers were largely precluded from purchasing farmland because agricultural land had to remain agricultural land and other rural dwellers had first opportunity to obtain land.

Permission to change land use was to be granted only in "exceptional cases." Stiff penalties existed to prevent rural land from being turned into urban land. If land was used without permission for purposes other than its originally stated purpose, the right to private ownership would be revoked, the land could be confiscated without compensation, and the seller was subject to a large tax.[122] In Kostroma Oblast, land shares that changed hands sometimes led to land use conversion, as witnessed by the construction of large private homes in rural areas and in the suburbs. In Kostroma Oblast, land officials said that land use changes were approved in undeveloped areas with relatively close proximity to the city of Kostroma.[123] In these cases, raion administration officials approved the changes in land use, and their decision was subject to approval by the oblast administration, but these occasions were relatively rare.

This limitation of the land market had a tremendous effect on its functioning. Relatively few members of the parent farm—the so-called collective—had the willingness or ability to purchase land shares of fellow members even though they were given preference in obtaining these shares. These restrictions

Table 23. Average Size of Private Peasant Farms in European Russia, 1991–1997 (total land holdings, in hectares)

	1991	1992	1993	1994	1995	1996	1997
Russian Federation	41	42	43	42	43	43	44
Northern region	30	23	24	22	22	22	22
Northwestern region	25	20	17	15	14	14	15
Central region	58	37	32	29	27	26	26
Volga-Vyatka region	43	28	25	23	25	26	27
Central black earth region	63	39	42	40	41	41	43
Volga region	116	88	75	81	86	90	91
Northern Caucasus region	14	20	21	19	20	20	21
Urals region	55	44	47	47	48	49	50

SOURCE: Private farm data from Goskomstat.
NOTE: All dates January 1 of respective year.

acted as a brake on land use conversion by preventing large-scale transformation of agricultural land into urban or suburban land. As a consequence, Russian land use has not changed dramatically, despite the ability to buy and sell land. Relatively little agricultural land has turned into urban-use land, which suggests that a lot of marginal land still is being used for agricultural production. Furthermore, the intent to keep agricultural land as agricultural land meant that perspectives are limited for urban growth and the development of Russian suburbs populated by single-family homes.

Furthermore, the fact that urban dwellers were largely precluded from the rural land market affected the nature of the land market. Generally speaking, rural dwellers and private farmers had little money with which to purchase rural land; and bank loans that could be used to purchase land were both expensive and hard to obtain. Therefore, because rural demand for land was rather low, and urban demand for land was precluded from the farmland market, two things happened. First, the size of land plots that were sold tended to be quite small.[124] Second, private farmers were not able to significantly increase the size of their farm, which would in turn increase production potential. As shown in table 23, since January 1994 average farm size has increased significantly only in the Volga region.

We can see the effects of a restricted land market when we consider specific oblast examples. Based on the provisions of Yel'tsin's October 1993 decree, by the summer of 1994 a land market had emerged. Urban and rural land sales were reported in a number of oblasts, especially in Moscow Oblast.[125] During all of 1994 in Moscow Oblast, more than 2,270 urban land deals were concluded, nearly 5,300 rural land deals, and more than 8,400 transactions encompassing abandoned land, for a total of more than 16,000 land deals involving nearly 10,500 hectares.[126] The picture was somewhat more complex than the numbers of transactions would imply, and the nature of the land market should be further clarified. According to unpublished data from the Committee on Land Resources in Moscow Oblast, during the first half of 1994 there were 3,751 land transactions in which just over 417 hectares were exchanged, the average size of a land transaction in Moscow Oblast being just 0.11 of a hectare as of July 1994.[127] For agricultural land, the "land market" in Moscow Oblast consisted primarily of the sale of private plots—*ogorody,* or land within a collective garden—in other words, very small plots of land.[128]

Data obtained from the Committee on Land Resources in Kostroma Oblast showed that land market patterns in Moscow Oblast were not unique to the country's most urban and populated oblast. Kostroma Oblast, which has a population density of less than fourteen persons per square kilometer,[129] displayed the same general characteristics: small plots of land being bought and sold and rural transactions dominating urban transactions. The primary difference between the two oblasts was that Moscow's land transactions were much more dispersed than in Kostroma, where transactions tended to cluster around the main city and oblast center, the city of Kostroma. In Kostroma Oblast, 1,057 land transactions had been registered from January 1, 1994, to July 1, 1995;[130] and 90–95 percent of those land transactions involved the sale of private plots, ogorody, dacha and dacha plots, or small individual holdings within collective gardens. Thus land plot sizes involved in land transactions were quite small.

Although regional variations in the land market exist, Kostroma shares with other regions the fact that the average size of a purchased land plot tends to be extremely small, and this pertains to both urban and rural land. For example, in Kostroma Oblast the average size of a rural land plot that was bought or sold was 0.43 hectares, and if Nerektskiy raion is excluded (which accounts for more than one-half of the total area of purchased land throughout the oblast), the average size for the other 17 raions was 0.17 hectares.[131] In Kostromskoy raion the average size of a purchased land plot was 0.07 hectares, and within the city of Kostroma the average size was 0.05 hectares.[132]

We should note in passing that these data do not necessarily capture the full extent of the "land market" in the broadest sense of the term. There were undoubtedly many transactions that were not registered, in order to escape land taxes and transaction fees. In other cases, land plots that had been privatized were exchanged for plots in other locales. Private plots could even be included in a package deal—for a bigger apartment, for example, trading a two-room apartment in the suburbs plus a land plot for a three-room apartment with a more central location in a nearby city. Anecdotal evidence suggests that land exchanges for apartments or as part of apartment exchanges were more common than outright sales of land, at least in Kostroma, owing to the high cost of land (by Russian standards) and the low purchasing power of the vast majority of Russian consumers.[133] Land reform officials register all transactions regarding land but do not keep statistics on exchanges regarding land, thus the frequency is difficult to quantify exactly.

The second way in which state legislation affected the land market involved rural demographics. In essence, rural demographics and the requisites for a land market clashed. It is necessary to remember that, because such a large percentage of the rural population was of pension age (about 23 percent of the Russian rural population), many recipients of land shares—new landowners—were pensioners. The age structure of the rural population definitely influenced who had land shares, and because such a large percentage of farm members often were at or near pension age they tended to possess most of the land shares that comprised the bulk of the land market.

The ramifications of the rural demographic situation were twofold. First, large percentages of farm members did not want to sell their land shares. A survey taken in Rostov Oblast in 1994 showed that 96 percent of land shareholders in collective farms preferred to turn over their shares to the farm; that is, land remained collective property.[134] Part of the reason behind these preferences was that social services were provided by the farms, and many farmworkers were unwilling to strike out on their own. This interpretation is supported by a survey conducted in Voronezh Oblast, which showed that pensioners who were willing to sell their land shares had either children or other relatives they could live with. Those who did not had no other place to go, and therefore the issue of dispossession not just of land but of place of residence was a key factor in the land market.[135] Thus, the first implication was that old farm members were not willing to give up their security by selling their land shares, and therefore relatively little land was made available for sale.

The second demographic implication was that relatively few rural persons (and even fewer rural men) of farming age were available as potential buyers of land shares. The most common age for private farmers was between thirty and forty, and men were overwhelmingly the ones who began private farms. However, in Kostroma Oblast persons (of both sexes) between those ages constituted only 15 percent of the rural population, and rural men comprised only 46 percent of rural dwellers. Kostroma Oblast is representative of the demographic situation in other non–black earth oblasts and even some black earth oblasts.[136]

Thus, for a variety of reasons the land market has not had the impact that was expected and has not led to the transformation of the Russian countryside.

A well-known proponent of reform, V. Uzun, summed up the situation when he noted that farm members who received land shares in Nizhniy, Orel, Ryazan, or other oblasts where farm privatization has been introduced

> treat [their land] with great care. Cases are very rare when a new owner agrees to sell his land allotment. . . . Also, it is not easy to find buyers deep inside Russia. No one wants to invest money in agriculture. As a rule, land allotments are purchased by agricultural enterprises with guaranteed life-long support of pensioners.[137]

By late 1997, a notable constraint on the land market was limited supply of and demand for agricultural land, as well as the inability to convert its usage. There was little to suggest that the Russian land market was likely to change significantly in the near future. The factors that had limited the market remained in place: an unfavorable demographic structure, difficulty obtaining loans to purchase agricultural land, the exclusion of urban demand, and state restrictions on land use. Despite some unduly optimistic reports in the Western press, the Russian land market remained restricted in both use and sale, and therefore predictions of significant rural transformation were premature.[138]

Conclusion

In this chapter we have argued that the best way to understand the development and operation of the private sector is to focus on state reform initiatives. The state influenced the environment and conditions in which agricultural enterprises and rural individuals operated. The state created incentives to which rural dwellers responded, and therefore the state was at the center of rural behaviors. In this way reform legislation was central to defining the nature of rural social policy in the emerging rural private sector. From our discussion of the impact of rural legislation we may conclude that state-sponsored changes in reform legislation created opportunities and incentives for the "right" people to become private farmers and opened up possibilities for the acquisition of good quality land as indicated by statistical data showing where farms were located. Those were beneficial aspects of reform legislation.

Reform legislation and the incentives that flowed from reform also fundamentally affected production and income potential. Among the most important aspects of reform was the limit on land that could be obtained free. Reform institutions were egalitarian, and they mandated small land plots as part of a deliberate policy. Small plots of land translated into limited income potential, making it difficult to afford farm equipment, to hire labor, or to expand operations; all these factors affected farm yields, which were lower than on other agricultural enterprises. Limited income potential exacerbated private farm debt, making it harder to pay off past debt, obtain new credits, or produce enough to satisfy collateral demands by commercial banks. The high cost of credit and farm operations prevented many farmers from being able to purchase additional land, even though farmers argued that small plots made it impossible to conduct effective commercial operations.[139] Thus, small land plots were at the heart of the desire to prevent rural "latifundia."[140]

Thus, reform legislation facilitated an egalitarian distribution of land in which small parcels of land were received by all, irrespective of ability. Because the farm population tended to be old, the distribution of land shares did not necessarily put land into the "right" hands. These old farm members had the willingness neither to start private farms nor to sell their land shares, which would mean giving up various aspects of social security provided by the parent farm.

Although the basis for a limited land market was established, continuing restrictions limited land use and precluded the radical transformation of the countryside. The impact of reform legislation, therefore, created conditions for the free sale of land, but other restrictions made it difficult to change land use and thus the land market was quite limited. The Russian land market, as it were, involves mostly transactions of small land plots. Unless land use laws are changed to allow rural land to become urban land easily (an unlikely prospect given the political situation), the potential to transform rural Russia will be severely affected. Many Russian cities are ringed by collective farms, their successors, or other types of agricultural land. Without the ability to change land use, it is unlikely that suburbs of individual family homes will spread, or that shopping districts can be decentralized away from the city center. In short, without fundamental changes in the valuation of land, the face of the Russian landscape will not change significantly.

6

Financial Levers and the Impact on Private Farming

The previous chapter demonstrated that state-defined reform legislation affected rural social policy in the private sector by influencing incentives and the legislative environment. We now turn to an analysis of financial levers and how their usage has affected the development and operation of private farming. The particular focus is on how management of state credits and interest rate policies affected the economic environment in which the private sector developed.

Although the contemporary Russian state is much more decentralized than its Soviet predecessor, financial flows to agriculture remain largely center-directed. All farming enterprises, including private farms, receive subsidies and credits from the center (some financial support is granted from oblast budgets, but these sums were relatively small particularly for private farms). Central to the economic environment of private farming were state financial levers. Given the lack of financial capital and the low levels of equipment that faced starting private farmers, the success of private farming in Russia could not and cannot be achieved without financial support from the state. The crucial economic issue, therefore, was how to use financial levers in the most efficacious manner possible.

In order to understand the use and impact of financial levers, in this chapter we shall address the following questions:

1. How were state financial levers toward private farms used?
2. How did state financial levers affect the development of private farming?

Financial Levers and Private Farming   183

3. What was the economic impact of financial levers on private farming?
4. What was the political impact of financial levers among rural liberals?

This chapter will highlight the conflict between economic and political goals. The way in which financial levers initially were used led to private sector dependence on state credits. The "solution" further complicated the situation and led to economic and political consequences. As macroeconomic policy led to fiscal tightening, a sharp downturn in the rate of farm creation and a skyrocketing rate of farm bankruptcies occurred. An unintended consequence of the effort to lower inflation and reduce deficit spending was the undermining of the private farm movement. Decreased financial assistance to the rural private sector also had political ramifications, as reform supporters in the rural sector began to distance themselves from the state, further eroding rural support for agrarian reform, which already was low. We shall now first survey the evolution of state credit policies toward private farming and then analyze the economic and political effects of those policies.

Credit Policy and the Private Sector

Once reform legislation made it easier for farm members to leave a parent farm and to receive land, the number of private peasant farms increased significantly. For example, during 1992 alone, the number of private farms in Russia increased from about 49,000 to over 183,000. But private farmers needed financial support. Estimates in 1991 of the start-up costs for peasant farms ranged from 60,000–100,000 rubles at the low end all the way to 500,000–1,000,000 rubles to begin and equip one farm.[1] Disagreements over which estimate was too low or how much start-up costs might have been inflated for political reasons miss the main point: beginning farmers did not have even 60,000 rubles in personal bank accounts; therefore access to cheap credit was essential to help nascent farmers establish a farm, obtain machinery, equipment, livestock, and to construct farm sheds.[2] Not only did prospective private farmers have little capital of their own, the effects of price liberalization led to extremely high inflation and a growing price disparity between industrial and agricultural goods. Therefore, to facilitate the development of a significant private agricultural sector, during 1992–1994, a number of resolutions were adopted that provided state financial support.

The origins of state support for the private farming sector date to October 1991 when Russian president Boris Yel'tsin placed land and farm privatization high on the list of his reform priorities. In his October 1991 speech Yel'tsin laid out his reform priorities and proposed that the Russian government, starting in 1992, allocate state credits for the development of private peasant farms. He further suggested the assignment of twenty-four thousand tractors and many other types of machinery.[3]

Following this proposal, at the end of January 1992, the Russian government adopted a resolution that formally created the Russian Farmer program.[4] This program was to be the backbone of state support for private farming. It was to provide credits at state-subsidized interest rates. The Russian Farmer program became the principal method of financing the creation of private farms: during 1993 some 83 percent of private farms received state-subsidized credits.[5] Through this program the symbiosis between the private sector and the state was born.

How a Farmer Received Credit

Under the Russian Farmer program, officials in the Moscow Russian Farmer office decided allocations of state credits to oblasts, krays, and republics within Russia. These credits were then allocated on a monthly basis by the Ministry of Finance through Rossel'khozbank, the Russian Agricultural Bank, which would channel the credits to state agricultural banks in the oblasts. Those banks then established special accounts for private farming within five days of the receipt of the funds.[6]

After central allocations were made, the actual responsibility for distributing funds to peasant farmers from the Russian Farmer program was given to oblast and raion offices of the Association of Peasant Farms and Cooperatives in Russia (AKKOR). A private farmer wanting to receive state credits submitted an application to the local branch of AKKOR, which in turn submitted a loan guarantee to the raion or oblast bank that distributed credits.[7] If AKKOR guaranteed the loan, the farmer went to the local Rossel'khozbank branch and opened an account in his name. Then credits were transferred to his account.[8] When AKKOR gave a loan guarantee on behalf of a farmer, relatively little information about the applicant was collected by the bank. The bank merely assumed that AKKOR had the resources in its account to back the loan, and the receipt of credit by individual farmers was virtually automatic. This credit procedure with AKKOR backing was explained to me in 1992 by a vice president, Mr. Zaitsev, of the Kostroma Rossel'khozbank, who characterized the process as a "political, not an economic decision."[9] He meant that loans were distributed with little—if any—information gathered as to how the credits would be used or the ability of the farmer to repay the loans; nor was collateral required. The loan process was extremely lax, and at one point Mr. Zaitsev indicated that the only people who were turned down for credits were those who showed up for their appointment drunk. He estimated that less than 5 percent of the applicants were denied credit.

Credit distribution through AKKOR offices was governed by guidelines established by the national AKKOR organization for its local branches at the oblast, kray, and raion levels. These recommendations provided guidance as to how federal credits were to be distributed to farmers. According to these recommendations, 25–30 percent of credit was to be assigned to the "growth in the number of farm and land area under cultivation," that is, to new farmers; 45–50 percent of credit was be divided among the number of existing private farms; and the final 20–30 percent was to be allocated to the development of agricultural cooperatives and interfarm enterprises.[10] There was evidence that local branches could change these percentages at the margins, if approved by the local leadership.[11]

As long as the local AKKOR branch had sufficient resources to guarantee credit applications the process was relatively simple. But this condition seldom existed. AKKOR officials' chronic complaint was that they did not have enough credit in their account. Shortages of credit in turn led to alleged corruption among AKKOR officials who were charged with distributing credit to friends and to persons who offered bribes. In theory, any private farmer was eligible to receive state credit no matter if he belonged to AKKOR. There is some anecdotal evidence, however, that members of AKKOR received preference. I interviewed private farmers who felt that they, being nonmembers, had been discriminated against. One can assume that access to cheap credits was a prime motivation for joining AKKOR, and the head of AKKOR in Salsk raion in 1993 indicated that "every" private farmer (over four hundred at the time) had joined his organization although there were other peasant organizations in existence.

If the local AKKOR branch did not have enough credit in its account, then farmers had to borrow at market rates from a commercial bank, where rates were much higher.[12] In such a case, AKKOR was not able to guarantee the loan. Legislation allowed the farmer to mortgage *(zalog)* his crops, property, and later his land to receive credit, or he could turn to an insurance organization, which also required some kind of guarantee and charged a small percentage of the loan as a fee.[13] In commercial banks, the application process was much more rigorous, and application forms from commercial banks in Kostroma typically ran to six or seven pages, asking for detailed information that allowed the bank to ascertain the applicant's ability to repay the loan. Anecdotal evidence from Kostroma and reports in agricultural newspapers indicate that commercial banks were very reluctant to extend credit to private farmers, even at higher interest rates, because private farms were considered uneconomical and high-risk ventures. As a consequence, in 1994 less than 10 percent of private farm applications for credit from commercial banks were approved. Whether the loan was AKKOR-guaranteed or not, the maximum length of the repayment period was five years, although most credits were short-term in nature, which meant repayment of principal was due within a year.

Credit Policy and State Support of the Private Sector in 1992

In order to facilitate the development of a private farming sector, various advantages were offered by state programs. For example, the children of private farmers were exempted from military service. Most state support, however, concerned material and financial resources. The January 1992 resolution that established the Russian Farmer program provided for many different types of state support, of which the most important category was financial. The January resolution stipulated that 6.5 billion rubles were allocated for assistance to peasant farmers during 1992; and because credits were indexed to inflation on a quarterly basis, the total for 1992 actually reached 94 billion rubles.[14] Most of this allocation was used to facilitate the creation of private farms, that is, unrestricted credits that could be used as the farmer saw fit: to purchase machinery or livestock, to build sheds for livestock, or to pay for other operational costs.

Another form of financial support to private farms was a variety of production subsidies and compensations that were offered to all food producers for food products sold to the state. With the low volumes of production during 1992 this particular form of state financial support benefited primarily the more established state and collective farms,[15] although private farmers did benefit from having a market for their produce. Furthermore, in order to provide an incentive for private farming, the state would pay for relocation costs for persons moving from urban to rural locales, or from one rural locale to another. In addition, beginning private farmers were given a onetime payment to help them get started: 75,000 rubles (about $330) to the head of the family and 15,000 rubles (about $66) to each member of the family in 1992, and these payments were indexed to inflation.[16]

Farmers were also given various tax advantages. For instance, peasant farmers were freed from the land tax for a period of five years (from the date of farm creation) for land that was distributed free, but they remained liable for taxes on land that was purchased. They were freed from income taxes on income derived from agricultural activities; and they were freed from taxes on enterprises that produced, processed, or stored agricultural products.[17]

A further type of state support concerned construction of rural infrastructure. The 1990 Law on Peasant Farming stipulated that the state had responsibility to construct rural infrastructure for private farmers. The January 1992 resolution stated that not less than 15 percent of state capital investments were to be used for the construction of infrastructure for private peasant farms.

Other types of support concerned the allocation of needed machinery. The state also promised to expend hard currency in order to purchase farm machinery for private farmers. Private farmers were to receive not less than 30 percent of the tractors, automobiles, and agricultural machinery and equipment allocated by the Ministry of Trade and the Ministry of Agriculture. Finally, there was educational and instructional support. The state also would pay for free training and instruction. In 1992, a program for education and preparation of farmers was begun and run by the Ministry of Agriculture and the Ministry of Education.[18]

About two weeks after the January resolution, it was announced that AKKOR and the Russian government had signed an agreement for mutual cooperation. The agreement stipulated that the state would provide 8 percent interest rates for credits and other types of financial and nonfinancial support if peasant farmers would "support the president and the government of the Russian Federation in working out and implementing a course of deep economic reform," and if peasant farmers agreed to sell up to 25 percent of their produce to the state through a "system of state contracts at market prices."[19] With this system of subsidized credits the government compensated banks for the difference between the state-guaranteed interest rate and the interest rate charged by the Central Bank. For example, when the Central Bank rate was 80 percent in 1992 and the state-guaranteed rate was 8 percent, the government compensated agricultural banks for 72 percent of the interest due on the loan. It is necessary to remember that inflation during 1992 was about 2,600 percent, so even 80 percent interest rates were negative real rates.

Credit Policy and State Support for the Private Sector in 1993

Building on the 1992 state support for private farms, other measures were adopted during 1993. Again, a range of financial and other types of assistance was offered. In early 1993 it was announced that the government planned to allocate some 225 billion rubles for the development of private farming (about $100 million, in December 1992 rubles) to be indexed to inflation.[20] Furthermore, a resolution adopted in January 1993 stipulated that peasant farmers who sold their output through state channels were eligible to receive a range of production subsidies and compensations spelled out in the resolution.[21]

In April 1993, a resolution for state support of the agroindustrial sector entitled "On Supplementary Measures for the Stabilization of the Agroindustrial Complex in 1993" indicated that financial support for peasant farms not only would be indexed to inflation but also would be tied to the growth in the number of farms. The resolution recommended that the RSFSR government as well as executive organs in krays and oblasts allocate not less than 15 percent of their capital investments to the construction of infrastructure that would benefit peasant farmers; that the government and organs of executive power in krays and oblasts provide electricity and build roads for agricultural enterprises; and that peasant farmers be compensated from the state budget for 50 percent of the cost of building livestock sheds (for not less than ten head of cattle).[22] Finally, persons migrating to rural areas to undertake peasant farming in 1993 were to be given a onetime monetary payment, which by this time had increased (in nominal terms) to 200,000 rubles (about $200) for the head of the family, and 50,000 rubles ($50) per family member for rural families; 300,000 rubles ($300) to the head of the family, and 75,000 rubles ($75) per member of the family for persons relocating from a city or from a republic other than Russia; and 500,000 rubles ($500) for the head of the family and 100,000 rubles ($100) per member for military personnel who had been demobilized.[23]

Private farmers also continued to receive subsidized credits. In 1993 the government increased the subsidized interest rate to 25 percent (plus a 3 percent margin) at a time when state and collective farms were paying 170 percent interest. Despite the increase in interest rates for 1993, private farmers were still able to obtain credits at real negative interest rates as the rate set by the Central Bank exceeded 200 percent annually. Continued state support for the system of subsidized credits for peasant farming was contained in a decree signed by Yel'tsin on July 27, 1993, entitled "On Urgent Measures for the Support of Peasant Farming and Agricultural Cooperatives," which pledged that subsidized credits through the Russian Farmer program would be retained during 1993–1994.[24] In addition, the decree allowed peasant farms to delay making interest payments on their short-term loans until January 1, 1994.

Problems in Receiving Credit

The creation of a new private farming sector confronted the Russian government with a number of dilemmas. On the one hand, a main purpose of agrarian reform was to make the agricultural sector less costly, specifically, to reduce budget outlays and subsidies. On the other hand, given the start-up costs for private farms, the increasing cost of farm equipment and machinery, the lack of preexisting capital held by prospective farmers, and high inflation, private farms required access to subsidized credits.

The need for state credits can be seen in several examples. A major impediment to private farming was the low level of equipment most farms possessed. As of January 1, 1992, there was one tractor for every two private farms, one truck for every five farms, one grain combine for every twelve farms, one plough for every five farms, one seeder for every seven farms, and one mower for every ten farms.[25] Equipment and machinery levels improved notably by

January 1, 1993, but still were far below levels found in the United States and other Western nations. The situation in 1992 led one author to write that "farmers approached the harvest without equipment and without hope . . . what is a farmer to do if he has no harvesting equipment? Where is he supposed to process, to store his production?"[26]

Thus, if private farming were to be successful, the state would have to finance the development of a new rural sector. As a result of the need for state-backed financing, it would be fair to say that private farms became dependent on state financial support for their operation and their financial survival. The growth and development of the rural private sector was directly linked to the provision of state credits. As one reporter wrote in late 1992, "Without credit the rise of private farming in the countryside is unimaginable today. Earnings from the sale of agricultural products are not sufficient to cover the costs of inputs, which are sharply more expensive."[27]

The importance of credit to private farmers is shown by data from one of the most productive agricultural oblasts in the nation. According to a survey of 1,570 private farms in Rostov Oblast during 1992, *not a single raion had earnings from farm production that exceeded expenditures*. This does not mean that individual farms were not profitable, but at the raion level private farm expenditures were considerably higher than income. (Credits were considered income, and therefore if the sum of income *[dokhod]* and credits *[kredit]* exceeded expenditures *[raskhody]*, a farm was considered profitable.) Overall, for the 1,570 farms that were surveyed, each private farm had average earnings of 362,811 rubles ($1,612) and an average expenditure level of 535,307 rubles ($2,379) per farm.[28]

The same general pattern was true for Russia as a whole. In each of the economic regions private farm expenditures exceeded farm revenues, and thus farms were dependent upon state-subsidized credit. This trend is illustrated in table 24.

The table shows that in every economic region farm expenditures exceeded income from production. In non–black earth oblasts, expenditures exceeded revenue by a factor of more than 2.5; whereas in the more productive southern regions, although farms on average did not cover expenses, at least the gap was narrower. Piecing together evidence from a number of different sources we can depict the financial situation with reasonable accuracy. First, we should note that most farm income was used to meet operational costs. Based on a 1991 survey of over twenty-nine thousand private farms in Russia (60 percent of the total at that time), on average about 83 percent of private farm expenditures went for farm operations, and less than 15 percent of farm expenditures went to wages of farm members or hired labor.[29] The percentage of farm income de-

Table 24. Private Farm Revenues, Expenditures, and Credits by Region, 1992 (on average for one farm, in thousand rubles)[a]

	Revenue	Expenditures	Credits Received[b]
Russian Federation	226.4	358.3	408.0
Northern region	80.3	200.9	217.2
Northwestern region	58.7	155.0	202.2
Central region	120.1	304.2	374.9
Volga-Vyatka region	124.3	312.9	345.5
Central black earth region	273.6	406.0	527.4
Volga region	408.3	483.3	573.2
Northern Caucasus region	229.2	293.8	338.0
Urals region	245.5	403.9	431.6

SOURCE: *Krest'yanskiye (fermerskiye) khozyaystva Rossiyskoy Federatsii (po dannym obsledovaniya na 1 Yanvarya 1993 goda)* (Moscow: Goskomstat, 1993), pp. 52–53.

a. The average ruble-dollar exchange rate during 1992 was 222.7 rubles to the dollar.

b. Credits include both short-term (to be repaid by the end of the year) and long-term (up to five years).

voted to wages most likely declined and the percentage used for farm operations increased because of the tremendous increase in input prices after 1991.[30] Where did a farm get its financial resources? Data for 1992 show that private farm revenues from production did not cover farm operational costs, and therefore state-subsidized credits helped to cover the difference between expenses and income and to provide some operating capital.

Thus, state credits to farmers were vital, not only to bridge the gap between earnings and expenditures but also to allow the farmer to obtain machinery, seed, fertilizer, and other necessary inputs. It was precisely the problem of the high cost of farm inputs that was cited by 73 percent of private farmers in Rostov Oblast as a key obstacle to private farming.[31] Nationwide, during 1992–1993 some 80 percent of private farmers surveyed indicated that the high cost of inputs was the primary factor slowing the growth of private farming.[32] In subsequent years, surveys of private farmers continued to demonstrate that economic and financial problems were the main impediments to the development of private farming.[33]

Despite the central importance of the Russian Farmer program, a host of problems plagued the program. A major problem for farmers was the amount of credit available. One article maintained that, during 1992, due to credit shortages private farmers could obtain on average less than 60,000 rubles per farm.[34] Given the tremendous price increases of farm machinery and equipment, this sum was not sufficient. In December 1992 one tractor cost at least 800,000 rubles, ranging to several million. By December 1993 the smallest, cheapest tractor cost 3 or 4 million rubles. Larger tractors started at 20 million rubles and ranged up to 150 million rubles.[35] High input prices, small farm sizes that limited production capacity and thereby farm income, and limited credit meant that private farms were debt-ridden and unprofitable except for state financial support.

Nor did the credit situation improve over time. In early 1994 the vice president of AKKOR, Yuriy Linin, noted that during 1993 the state had allocated an average of 950,000 rubles per peasant farm, but in constant 1991 rubles this equaled only 1,000 rubles.[36] According to AKKOR data, in constant rubles financial support per private farm averaged 30,000 rubles in 1991, 4,500 rubles per farm in 1992, and only 1,000 rubles per farm in 1993.[37] Conservatively speaking, state support was thirty times less in 1993 than in 1991, although one calculation put the difference as high as fifty times.[38] More generally, the trend in state support for private farming declined in subsequent years. When calculated in dollars, state support for private farming declined from 42.3 million dollars in 1992 to 6.9 million dollars in 1995 (not including production subsidies and compensations). Thus, federal support for private farming declined as time went on. The trend since 1992 has been to reduce federal expenditures, putting more of the budgetary burden on local budgets.

Another key problem was that the allocation of credits to agricultural banks was very slow in arriving. As early as 1992, AKKOR officials were complaining that credits were not distributed to meet needs. For example, in October 1992 the president of AKKOR, Vladimir Bashmachnikov, stated that only 15 percent of state credits had been dispersed during the first nine months of the year.[39] The problem was that the Ministry of Finance often did not compensate agricultural banks for many months with the difference between the market rate of interest and the state-guaranteed rate. When state-subsidized credits were not available, private farmers had to pay commercial rates for loans. Even when subsidized credit was available, it often was only a short-term loan, meaning it had to be repaid by December 1 of that year. During 1992, for example, about three-quarters of credits were short-term,[40] and during 1993 between 50 and 60 percent of available credit was for a short-term duration.[41]

A related problem was that often credits were not distributed at all, even after having been included in the budget. From the earliest days of the Russian Farmer program there were reports that credits allocated in the budget were not reaching private farmers. For example, of the 94 billion rubles that were assigned to facilitate private farm creation in 1992, only 54 billion were in fact distributed.[42] During the early fall of 1992, minister of agriculture Khlystun admitted that the credit program to farmers was not being fulfilled.[43] During 1993, the situation improved, but still only 80 percent of budgeted rubles were distributed.[44]

Unpublished data from AKKOR in Kostroma Oblast illustrated the critical situation that farmers faced in obtaining credit. Although data are not complete for all of 1993, what does exist is very instructive. In February 1993, Kostroma Oblast received 105 million rubles for distribution from the Russian Farmer program. However, during that month, a total of 57 million rubles were distributed, or just a little over half the sum that had been allocated. In March 1993, the story was much the same: out of 115 million rubles assigned by the Russian Farmer program, again about one-half of the total sum was distributed. In April 1993 almost the entire amount was distributed, but in May 1993 distributions declined again.[45]

From these data it is obvious that credits were not reaching private farmers in the amounts intended. Although the data from Kostroma did not indicate how many farmers in each raion received credits, the data did show the total amounts of credit issued per raion. Usually, for any given month during 1993, only about one-half of Kostroma's twenty-four raions received 2–3 million rubles in credit ($2–3,000); others would get nothing. In a few rare cases a raion would receive 5, 6, or 8 million rubles, and in one case a raion (Nerekhtskiy raion in February 1993) received 10 million rubles in credits. In February 1993 Nerekhtskiy raion had forty-nine private farms, so if we assume that every private farmer was given an equal share of credit, each farmer received just over 200,000 rubles ($200). Thus, even when credits were available they were in amounts that were quite small relative to input and production costs. With farm start-up costs estimated at 20–30 million rubles in 1993, often the average line of credit was insufficient.

Where did the money go if not to farmers? There is ample evidence of shady behaviors, if not outright corruption, in the Russian Farmer program. The availability of "free money" led to abuses. Persons obtained land and became "a farmer" simply to be eligible to receive cheap credits, even if they never planted a single crop. Mr. Berulin, the head of the Kostroma Oblast branch of AKKOR, claimed that his organization had been hurt by a number of people who had obtained state-guaranteed credits but who used them for purposes other than agricultural production. For example, he noted that some "farmers" were building spacious private homes on the banks of the Volga River.[46] As a result of abuses of credit, the AKKOR branch in Kostroma Oblast during July 1993 was "virtually bankrupt," and so private farmers had to turn to commercial banks, at that time facing interest rates of at least 150 per annum. Other published examples of abuses included the buying of cars or shares of stocks. In addition, published accounts reveal that many regional and raion peasant association leaders assigned credits based on personal relations. In other cases state credits were not assigned to peasants at all but remained in local budgets,[47] or even in the local AKKOR's account.[48] For all these reasons there were numerous reports of farmers facing difficulties in receiving credit.[49]

In mid-July 1995 the results of a government audit *(proverka)* as to where the money to agriculture was going were published. In the published report, the government blamed commercial banks for withholding financial resources for anywhere between one and nine months.[50] Although commercial banks were by no means blameless, they in fact accounted for the smallest percentage of credits to private farmers. Data reveal that, among private farmers who received credits in 1993, about two-thirds received credits from the Central Bank or a branch of the state agricultural bank, Rossel'khozbank. Thus full blame cannot be laid on the commercial banks. In fact, the nondistribution of credits and other financial resources to the rural private sector was simply part of a larger pattern of nondistribution to the agricultural sector at large.

To conclude, financial support to private farmers was critically necessary, but financial resources were not distributed in an efficacious manner. Negative real interest rates led to excessive demand for credit—which the state could not meet—and set off conflicts among farmers.[51] Credits were distributed in a manner that rewarded the creation of numerous small unprofitable private farms that were credit-dependent and not commercially viable. Furthermore, the need to build rural political support led to lax oversight of credit distribution, which in turn led to abuses and corruption.

One alternative would have been to provide below-market interest rates, but to charge a higher interest rate to discourage waste and fraud. In the summer of 1994 this is precisely the recommendation made during an interview with the deputy director of the Peasant Party of Russia, Aleksandr Khokhlov, who suggested that the state provide subsidized credits at 60 percent interest to private farmers at a time when the Central Bank was charging about 130 percent per annum.[52] AKKOR itself adopted a similar position in resolutions and open letters to the government.[53] Moreover, credit distribution should have prioritized the creation of fewer—but larger—farms that could become commercially competitive. However, this latter course was not followed.

During 1992–1993 the state pursued a land and farm privatization campaign, augmented by a program of financial support, that expanded the number of private farms without concern for the fact that private farms were nearly completely dependent upon state support to survive and operate, a reality that led some Western analysts to characterize private farms as "state peasant farms."[54] Furthermore, once credits were indexed to inflation and tied to the growth in the number of farms, a vicious self-sustaining cycle was formed that ultimately the government could not afford to sustain. The state's "solution" to the situation it had created undermined the centerpiece of its reform strategy and devastated private farmers who were the most economically vulnerable segment of the agricultural sector.

Patterns of Credit Distribution to Private Farms

The Russian Farmer program, combined with changes in land reform laws, led to an explosion in the registration of peasant farms: from 49,000 at the beginning of 1992 to more than 183,000 by January 1, 1993. By October 1993, when subsidized credits ended, there were more than 265,000 peasant farms registered throughout Russia. More important than the overall number of peasant farms was that the "right" areas had experienced the fastest growth. Early in the land reform process the largest percentage of peasant farms had been established in non–black earth zones with relatively poor soil, owing in large part to the way legislation governing the reform process had been defined. However, between January 1992 and January 1993, the percentage of peasant farms found in areas with the best soil in Russia—the Central black earth region, the Volga region, and the Northern Caucasus region—increased from 36 to 43 percent of all peasant farms,[55] and this trend continued throughout 1993.[56] Linked to this process was the provision of subsidized credit to beginning peasant farmers.

It is therefore clear that farm credits from the Russian Farmer program played a crucial role in facilitating the development of private farms throughout Russia. Nonetheless, one of the major reasons for the stopping of subsidized credits was the argument that low interest rates led to the overconsumption of credits, particularly in poorer agricultural regions (and here we should note that negative real interest rates probably contributed to excess demand in all regions). Excess demand for credits precluded scarce credit resources from reaching more productive areas, although farmers in all regions often had difficulty obtaining the level of credit they desired. Areas where agricultural production was highest and where productivity gains were likely to be greatest from land privatization were actually shortchanged in credits. Criticism about the method of credit distribution was voiced even within AKKOR. For example, at AKKOR's fourth congress in February 1993, some argued that it would be preferable to allocate fewer farms more credit, which would enable them to become larger and expand production, rather than distribute small sums of credit to many farms, essentially an egalitarian mode of distribution.[57]

Table 25. Summary of Private Farm Indicators: Rank Order of Regions, 1992

Volume of Credits by by Region	Gross Grain Production	Average Grain Yield (per hectare)	Percentage of Private Farms
Volga region	N. Caucasus	N. Caucasus	N. Caucasus
N. Caucasus	Urals region	Central black earth	Volga region
Urals region	Volga region	Volga-Vyatka	Urals region
Central region	Central black earth	Central region	Central region
Central black earth	Central region	Volga region	Central black earth
Northwestern	Volga-Vyatka	Urals region	Northwestern
Volga-Vyatka	Northwestern	Northern region	Volga-Vyatka
Northern region	Northern region	Northwestern	Northern region

SOURCES: Author's calculations based on *Rossiyskiy fermer*, no. 1, February 9–16, 1993, p. 4; *Krest'yanskiye (fermerskiye) khozyaystva Rossiyskoy Federatsii na 1 Yanvarya 1993 goda* (Moscow: Goskomstat, 1993), pp. 2–4; and *Itogi khozyaystvennoy deyatel'nosti krest'yanskikh (fermerskikh) khozyaystv Rossiyskoy Federatsii v 1992 godu* (Moscow: Goskomstat, 1993), pp. 23–24.

An Interregional Analysis of Credit Distribution

The logic of the Russian Farmer program suggested two main objectives: first, to support private farm development in general, and second, to support private farm development in the most productive areas. The first objective was fulfilled, and the tremendous increase in the number of registered farms coincided with the provision of subsidized credits. Regarding the second objective, published data on regional distributions of credit allow an assessment of the program to direct resources to the most agriculturally productive areas. During 1992 the distribution of credits flowed in the greatest volumes to the Volga region, the Northern Caucasus region, and the Urals region, in that order. These three regions also had the largest percentage of private farms, with the Northern Caucasus first, followed by the Volga region and then the Urals region. Table 25 presents a rank ordering of volume of credits, gross grain production, yield per hectare, and percentage of private farms during 1992.

This table indicates that the distribution of credits during 1992 was "right" in general, but not always in particular. At a general level it is clear that the three regions with the most private farms and the highest gross grain production received the most credits.[58] However, when considering the exact rank ordering of regions, the Volga region received the most credits even though it ranked third in grain production, fifth in yields, and second in the percentage of farms. The Northern Caucasus region ranked first in gross grain production, yields,

Table 26. Summary of Private Farm Indicators: Rank Order of Regions, 1993

Volume of Credits by by Region	Gross Grain Production	Average Grain Yield (per hectare)	Percentage of Private Farms
Volga region	N. Caucasus	N. Caucasus	N. Caucasus
N. Caucasus	Volga region	Central black earth	Volga region
Urals region	Urals region	Central region	Urals region
Central region	Central black earth	Volga region	Central region
Central black earth	Central region	Volga-Vyatka	Central black earth
Volga-Vyatka	Volga-Vyatka	Northwestern	Northwestern
Northwestern	Northwestern	Northern	Volga-Vyatka
Northern	Northern	Urals region	Northern

SOURCES: *Rossiyskiy fermer,* no. 34, September 28–October 5, 1993, p. 2; *Itogi khozyaystvennoy deyatel'nosti krest'yanskikh (fermerskikh) khozyaystv Rossiyskoy Federatsii v 1993 godu* (Moscow: Goskomstat, 1994), pp. 45–46; Goskomstat private farm data; and author's calculations.

and percentage of farms but received the second most credits. Other anomalies existed for the remaining regions as well. For example, although the Northwestern region had the lowest grain yields per hectare in European Russia, that region received more credits than did Volga-Vyatka, which had the third-highest grain yield per hectare. Overall, there was no discernible consistency or pattern that matched credit distribution to output by farms during 1992.

For these reasons there are valid reasons for criticizing the system of credit distribution, and criticism arose at the end of 1992. As a result, more effective use of credits was promised for 1993. Once again the Volga, the Northern Caucasus, and the Urals regions received the most credits. However, among the top three regions, the Volga region was the big winner this time, receiving over 19 percent of the credits distributed while accounting for less than 14 percent of all private farms. The relative losers were the Northern Caucasus and the Urals regions. The Northern Caucasus region had nearly 25 percent of all private farms but received only 17 percent of the credits; and it ranked first in gross grain production and yield per hectare. The Urals region received 15 percent of state credits but had less than 13 percent of all private farms.

Summary indicators for 1993 showing the rank order by region for volume of credits, gross grain production, grain yields, and percentage of farms are illustrated in table 26. The table indicates that many of the same anomalies continued in the distribution of credit during 1993. As in 1992, the Volga region received the most credits, with the Northern Caucasus region second, although the Northern Caucasus region had the largest number of farms, the highest level of gross grain production, and the highest average grain yields per hectare. Again there appears to be no discernible pattern that would reflect a consistent method for decision making about credit distribution, which raises questions about how decisions were made and for what purposes. A comparison of rank ordering by region for 1992–1993 of credits received and private farm grain production is shown in table 27.

The table indicates that very little changed in patterns of credit distribution between 1992 and 1993. With the exception of the Volga-Vyatka region moving up from number 7 to number 6 in amount of credit received during 1993, the table depicts basic continuity of 1992 patterns. This trend means that, despite promises to correct credit distribution patterns, credits during 1993 were not linked to gross output or yields. All private farms, regardless of location, were underequipped and suffered from poor infrastructure. However, farms with "good" soils and climates needed even more state support because their production was higher and their operating and production costs were thus greater.

Yet, credit data reveal that the largest producers were not necessarily the recipient of the most credits. During 1992, the losers included the Northern Caucasus, the Urals, the Central black earth, and the Volga-Vyatka regions ("loser" being defined as a region ranking lower in production than in receipt of credits). During 1993 the losing regions included the Northern Caucasus and the Central black earth regions, with the Urals and Central regions the only winners. A basic conclusion from these data is that, because of the manner in which credit distribution was implemented, the most productive areas did not

Table 27. Comparison of Rank Order of Regions, 1992 and 1993

Credits per Region, 1992	Credits per Region, 1993	Gross Grain Production, 1992	Gross Grain Production, 1993
Volga region	Volga region	N. Caucasus	N. Caucasus
N. Caucasus	N. Caucasus	Urals region	Volga region
Urals region	Urals region	Volga region	Urals region
Central region	Central region	Central black earth	Central black earth
Central black earth	Central black earth	Central region	Central region
Northwestern	Volga-Vyatka	Volga-Vyatka	Volga-Vyatka
Volga-Vyatka	Northwestern	Northwestern	Northwestern
Northern region	Northern region	Northern region	Northern region

SOURCES: Tables 25 and 26 and sources cited therein.

necessarily receive the most credit. Therefore, the Russian Farmer program did not fulfill its second task, that being to support the most productive regions with the highest level of state credits.

The Great Turn

Despite the centrality of subsidized credits to the development of private farming, the regime faced a critical choice between helping private farming succeed and pursuing a tight monetary policy to meet IMF guidelines on deficit reduction. The choice was made to reduce government spending by curtailing subsidies and credits to agriculture in general, which included private farming. Already in early 1993 there were reports that former vice premier Yegor Gaydar wanted to shift more of the burden of state support for private farmers onto oblast budgets.[59] By the end of August 1993, there were signs that the government was reconsidering its policies of financial support to private farmers, perhaps because the government had already expended the budgeted allotments for 1993 during the first eight months.[60] As a forewarning of what was to come, a short Agrofact article at the end of August 1993 reported that the government would retain subsidized credits to private farmers at least through the third quarter of the year. The article noted that "there is no reason for concern. In any case until October."[61]

The most drastic move of overall fiscal tightening came in late September 1993 when Prime Minister Chernomyrdin signed resolution number 975, which effectively ended subsidized credits to private farmers.[62] This resolution ended access to the below-market interest rates on credits that private farmers had enjoyed during 1992 and most of 1993 (state and collective farms also received below-market interest rates, but their rates were much higher than those offered private farmers). Instead, private farmers would be subject to interest rates established by the Central Bank of Russia, which were 210 percent a year from Rossel'khozbank and 420 percent from commercial banks at the time the resolution was signed.[63] Interest rates remained at those levels until about mid-1994 before dropping somewhat, to between 130 and 150 percent from Rossel'khozbank.[64] In some regions as early as the summer of 1993 subsidized credits had already "disappeared" according to the vice president of AKKOR.[65] In other regions the Central Bank had begun to charge commercial banks a nonsubsidized interest rate prior to September, whereas subsidized credits continued in some parts of the country until November to allow completion of the harvest.[66] Essentially, however, the September resolution marked an end to the era of state-subsidized credits to private farmers.

Why did the state end a policy that was so clearly central to the development of private farms and the centerpiece of its agrarian reform program? In an article published subsequent to resolution 975, Aleksandr Kalinin, head of the department of the agroindustrial complex apparatus within the Council of Ministers, shed light on the reasons for the government's decision. While admitting that "no one can doubt the fact that agriculture needs a special system of credits," he also argued that the previous course of financial policy had led "our government to the edge of a financial catastrophe." Kalinin argued that "the present year has shown that we spent a lot of money on agriculture, but ineffectively."[67]

Other pressures on the Russian Farmer program came from the burden of the deficit on the budget. With Western institutions pressuring the central government to reduce its spending in order to qualify for loans, the government tried "to protect itself from direct responsibility for the development of the farmer movement."[68] As a consequence, more of the burden of supporting private farms and agriculture in general was shifted to oblasts. Some regions were able to provide adequate support for private farmers (Rostov on-the-Don, Volgograd, Samara, and Belgorod Oblasts, and Stavropol and Krasnodar Krays are prime examples). With the exception of those few regions, most local budgets simply did not have the resources to provide sufficient support to private farmers. In 1995 for example, budgetary allocations from regional budgets totaled 140 billion rubles, or 500,000 rubles per farm—$100 per private farm. This level of financial support was 10–15 times less than recommended by government experts. Some regions, such as Tartarstan and Bashkortostan did not allocate any financial support to private farmers at all.[69]

This strategy in turn had ramifications, as fieldwork suggested that oblasts often did not have the means to support private farmers to the degree they needed, and thus the overall movement suffered. In Kostroma Oblast, for example, the 1994 budget envisioned 49 billion rubles to the agricultural sector as a whole.[70] However, the oblast budget allocated only 1.5 billion rubles for assistance to private farmers, but as of the end of September 1994 only 69 million rubles, or 4.5 percent, had been distributed.[71]

Shortly after the September 1993 resolution that ended subsidized credits, top government officials tried to soften the blow by indicating that state support would continue for private farmers. Former vice premier Yegor Gaydar, for example, maintained that "the priority sector for us will be the sector of private farms." However, he was very clear that future support would not be in the form of subsidized credits. Gaydar indicated that support would be forthcoming in the area of trade, where export tariffs would be lowered and the monopoly grain-purchasing agent Roskhleboprodukt would be reorganized and privatized. Furthermore, he expressed support for the development of small investment projects "that would allow farmers to escape the *diktat* of [food] processors," and toward this end he advocated the shifting of financial support from the state enterprises to private entrepreneurs and enterprises.[72]

Following the September 1993 resolution, it would not be fair to say that state support for private farming ended completely. Certain tax breaks for peasant farmers remained in place, which freed them from various taxes. About two months after the September resolution, another governmental resolution was adopted entitled "On Measures of State Support of the Agroindustrial Complex in 1993–1994," which specified the types of support that the state intended to provide for farmers.[73] The resolution stipulated that subsidies would be continued for production sold through state channels, and the state would pay 50 percent of the transportation costs for feed during the fourth quarter of 1993 and the first six months of 1994. These forms of support, however, were unreliable, as AKKOR officials charged that "the state withholds payment for products that were delivered," which drove many farmers out of business.[74] State suppliers could not be counted on to deliver the fuels that farmers needed to conduct farming operations.[75] In addition, during 1994 only about 50 percent of the monies intended to compensate banks for subsidized credits received during 1992–1993 were distributed; only twenty oblasts or krays actually paid persons who relocated to the countryside to undertake private farming; and only ten oblasts or krays in the entire country provided support for the development of rural infrastructure and rural social construction.[76]

After 1993, the state continued to provide directed assistance to private farmers. In 1994, the government no longer provided subsidized credits, but deferred until December 1994 payment of the interest due on credits given to farmers during 1992–1993. The government also continued the benefits regarding relocation and land taxes, and it indicated it would cover 50 percent of the costs associated with the construction of animal sheds and barns for private farmers who had at least ten head of livestock.[77] In addition, in June 1994 a government resolution allocated 850 billion rubles to create a fund of agricultural machinery that could be leased on a long-term basis.[78] A few days later, Prime Minister Chernomyrdin signed a resolution that created a federal program for leasing agricultural equipment to food producers. While any food producer was eligible to use this fund, private farmers stood to gain, since they were the most undercapitalized.[79]

During 1995, the aforementioned policies were continued. The government also allocated 314 billion rubles to pay the interest on private farmer loans taken out before November 1, 1993, and continued to finance the publication of specialized literatures to help educate farmers.[80] In addition, in May 1995 the federal law "On State Support of Small Entrepreneurs in the Russian Federation" was adopted that provided for a series of measures of state support for small enterprises, which for agricultural operations was defined as fewer than 60 persons. In other words, the law specifically applied to private farms, which had on average 3–4 persons, and newly reorganized farms according to the Nizhniy Novgorod model, which also commonly had less than sixty employees.[81]

In 1996, the aforementioned policies were continued and a number of additional measures were enacted. First, in February 1996 the government allocated 240 billion rubles in subsidies and production support to private farmers—2.1 million rubles per farm (about $420). Another 360 billion rubles was allocated for the special leasing fund that would allow private farmers to lease agricultural machinery. In April 1996 President Yeltsin signed a decree allocating 800 billion rubles to private farmers to allow them to obtain land. Another point of the decree allowed "organizations of the agroindustrial complex" to obtain credits at not more than 25 percent of the interest rate set by the Central Bank.[82] In December 1996 the "Federal Program for the Development of Private (Peasant) Farms and Cooperatives during 1996–2000" was adopted by the Russian government and officially entered into force upon the issuance of resolution number 1499 by Prime Minister Chernomyrdin on December 18, 1996. This special program reflected an ambitious effort to increase both the number of private farms and their production. Officially, the goal of the program was to ensure "steady development and increase the effectiveness of the peasant farming sector as a component part of the diverse agrarian economy, [and] to provide social defense of the peasantry in a market environment."[83] Table 28 compares the quantitative goals for the program by the year to 2000 to trends during 1994–1996.

The Ministry of Agriculture and Food was put in charge of implementing the program. It would be achieved in two stages. The first stage, 1996–1998, would see the "perfection" of the legal basis for private farms and cooperatives. The second stage of the program is to run 1999–2000, during which time state support for private farming is to be "strengthened," the number of farms and the area cultivated is expected to increase, and the process of farm specialization is to accelerate.[84] Various types of state financial support were envisioned in the program, including subsides, compensations, and special programs to aid private farmers. In the page-and-a-half list of support—some of which are loans or credits that have to be repaid, we should note that no new measures were proposed. While the program admits that deficiencies are severe in rural infrastructure, farm buildings, storage, and other means central to the food production process, what is most interesting is the expected source of financing capital investment. According to the program, in 1995 private farmers provided over 72 percent of total capital investments on their own (844.8 billion rubles out of capital investments of 1.16 trillion rubles), a sum that was expected to decline to 63 percent in 1996 (1.4 trillion out of investments of 2.2 trillion). By the year 2000, capital investments are forecast to increase fourfold (to 9.36 trillion rubles), but private farmers are expected to provide financing for 5.92 trillion (63 percent) of capital investments.[85] In short, much of expected capital improvements are to be financed by farmers themselves.

For the program's forecasts to be fulfilled, private farmers would have to increase the level of capital investments by more than 1.1 trillion rubles *annually* during 1997–2000 (in 1995 rubles). It is difficult to imagine where farmers will find these resources. In 1995, an estimated 10 percent of private farmers were profitable, and since that time there has been little if any evidence to suggest that their financial position has improved, indicated by the sharp rise in bankruptcies. Thus, even if we assume the best-case scenario that no deterioration has occurred since 1994, a small minority of private farmers would have to contribute the bulk of capital investment increases. An additional factor is that land tax abatements that were offered to private farmers in 1992 and 1993 will expire in 1997 and 1998. Unless the abatement is extended, an idea considered in the summer of 1996 with no apparent decision, private farmers who have survived will have to begin to pay land taxes.

The types of support envisioned by the program coming from federal and

Table 28. Development of Private Farming and Goals of Federal Program for Private Farming, 1996–2000

	1993	1994	1995	1996	2000 (forecast)
Number of private farms (end of year)	270,000	279,200	280,100	278,600	350,000
Area of agricultural land (end of year, mil. hect.)	11.3	11.8	12.01	12.2	16
Area per one farm	42	43	43	43	46
Grain production (mil. tons)	5.15	4.16	3.01	4.29	8.9
Sugar beets (thous. tons)	1,000	487	669	720	1,680
Sunflower seed (thous. tons)	273	260	519	450	660
Potatoes (thous. tons)	376	302	363	360	780
Vegetables (thous. tons)	102	111	148	185	440
Milk (thous. tons)	495	568	576	645	1,950
Wool (tons)	2.9	3.3	4.2	4.0	6.8

SOURCES: *Rossiyskiy statisticheskiy ezhegodnik* (Moscow: Goskomstat Rossii, 1996), p. 555; "O federal'noy tselevoy programme razvitiya krest'yanskikh (fermerskikh) khozyaystv i kooperativov na 1996–2000 gody," *Sobraniye zakonodtel'stva Rossiyskoy Federatsii*, no. 1 (January 6, 1997); annual Goskomstat data on private farms.

Table 29. Financing the Development of Private Farming, 1996–2000

From federal budget, financing will be provided for:
1. Delivery of equipment and pedigree livestock through leasing funds
2. Development of cooperatives
3. Land surveying
4. Working out documentation for protection of land
5. Publication of specialized literature
6. Information-consultation services
7. Construction of housing for farmers

From local budgets, financing will be provided for:
1. Technical services
2. Development of production infrastructure
3. Construction and repair of animal sheds and other agricultural objects used for production and housing
4. Obtaining farm equipment
5. Development of cooperatives
6. Subsidies for relocation to rural areas to begin private farming
7. Support for beginning farmers and rural orphanges
8. Land surveying
9. Documentation for protection of land
10. Subsidies for production of certain agricultural products and other goals
11. Allocation of 15 percent of local budgets to be used for acquring farm machinery, construction of housing, and other objects of production and social infrastructure

From federal and local budgets, financing will be provided for:
1. Construction of projects related to agricultural production
2. Credits for seasonal production work
3. A special credit fund from which farmers may borrow at not more than 25 percent of the interest rate set by the Central Bank

SOURCE: "O federal'noy tselevoy programme razvitiya krest'yanskikh (fermerskikh) khozyaystv i kooperativov na 1996–2000 gody," *Sobraniye zakonodtel'stva Rossiyskoy Federatsii*, no. 1 (January 6, 1997): 213–14.

local budgets are indicated in table 29. Because state support for private farming is such a critical issue for the future, the question is how much credence to put in a federal program from a government that has consistently demonstrated an inability to carry through on its promises. This question becomes especially acute in the wake of drastic budget reductions in May 1997. Those cuts call into question the program's ability to reach its goals to develop private farming.

In 1997, the list of compensations and subsidies from previous years was continued. In addition, monies allocated to the leasing fund were increased to 2.8 trillion rubles. In May 1997, two special funds were announced to support private farmers: 10 billion rubles to partially finance spring sowing and a 100-billion-ruble fund that allowed a farmer to borrow up to 1.5 billion rubles in order to purchase farm machinery, equipment for processing, pedigree livestock, and construct animal sheds. The interest rate could not exceed the rate set by the Central Bank, and the loan was repayable after three years. The loan would be secured through mortgages on production, land owned by the farmer, or by other property held by the farmer.[86]

Despite the continuation of certain types of state support after September 1993, the bulk of state assistance to private farmers went from unrestricted financial support to restricted support. After September 1993, private farmers were eligible to draw from special purpose funds, but these funds restricted usage and were not always reliable. The ending of state-subsidized credits to private farming was a crucial blow and severely affected the fortunes of the private farmers.

The Economic Impact of Ending State-Subsidized Credits

Despite continued state support for private farmers, after the September 1993 resolution the problem was that continuing state financial support was both ineffective and not directed to the primary needs of farmers. The types of financial support that were offered could not replace the subsidized credits that farmers needed to fund operations and equip their farms. For example, post–September 1993 state support came in the form of liberalized trading conditions, production subsidies, and price supplements, but this support definitely presumed the ability to produce and harvest. Commercial banks were unwilling to fill the void as they considered private farmers to be high-risk. As a result the private farm movement as a whole was severely affected by the ending of state-subsidized credits.

Moreover, as the state moved from unrestricted to restricted forms of financial support, private farms did not have sufficient resources to address and rectify infrastructural and operational deficiencies. Because the state never fulfilled its legal obligation to construct rural infrastructure, farmers themselves were forced to assume this financial burden, and their ability to meet production and infrastructural costs was limited. Former deputy prime minister in charge of agriculture, A. Zaveryukha, revealed that during 1993 more than one-half of private farmers did not have running water, one-third had no electricity, and almost one-half did not have suitable roads.[87] During 1993 one farmer in fifty had begun construction on a house; and only one in fifteen private farmers was able to afford construction of animal sheds.

Similar to most state programs, there was a huge gap between state promises and reality. Even restricted state support that existed on paper often was not provided. In March 1995, at the sixth congress held by AKKOR, it was revealed that only 61 percent of intended state support for private farmers had actually been distributed during 1994.[88] Furthermore, although private farms received 10 percent of all production subsidies allocated to agriculture during 1994, between 60 and 70 percent of those subsidies was concentrated in a handful of oblasts, so that most private farmers received very little.

In real terms, private farmers not only lost their key source of finance when subsidized credits ended, but state support supposed to replace the credit was ineffective and insufficient. Responsibility to fund private farmers fell to the oblasts, but (as noted above) most obasts were often not in a position to offer much financial assistance. The loss of state-subsidized credits and unfulfilled promises of other types of support had three discernible effects on private farmers.

Rural Preferences and Private Farm Creation

The first impact of state financial policy toward the private sector was on rural preferences and rates of private farm creation. It is necessary to remember that private farms competed with other agricultural land users, with small-scale recreational and large-scale collective enterprises, with urban users of land, with agricultural subsidiary operations of enterprises, and with forestry and lumbering interests. Farm members and rural dwellers were offered options as to the form of landownership and the type of farming they wished to engage in. Patterns of land privatization, therefore, should be seen as an expression of rural preferences within the context of this competitive environment. What were rural preferences? Private farming was never tremendously popular as a rural preference in comparison to other forms of agricultural activity, and it became increasingly less popular after state-subsidized credits ended.

Even though the number of private farms officially registered grew rapidly during 1992–1993, these numbers still paled in comparison to the number of small privatized plots. Data at the national level as well as from individual oblasts consistently show that Russians preferred small-scale agriculture—that is, personal plots, ogorody, fruit and vegetable collective gardens—since the time that land reform was first undertaken. For example, at the national level during the first full two years of land reform, the number of Russian families with a collective fruit or vegetable garden increased from 13.6 to 21.4 million.[89] Trends in rural preferences for land privatization in two non–black earth oblasts are shown in table 30.[90]

It is clear from this table that not only were private farms less preferred on a relative basis, but future desire—measured by outstanding applications—also was considerably below that expressed for other forms of private endeavors. In contrast to the existence of about 280,000 private farms, by the end of 1996 over 16 million land plots were privatized for personal plot usage, as were

Table 30. Land Privatization Patterns in Kostroma and Moscow Oblasts as of July 1, 1994

	Kostroma Oblast	Moscow Oblast
Number of private farms	1,151	6,100
Number of hectares of private farms	42,568	58,368
Outstanding applications for private farms	3	1,709
Number of personal plots	155,025	518,959
Number of hectares of personal plots	99,499	87,040
Outstanding applications for personal plots	12	41,412
Number of collective fruit gardens	63,508	1,190,844
Number of hectares of collective fruit gardens	5,238	100,851
Outstanding applications for collective fruit gardens	0	319,835
Number of collective vegetable gardens	65,502	559,432
Number of hectares of collective vegetable gardens	3,314	31,384
Outstanding applications for collective vegetable gardens	0	146,514

SOURCES: Unpublished information from Committee on Land Resources and Land Surveying, Kostroma Oblast and Moscow Oblast.

more than 20 million plots in collective gardens, dacha plots, and other small-scale individual farming activities. We are able to understand rural preferences by looking more closely at Kostroma Oblast. In Kostroma Oblast the three raions with the best bioclimatic location and proximity to the city of Kostroma were Kostroma raion, Krasnosel'skiy raion, and Nerekhtskiy raion. These three raions were also among the most agriculturally productive in the oblast. Prospective private farmers fared poorly in the competition for land, as personal plots were by far the most popular option for land privatization in these raions. As of July 1995, Kostroma raion had 18,881 personal plots and 75 private farms; Krasnosel'skiy raion had 9,098 personal plots and 48 private farms; and Nerekhtskiy raion had 9,702 personal plots and 71 private farms.[91] Thus, there were vast differences in rural preferences for small-scale, low-risk, low-debt ventures and preferences to undertake private farming. The revival and resurgent popularity of personal plots occurred not just in Kostroma Oblast, but throughout Russia.[92] As a result, the percentage of land used nationwide by private farmers remained at about 6 percent of agricultural land from 1994 onward. The percentage of agricultural land used by private farmers in the Central region is shown in table 31.

Financial Levers and Private Farming   207

What explains these patterns? They are at least partially linked to the demographic characteristics of the countryside. The Soviet countryside experienced decades of rural out-migration, primarily among the young, skilled, and women. Because of rural migratory trends, the inadequacy of implementing programs to revitalize the Russian countryside, and a high rural mortality rate per thousand persons, the population of the Soviet countryside became older as time progressed. By the early 1990s in the Central economic region, over one-quarter of the rural population was too old to work, and this situation was mirrored throughout Russia. By the early 1990s many oblasts had over 30 percent of their rural populations too old to work.[93] It is logical that an older rural population would be less likely to engage in a high-risk and high-debt venture such as private farming, let alone be able to put in the labor necessary to make a private farm successful. High risk and high debt were not attractive options for an older population, and for these reasons the most common age bracket for private farming was from thirty to thirty-nine.[94] However, in Kostroma Oblast, rural and urban men in this age bracket comprised 18 percent of the oblast's male population, whereas men aged fifty and above accounted for 25

Table 31. Percentage of Agricultural Land Registered for Private Farms by Oblast, Central Region, as of October 1, 1994

	Area Registered for Private Farms (thous. hectares)	Percentage of Agricultural Land Used for Private Farms
Central region	849.5	4.2
Bryansk	43.2	2.3
Vladimir	36.7	3.6
Ivanovo	23.9	2.7
Kaluga	48.6	3.5
Kostroma	42.6	4.1
Moscow[a]	58.4	3.3
Orel	90.5	4.4
Ryazan'	82.4	3.3
Smolensk	166.6	7.6
Tver'	106.6	4.4
Tula	101.4	5.2
Yaroslavl'	48.8	4.2

SOURCES: Calculated from private farm data and *Moskovskaya oblast' i eyo mesto v Tsentral'nom Rossiyskom regione (ekonomicheskoe razvitiye)* (Moscow: Goskomstat Rossiyskoy Federatsii, 1993), p. 81.
a. Moscow Oblast excludes the city of Moscow. Numbers have been rounded off.

percent of the oblast's male population.[95] These percentages are representative of demographic patterns throughout the Central economic region and more generally for the Russian non–black earth.

Because the rural population cohort for private farming was relatively small, this left a large percentage of the rural population disposed to small-scale agricultural activities such as personal plots and dacha plots. Private farming was not a prospect that was attractive to an older cohort. However, rural demographics alone do not explain rural preferences. Another factor was the impact of state financial policy toward private farming.

Even when subsidized credits were available to private farmers (during 1992 and the first nine months of 1993) private farming was a less preferred option. After October 1993, private farms faced annual interest rates in excess of 200 percent from the Central Bank and more than 400 percent from commercial banks (and this assuming that credit could indeed be obtained). Interest in becoming a private farmer showed a sharp decline, according to one survey falling from 12 percent in 1991 to 5.8 percent in 1992 to 1.4 percent in 1993.[96] The importance of state financial policy is evident when we view the rate of private farm creation. At the beginning of 1997 there were less than 280,000 private farms in Russia, far from the 400,000–500,000 farms predicted early in the reform process. The number of private farms had essentially stagnated since the ending of subsidized credits, adding only a net of 11,000 private farms from January 1, 1994, to July 1, 1997. During the heyday of land privatization the same number of farms were created in one month.

The chart below depicts a rapid increase in the number of private farms during 1992 and during the first half of 1993, the time when state-subsidized credits were available. The chart shows that the first and second quarters of 1992 and 1993 experienced roughly similar rates of farm creation. Beginning in the third quarter of 1993, however, the rate of farm creation slowed greatly, falling from 20,800 farms during 1992 to 10,500 in 1993. A comparison of farm creation rates over time is indicated in figure 3.

How can we explain the dramatic slowdown in the rate of farm creation? The answer is that it was directly linked to state financing. As subsidized credits for private farming declined and then ended altogether, and as financial prospects for success worsened, the rate of farm creation fell sharply. This decline continued after 1994. At the national level, from mid-1994 to the beginning of 1995 there was a net reduction of 6,600 private farms for the first time since 1990, although during 1994 the net number of private farms increased by just over 9,000. During 1995, the net number of private farms increased during the first half of the year but again declined during the second half, so the net increase was only 917 farms. During 1996, a net reduction of nearly 1,400 pri-

Figure 3. Rate of Private Farm Creation, 1991–1996

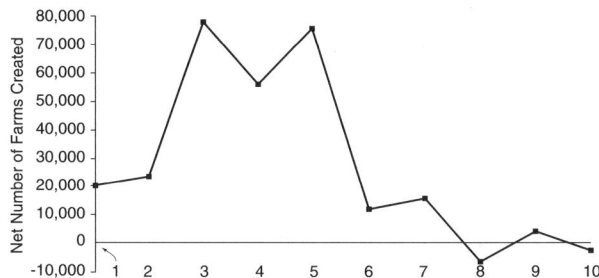

vate farms occurred. In general, starting in 1994 a common pattern was evident: the number of private farms increased during the first half of the year, but then the number decreased in the second half, resulting in a marginal net increase or even a net decrease (due to farm closures and bankruptcies, which in turn were linked to high farm costs, problems selling their production, input-output price disparities, and difficulties obtaining financing).

The impact of a deteriorating economic environment for private farmers after 1993 is clearly reflected in the rate of net farm creation at the regional level as well. These trends are indicated in table 32. The table shows that in 1993 all economic regions were creating a net of private farms. At the low end, some regions added only a hundred or so per month, whereas the Northern Caucasus was creating more than two thousand per month. Regions such as the Volga and Ural regions added just under a thousand net farms per month. During 1994, all regions experienced a significant decline in the net number of farms created, and for the first time two regions experienced a net decline (the Volga and Central black earth regions). During 1995 and 1996, five of the eight European regions experienced net declines in the number of private farms. The increase in the number of private farms in the Northern Caucasus region is at least partially explained by migratory trends into Russia from Central Asian states by ethnic Russians who were fleeing discrimination in those states. Evidence suggests that perhaps as many as one-half of migrants from the near abroad are settling into rural regions, due to a lack of urban housing and in pursuit of land and secure food supplies.[97] Only the Northern Caucasus region retained a significant growth rate of net farms.

One could argue that a process of weeding out economically weak farms was necessary in order to construct a more productive and efficient private farming sector. However, we should note that this shakeout has several disturbing aspects for the success of private farming. First, farm success—and conversely farm failure—correlated with the date a farm was created. Farms begun before price liberalization were the best equipped and the most productive.

Farms created in 1993 or after failed at a rate more than three times higher than those begun in 1992.[98] Thus the earlier a farm was started the better its chances for survival.[99] This fact is significant because we know from survey data that between two-thirds and three-quarters of private farms begun during 1991–1992 were started by persons who were not former farmworkers; often they were former urban dwellers or former white-collar farm employees who used "connections" to obtain the best land, the best equipment, and the best farm machinery.[100] During 1993 the composition of private farmers changed rapidly as more former farmworkers became private farmers, so that by the end of 1993 most private farmers were former farmworkers. Thus, there is the possibility that those who were shaken out during 1993 and after were those with the most agricultural experience. Ironically, farms begun by former urbanites had the best chance of survival. But it was precisely farms begun by former urban dwellers that government leaders criticized for being subsistence farms and not able to produce for a commercial market.

Thus, the decline in farm creation is linked directly to the ending of state-subsidized credits. From late 1993 on, the rate of private farm creation plummeted and the rate of private farm closures increased significantly. We shall examine bankruptcy trends later, but for now we should note that the decline in farm creation has meant that the number of private farms per thousand persons has remained extraordinarily low. The number of private farms per thousand persons in European Russia from January 1992 to January 1996 is shown in table 33.

Table 33 shows that only two economic regions had more than two private farms per thousand persons, and only one region—the Northern Caucasus—had more than four. Moreover, from January 1994 onward, only the Northern

Table 32. Private Farm Creation Rate by Region, European Russia, 1993–1996 (net number of farms created)

	1993	1994	1995	1996
Russia	87,094	9,223	917	-1,387
Northern	868	121	-452	12
Northwest	2,864	1,239	196	117
Central	7,636	1,608	-336	-236
Volga-Vyatka	2,423	514	205	-159
Central black earth	3,571	-1,338	-461	-768
Volga	10,027	-784	-813	-445
N. Caucasus	27,231	5,634	5,729	4,882
Urals	10,855	277	-1,439	-1,466

SOURCES: Author's calculations from Goskomstat private farm data.

Table 33. Number of Private Farms per Thousand Persons by Region, European Russia, 1992–1996

	1992	1993	1994	1995	1996
Northern	0.22	0.59	0.61	0.75	0.68
Northwest	0.28	0.89	1.22	1.38	1.41
Central	0.19	0.70	0.95	1.01	1.00
Volga-Vyatka	0.21	0.68	0.98	1.03	1.05
Central black earth	0.36	1.38	1.84	1.67	1.61
Volga	0.38	1.67	2.27	2.23	2.18
N. Caucasus	0.55	2.50	4.0	4.39	4.73
Urals	0.32	1.14	1.70	1.69	1.62

SOURCES: Census data in *Vestnik statistiki*, no. 3 (March 1990): 74; and author's calculations from Goskomstat private farm data.

NOTE: The entire population of a region was used in the calculations because land can be obtained by urban or rural individuals for private farms. The dates are January of the relevant year.

Caucasus region experienced a significant increase in the number of farms per thousand persons, while three economic regions had net declines. We are able to understand rural preferences and the competition for land by looking more closely at Moscow Oblast.

Rural Preferences in Moscow Oblast. The relative unpopularity of private farming is evident when considering Moscow Oblast. If private farming were to be successful anywhere in the Central region, one would expect it to be in Moscow Oblast. Unlike most other oblasts in the non–black earth, Moscow Oblast experienced a net population inflow into its rural areas during the decades other oblasts experienced net outflow. As a consequence, the age structure of Moscow's rural sector was not as unfavorable as in other non–black earth oblasts.[101] In addition, because Moscow Oblast is heavily urbanized, one might have expected popular opinion to influence the course of agrarian reform. A survey by the Agrarian Institute and All-Russian Center for Public Opinion in 1993 indicated that the "most active" support for land reform came from urban residents.[102] In Moscow Oblast there is also a huge food market represented by the city of Moscow, and a relatively good road and infrastructural system. Moscow Oblast therefore possessed a favorable political, economic, and demographic climate for private farming.

At first glance it does appear that private farming in Moscow Oblast was more successful than in any other oblast in the Central region. Since September 1991, Moscow Oblast has had the largest number of private farms in the Central region. In January 1991, Moscow Oblast had only twenty-one farms (just under 5 percent of the total for the region), and as late as May 1991, it had three hundred farms (representing almost 13 percent of all farms in the region), which placed it behind Tver' and Yaroslavl' Oblasts. By September 1, 1991, Moscow Oblast took the lead in the number of farms, with 739 private farms, and since January 1992, it has accounted for roughly 19 percent of the total number of private farms in the Central region.[103]

Even though Moscow Oblast had the largest number of private farms in the Central region, this number is somewhat misleading because it also had the largest overall population and by far the largest rural population in the Central region. The rural population in Moscow Oblast was almost three times larger than in Tver' Oblast, which had the second-largest rural population.[104] Although Moscow Oblast had the largest number of private farms, its degree of farm privatization—measured through the creation of private farms per thousand persons—was the second lowest in the Central region. Comparing Moscow Oblast with other Central region oblasts, table 34 illustrates the number of private farms per thousand persons as of January 1, 1996.

Within Moscow Oblast, most raions had very few farms per thousand persons. By mid-1994 only two raions had more than ten farms per thousand persons, and six raions averaged more than five farms per thousand persons. At the same time, twenty-seven raions had fewer than three farms per thousand persons. The two raions with the highest number of farms per thousand persons were located far from the city of Moscow: Volokolamskiy raion in the west of the oblast, about 120 kilometers from Moscow; and Zarayskiy raion in the far southeast corner of the oblast, more than 120 kilometers from the outskirts of Moscow. These lands were not only far from Moscow but also generally of poor quality.

Thus we see that, even in an oblast that should have been among the success stories, private farming was relatively unpopular. Whereas the rate of private farm creation per thousand persons was less than 1.0 in July 1994, the rate of personal plot operation was 78 per thousand persons and for collective gardens (both fruit and vegetable) 179 per thousand persons.[105] These findings replicate patterns in other regions of Russia, including rural oblasts to the north and southern oblasts with good quality soil. These data substantiate the conclusion that private farming was much less popular than other forms of individual farming on a standardized statistical basis.

Private farming in Moscow Oblast was also squeezed by competition from urban land users. In Moscow Oblast, about two-thirds of land for privatization was set aside for residents from the city of Moscow.[106] But demand for land outstripped supply, forcing the city to turn to neighboring oblasts for land for Muscovites. During the winter of 1991 the RSFSR Council of Ministers adopted

Table 34. Number of Private Farms per Thousand Persons in the Central Region on January 1, 1996 (using population data from 1989 census)

	Urban Population (thous.)	Rural Population (thous.)	Number of Private Farms	Private Farms per Thousand Persons
Bryansk	992	483	1,479	1.0
Vladimir	1,310	344	2,463	1.5
Ivanovo	1,075	242	1,058	0.80
Kaluga	735	332	1,817	1.7
Kostroma	555	255	1,102	1.4
Moscow[a]	5,311	1,383	6,480	0.97
Orel	555	336	1,450	1.6
Ryazan'	885	461	2,370	1.8
Smolensk	788	370	2,778	2.4
Tver'	1,194	476	3,803	2.3
Tula	1,513	354	3,212	1.7
Yaroslavl'	1,200	270	2,252	1.5

SOURCES: *Vestnik statistiki*, no. 3 (March 1990): 74; and author's calculations from Goskomstat private farm data.
  a. Moscow Oblast excludes the city of Moscow.

Resolution No. 111, "On Priority Measures for Providing Residents of Moscow with Land Plots for the Organization of Collective Gardens and Vegetable Gardens."[107] Flowing from this resolution, during the spring of 1994, the government of the city of Moscow concluded a series of agreements with neighboring oblasts: Tver', Vladimir, Kaluga, Ryazan', Smolensk, and Yaroslavl'.

Published documents show that land was to come from specific raions within each oblast; by summer 1994, fourteen specific areas had been identified where land would be made available.[108] From these oblasts it was envisioned that almost 20,000 hectares would be made available for collective gardening by Moscow residents.[109] Each administrative district within the city of Moscow would be allocated a certain quantity of land. For example, from Gagarinskiy raion in Smolensk Oblast, 250 hectares would be assigned to Moscow residents, distributed to urban raions in the city of Moscow. Published documents show that, in addition to the land from Smolensk, in the early stage Muscovites would receive 450 hectares from Zubtsovskiy raion in Tver' Oblast and 200 hectares from Kol'chuginskiy raion in Vladimir Oblast.

Due to the competition for land and the nature of rural preferences, the closer the raion was to the city of Moscow the smaller the percentage of agricultural land used by private farms. By mid-1994, the eight raions with the best proximity to the Moscow food market had just 1,283 hectares of land assigned to private farms, representing 2.2 percent of private farmland in the oblast. In comparison, prior to farm reorganization in 1992, forty-five state farms and twelve collective farms located in these same eight raions had 103,859 hectares, or 8.2 percent of the oblast's agricultural land.[110]

The small amount of land used by private farmers in these close-in raions was not solely because they were urban raions. Podol'skiy, Odintsovskiy, and Leninskiy raions had predominantly rural populations (after excluding the raion center). In two other raions, Krasnogorskiy and Balashikhinskiy, about half the population was rural. Instead, the explanation likely was because of popular preferences, that is, persons acquiring land preferred farming operations that were risk and debt averse.

Production Performance

The second economic impact of ending state-subsidized credits affected the production performance of private farms. We can measure this impact in two main ways. First, private farms remained small, which limited income and production capacity. Russian academicians point out that small farms suffer from an economies-of-scale disadvantage and we know from world experience that large farms produce more.[111] Even with their smaller size, private farms' yield per hectare was considerably below the average yields compared to other forms of farming for grains, vegetables, and potatoes during 1991–1994.[112]

At the beginning of 1997, nearly 80 percent of Russia's 280,000 private farms were less than fifty hectares, and less than 6 percent were more than a hundred hectares throughout Russia.[113] Although it is true that a large number of small farms contributed to the organization of cooperatives and the sharing of farm equipment among farmers (leading to the creation of "MTS from below"),[114] small farms had a negative effect, in that excessive demand was placed on farm equipment and input supplies, which had decreased significantly since 1991.[115]

The ending of state-subsidized credits also made it difficult for most private farmers to expand landholdings by purchasing additional land. In Kostroma Oblast, for example, private farmers had less land under cultivation on July 1, 1997, than they did on January 1, 1993.[116] In European Russia, on January 1, 1997, the Northern, Central, Central black earth, and Ural economic regions had less farmland held by private farmers than on January 1, 1995. Furthermore, since January 1993, average farm size increased significantly only in the Volga region among all economic regions in European Russia (see table 23).

A second way in which the ending of state-subsidized credits affected production was that private farms had to curtail production, forgo expansion of production capacity, or close altogether.[117] In Kostroma Oblast, for example, by early 1995 only one-fourth of private farms were producing food for commercial purposes, and these were farms that had been established during 1991–1992. Lacking capital and the access to affordable credits often meant that private farmers were unable to purchase or to increase holdings of cattle or other livestock. In Sudislavskiy raion (Kostroma Oblast), thirteen of thirty-one private farms had no cattle at all.[118] In general throughout Russia private farms that raised livestock tended to have but a few head. Nationwide, by January 1, 1995, private farms had an average of two head of cattle, one pig, four goats and sheep, and six head of poultry, or essentially the same holdings as in January 1992.[119] During 1996 livestock herds held by private farmers continued to decline. By late 1996 private farmers had 92 percent of the beef cattle, 94 percent of the dairy cows, 97 percent of the hogs, and 88 percent of sheep and goats in comparison with 1995.[120] The economic effects of state financial policy also influenced the economic environment of private farms in the most agriculturally productive regions of the country. To illustrate we turn to the Volga economic region.

Private Farming Performance in the Volga Region. The Volga region is one of the most important agricultural regions in Russia. During 1991–1993, this region produced about 20 percent of the nation's grain, 15 percent of the meat, 13 percent of the milk, and 25 percent of the wool. Thus, the Volga region was a rich agricultural region where private farmers could be expected to fare quite well if they were financially supported.

Although the Volga region has had the largest average private farm size, a significant percentage of private farms are small: throughout the entire region 15 percent of private farmers had less than ten hectares by early 1992, and 32 percent had between eleven and fifty hectares. In Saratov Oblast, which is in the Volga region, 31 percent of private farmers had less than thirty hectares, and another 45 percent had between thirty-one and one hundred hectares. One economist summed up the situation by stating "naturally, high effectiveness of production on small land plots, especially in the steppe zones of the Volga region, is impossible."[121]

Private farmers in this agriculturally rich region depended upon state-subsidized credits to help them meet operational costs and to compensate for disadvantageous terms of trade. In the Volga region during 1991, only 5 percent of the private farms had sufficient funds of their own to equip and operate their farm. More than 50 percent used state credits and budget subsidies, and 30 percent borrowed from state and collective farms.[122] According to analyses by Russian academicians, private farming was damaged the most by the terms of trade, which turned drastically against food producers, and by the ending of state-subsidized credits, which made acquiring operational capital prohibitively expensive.[123] In Saratov Oblast, where 35 percent of the Volga region's private farms are located, prices for industrial goods, farm inputs and services increased 1.9, 1.7, and 1.2 times faster than purchase prices of agricultural products during 1991–1993, respectively.

As input prices outstripped agricultural purchase prices, state-subsidized credits became ever more important. When these credits ended, private farmers could not afford market interest rates for loans, and the ways they responded affected production and income potential. One response was to curtail fertilizer purchases, which "sharply lowered the fertility of the soil and quality of grain." Another response was to reduce livestock holdings.[124] Squeezed by high input prices (particularly feed), low demand for expensive meat products, reduced subsidies, and a lack of subsidized credits, private farmers in the Volga region during 1994 saw the number of cattle decline over 12 percent in comparison with 1993. In Saratov Oblast, which had the most head of cattle in the region, the decline was over 24 percent.[125] Furthermore, gross agricultural production and productivity declined. During 1994 grain output in the Volga region decreased 30 percent, and yields dropped 17 percent in comparison with 1993. In Saratov Oblast, which accounted for the largest grain harvests in the region, yields per hectare of arable land were 1.6 times lower than on large agricultural enterprises in 1994.[126]

Thus, we see that the economic environment of private farmers in both poor and rich agricultural regions was affected by the ending of state credits. Without state credits, private farmers found it difficult to meet production and operational costs. Private farmers responded by curtailing the use of inputs, which affected yields, and by curtailing production, which further reduced their farm revenues.

Farm Profitability and Farm Closures

A third effect of ending state-subsidized credits was the deterioration in the financial condition of private farms. With access to state-subsidized credits in 1992, nearly all private farms were "profitable."[127] In 1993, rising input prices and the ending of state-subsidized credits meant that only 43 percent of all private farms received a profit, and even then the average profit was only 46,000 rubles ($46). By the middle of 1994, only 20 percent of private farms were profitable. Throughout the nation, private farms experienced average losses totaling 1.3 million rubles ($590) per farm. The percentage of private farm expenditures used for debt service rose from 7 to over 34 percent from 1992 to 1994.[128]

As the financial condition of private farms worsened, the number of private farms that closed due to bankruptcy greatly increased. By the end of 1993, an estimated 52 farms were failing for every 100 that were begun, up from 4 per 100 in 1992 and 5 per 100 during the first quarter of 1993.[129] The difficulty in obtaining credit and the high cost of credit meant that, by the fourth quarter of 1994, for every 100 farms created, 103 were stopping operations.[130] Overall, during 1993–1994, nearly 40,000 private farms closed (14,000 in 1993 and 26,000 in 1994).[131] During 1995, another 25,300 private farms closed, just under 10 percent of the total number of farms in existence in January 1996.[132] During 1996, more than 25,000 private farms ceased operations. During 1995, twenty-five regions of Russia experienced a net decline in the number of private farms; in 1996 that number rose to forty-nine.

After state credits were ended, farm failures hit the more productive agricultural regions no less hard than in non–black earth zones (with the exception of the Northern Caucasus region), as shown in table 35. The table clearly shows that the number of farm failures in each region increased significantly during 1993–1995. Farm failure data do not indicate a clear pattern of weeding out private farms in less productive areas. With the exception of the Northern Caucasus region, the percentage of farm failures to farms created in more productive agricultural areas (Volga, Central black earth, and Urals) were equal to or exceeded the percentage in poorer non–black earth regions. These data therefore suggest two important points: first, that ending subsidized credits hurt the "good" regions at least as much as the poorer agricultural regions. Second, the "market" was not sufficient to sustain high rates of private farm creation, and the economic environment was hostile to their operation. Only through state financial interventions and subsidized credits were large numbers of farms created, and the removal of subsidized credits led to massive numbers of farm bankruptcies.

Two ramifications flowed from the trends reflected by the data presented above. First, since 1993, when former farmworkers and rural dwellers started to become private farmers in greater numbers, the financial environment and economic conditions became more hostile, which made it more difficult for the "right" people to succeed. Second, as anyone who has visited a typical private farm can attest, private farmers were often quite poor, and their "economic stratification" was best measured in degrees of poverty. The high cost of building materials meant that buildings for livestock were a priority. Although some farmers managed to build small dwellings for themselves, some farmers either lived in the buildings with their livestock or lived in the back of a truck outdoors. The ending of subsidized credits made it all the more difficult for farmers to raise their material well-being.

Table 35. Private Farm Failure by Region, European Russia, 1993–1995

	Private Farm Closures in 1993	Ratio of Farms Created per Farm Closure, 1993	Private Farm Closures in 1994	Ratio of Farms Created per Farm Closure, 1994	Private Farm Closures in 1995	Ratio of Farms Created per Farm Closure, 1995
Russia	14,063	6.2	26,761	0.34	25,303	0.04
Northern	297	2.9	516	0.23	806	-0.56
Northwestern	405	7.1	945	1.3	934	0.21
Central	1,684	4.5	2,493	0.65	2,635	-0.13
Volga-Vyatka	725	3.3	1,047	0.49	732	0.28
Central black earth	1,292	2.8	2,644	-0.52	1,474	-0.31
Volga	2,138	4.7	3,430	-0.23	2,677	-0.30
N. Caucasus	1,315	21.0	3,029	1.9	3,614	1.6
Urals	2,302	4.7	4,593	0.06	4,169	-0.35

SOURCE: Table 32, Goskomstat private farm data, and author's calculations.
NOTE: Numbers have been rounded off.

The Political Impact of Ending State-Subsidized Credits

The economic impact on private farmers was not the only consequence of ending state-subsidized credits. A second important consequence concerns political support for reform and political alliances. Rural support for reform was important because the goals of agricultural reform cannot be achieved with only an urban constituency. A key ramification of ending state-subsidized credits and the general insufficiency of state support was the alienation of proreform rural groups from the Yel'tsin government. The erosion of political support for Yel'tsin was of extreme importance because, from the start, support for agrarian reform was weak in rural areas. The government badly needed rural allies to advocate reform in the countryside. In the larger political picture, the government also needed rural support in order to counter conservative political organizations and to rally political support (votes) that would keep a reformist regime in power.

Because agrarian reform did not originate from peasant pressures for social equity as in some Third World states, rural support for reform would have to be built. The task of building rural support was difficult from the beginning for two reasons. First, the nature of reform policies—with their emphasis on farm privatization, decollectivization, and a land market based on private property—dichotomized the Russian countryside. The political division of the countryside was neither surprising nor unusual, given the scope of change that was being attempted. What it meant was that ideological differences arose over the course of reform. Conservative agrarian interests opposed privatizing land at the expense of state and collective farms. At the Second RSFSR Congress of People's Deputies during November–December 1990, conservatives supported land reform legislation and private property only in return for promises of substantial state financial assistance to agriculture. On the whole, however, raion officials and farm managers resisted reform efforts to decollectivize farms and privatize land.

Second, the rural population is older and more conservative than the urban population, owing to decades of out-migration by the rural young. The age structure and conservative nature of rural Russia meant that rural support for reform would be difficult to obtain under any circumstances. Surveys have consistently found that rural residents opposed decollectivization and favored limits on land sales at much higher rates than did urban residents.[133] Voting results since 1993 have likewise shown a distinctive urban-rural split in policy preferences and candidates. For example, Yel'tsin himself did not have much rural support, as shown by the voting results in the April 1993 referendum, the December 1993 and December 1995 elections, and the 1996 presidential runoffs.[134]

If agrarian reform were to be successful, it was necessary to build support in the countryside in order to promote reform policies and to counter conservative organizations. Because reform policies directly threatened the interests of vested agrarian interests, they were not stable coalition allies and efforts at building support among those ranks would be futile. In order to promote reform policies and to defend the interests of newly created private farmers, AKKOR was founded in January 1990 and was conceived as a voluntary organization to defend leaseholders and peasant farmers against infringements of their sovereignty by local officials and farm managers, even though its functions were primarily economic and technical.[135]

AKKOR became the government's proreform rural representative. From the beginning, AKKOR was little more than a state appendage, being created with state power and intimately linked to the state. Its offices are housed within the Ministry of Agriculture; its origins were found in the State Committee on Land Reform; and during 1992–1993 it served as the guarantor of the Russian Farmer program credits channeled to private farmers. The minister of agriculture, V. Khlystun, and the president of AKKOR, V. Bashmachnikov, are also personal friends, which gave AKKOR privileged access to governmental policy making.[136]

AKKOR was afforded a privileged position vis-à-vis other rural groups by way of offering "selective incentives" to rural liberals.[137] As a result, AKKOR supported the early reform initiatives of the regime as part of a corporatist relationship. The strategy of giving various financial inducements to private farmers in order to build rural support for agrarian reform appeared to work. During 1992, for example, while numerous grain strikes and peasant protests swept the countryside, AKKOR abstained from joining the protests and private farmers continued to sell their grain to the state. Furthermore, AKKOR and most private farmers supported Yel'tsin in his showdown with the old Soviet parliament during September–October 1993.[138]

Despite ideological similarities, even from the earliest days, there were tensions in the relationship between AKKOR and the state, tensions that revolved around the perception that the state was not doing enough to help private farmers and was not fulfilling its promises.[139] By the fall of 1992, AKKOR officials were already complaining that private farmers had lived up to the agreement to sell 25 percent of farm production to the state, but the state was "sabotaging" land reform by not distributing credits to private farmers. AKKOR officials charged that only 15 percent of budgeted credits for private farmers had been distributed during the first nine months of 1992.[140] Other disagreements existed as well, as AKKOR lobbied the government to channel state investments and production credits through AKKOR branches, not state banks. Even though AKKOR did not get its wish, as long as the state provided the private sector with subsidized credits and provided other financial advantages to private farmers AKKOR supported the Yel'tsin regime.

Because AKKOR advocated land privatization and an unregulated land market, it did not represent a majority of rural opinion. Whereas rural dwellers largely opposed land privatization and a free land market (one survey by the Agrarian Institute placed rural support for these proposals at 20 percent), urban reform elites pushed these ideas vigorously. Because AKKOR's attitudes toward reform were similar to those of urban elites, AKKOR was linked to urban and state interests and therefore did not have much rural political support outside of private farmers. Thus, AKKOR was comprised primarily of private farmers—a relatively thin stratum of rural persons.[141] Even among private farmers, support for AKKOR was often linked to the monetary benefits to be gained from such support. After subsidized credits were ended and AKKOR lost its privileged position, it became much less popular among private farmers.[142]

AKKOR and the Peasant Party supported the Yel'tsin regime even though this support put them in an awkward position. Liberal rural groups, who represented primarily private farmers, found little support among rural dwellers and therefore had to rely on alliances with urban liberal groups to have any significance at all. But urban groups advocated policies that contravened rural interests, which further alienated the majority of rural dwellers from rural liberal groups.

The dilemma that liberal rural interests faced was made clear in an interview with the deputy director of the Peasant Party of Russia in August 1994. The Peasant Party of Russia is the oldest agricultural political party in Russia, having held its first congress in March 1991 at which the writer Yuriy Chernichenko was elected chairman of the party. (The party was officially registered in April 1991, at which time it was considered a party in opposition to mainstream state rural policies.) In its early days the party represented primarily the interests of private farmers, with some representation of food-processing workers and farm chairmen. Over time, as food-processing interests joined the Agrarian Union and later the Agrarian Party of Russia, the Peasant Party became the political party representing liberal (pro–land market and privatization) rural views and worked in close coordination with AKKOR. Even though the party claimed to "defend the political, social, and economic interests of peasants, farmers, and all agricultural producers," by its third congress in June 1993 the party was primarily a vehicle to promote private farm interests, land reform, and a free land market, without which "the food problem could not be solved."[143]

In October 1993, Chernichenko wrote Yel'tsin a personal letter asking him to allow private ownership of land without restriction on sales, purchases, leasing, mortgaging, inheritance, and so on. He asked Yel'tsin to protect the rights of citizens in obtaining land and property shares from state and collective farms, and to forbid "old agrarian structures" from withholding shares. Land vouchers were suggested for persons who did not have the right to land shares from state and collective farms. He concluded that "if we are not successful today, we will never be successful."[144] Whether coincidental or not, many of the planks in Chernichenko's letter were included in Yel'tsin's decree on land relations issued at the end of October 1993.

Despite a corporatist relationship between rural liberal groups and the government, by the summer of 1994 the political weakness of liberal rural views was apparent: the Peasant Party had established local organizations in only forty-one oblasts and krays, and in August 1994 the party had only 16,000 registered members.[145] In contrast, the Agrarian Party claimed more than 300,000 members. The weakness of the Peasant Party forced it into a political alliance with Gaydar's party, Russia's Choice.

In August 1994, I asked the deputy director of the Peasant Party, Aleksandr Khokhlov, in his Moscow office why his party supported Gaydar's party even though Gaydar advocated a number of antirural policies and had been a driving force in ending subsidized credits and reducing financial outlays to agriculture. Khokhlov admitted that the state's rural reform policies had made rural conditions worse and made it "difficult" at times to remain in the alliance. He appeared uncomfortable at the antirural bias of urban liberals and admitted that he wished they would show more support for food producers, although he expressed opposition to "throwing money at the problem." His main explanation for the alliance was that the Peasant Party was small and weak in the countryside. It needed a parliamentary ally, and Russia's Choice espoused views concerning democracy, a land market, and market reforms that were closest to the beliefs of Chernichenko and the party. Even though the antirural policies of Gaydar's party disturbed him and he wished for more moderate proposals, ultimately "there were no political alternatives" for rural liberals.[146]

As long as the state financially supported the private sector, AKKOR supported the Yel'tsin regime even though this meant that rural interests were subordinated to urban ones.[147] A crucial political turning point in the political relationship between AKKOR and the government came in September 1993 with the ending of subsidized credit to the private farming sector. Although the move was taken for economic reasons, politically this move undermined its primary source of rural support in the countryside. The impact was twofold. First, the popularity of AKKOR in the countryside became even weaker as farmers had no reason to join or remain members. Among private farmers interviewed in Kostroma Oblast during the summer of 1994, none indicated that any help whatsoever had been forthcoming from the local AKKOR and all felt that the organization was essentially "useless."[148] One farmer even charged that the AKKOR leadership was engaged in "illegal" behaviors and was interested only in helping themselves and friends.

Second, as financial conditions worsened for private farmers, AKKOR began to distance itself from the Yel'tsin regime by advocating subsidies and credits for agriculture, in effect becoming part of the so-called agrarian lobby.[149] In April 1994 the deputy head of the Founders' Council of AKKOR published an open letter in which he stated that "today relations between [land]owners and the state have become sharply strained," and he specifically criticized the financial policies of the state.[150] Articles in liberal, proreform agricultural newspapers openly questioned the nature of the relationship between private farmers and the government.[151]

The disaffection of rural liberal support culminated in December 1994 with the creation of a new economic-political organization, spun out of AKKOR, in fact led by V. Bashmachnikov (the president of AKKOR) and Yu. Chernichenko. This new organization—the Union of Landowners—advocated the same reform goals in the countryside as AKKOR, but the new union was different in that it was a political organization that would also fulfill economic functions. The creation of this organization was evidence that rural liberals did not feel their interests were being adequately represented or addressed through alliances with urban liberal groups.

The Union of Landowners was formed in an attempt to unify rural dwellers and to gain more broad-based rural support for land reform. Due to the political strategy that AKKOR had followed, there had been an erosion of support even among private farmers. Some private farmers had supported the reform planks of AKKOR and some had supported the policies espoused by the Agrarian Party simply because the Agrarian Party was more assertive in defending the financial, socioeconomic, and political interests of the rural sector.[152] By advocating a unified rural movement, Bashmachnikov implicitly admitted that dependence on state support had undermined rural support for proreform interests.

According to statements made at its founding congress, Union leaders wanted to distance the organization from the state. A distancing was necessary because close ties with the Yel'tsin regime had not yielded the needed or expected levels of financial support for the private rural sector, and AKKOR had suffered in popularity because it was associated with the antirural policies pursued by the Yel'tsin regime. Resolutions adopted at a meeting of AKKOR leaders in December 1994 implicitly blamed government policies for difficulties in the private sector. Specifically, these difficulties included a lack of legal protection, insufficient financial support, and an ineffective credit policy.[153]

A few months later, reflecting that rural liberals were politically weak in the countryside and needed urban alliances, Bashmachnikov suggested that the Union of Landowners join an electoral bloc with Prime Minister Viktor Chernomyrdin's party, Russia Is Our Home.[154] This move was bitterly contested within the union, and opponents argued that the party should operate independently, not as part of an urban alliance. Eventually, an "overwhelming majority" of delegates in the union voted to join the electoral bloc, but only after it became clear that the union lacked the money to operate an independent campaign.[155] The ratification of this electoral alliance was an implicit admission that, despite the desire for independent rural organizations, rural support for the political program of the Union of Landowners was insufficient to mount much of a campaign for independent seats. Bashmachnikov was elected as one of the delegates of Chernomyrdin's party, but he was virtually alone in terms of representing rural interests.

As financial trends for private farmers continued to deteriorate, so too did the relationship between AKKOR and the government. In early 1996 Vladimir Bashmachnikov charged that the minister of agriculture and the minister of finance were involved in a "secret agreement against private farmers."[156] In early 1997 Bashmachnikov was even more forceful: "Even the procommunist majority in the Duma annually favors a review of state support in the draft budget for farmers who are standing on their feet, but 'democrats' in the Ministry of Finance, with enviable tenacity, annually do not fulfill the law on the budget. It is solely for this reason that every year thousands of private farmers are destroyed."[157]

Thus, support for the government from rural liberals seriously eroded over time. Despite the need for rural support, state policy diverged from rural preferences on a number of core issues, and these differences were evident for both rural liberals and rural conservatives. For rural liberals, differences between policy preferences and state policy were evident in the following policy areas:

1. A guaranteed "market" (state purchases) for privately produced food that otherwise cannot be sold: Rural liberals maintained that the state should provide a guaranteed market, that is, to purchase private farmers' products if private farmers were not able to sell them on the domestic market due to continued trading monopolies and depressed consumer demand.

2. The issuance of state-subsidized credits to private farmers: AKKOR argued that the government should provide start-up capital for new private farmers because commercial banks often were unwilling to lend to private farmers. They also complained that market interest rates were high and repayment terms short, and insurance companies or banks had collateral requirements that private farmers could not meet. Cooperative rural banks and farmer association banks were still underdeveloped and therefore could not fulfill farmers' demands for loans.

3. An increase in state budgetary allocations to private farming (in constant rubles): At AKKOR's sixth congress in March 1995, AKKOR officials complained that, although the government promised to allocate 10 percent of federal monies budgeted for the agroindustrial complex to private farmers, during 1994 private farmers in fact received less than 6 percent. Furthermore, monies that were budgeted to the rural private sector often were not actually distributed. In October 1995, the president of AKKOR complained that during the first nine months of 1995 only 22 percent of federal monies had been distributed to private farmers, and almost all of those monies were used to make interest payments to banks from loans made during 1992 and 1993—there were no federal allocations to help beginning farmers, or to assist existing farmers obtain equipment and machinery.

4. Increased federal funding for rural infrastructure affecting private farmers: At AKKOR's sixth congress in March 1995, AKKOR officials complained that during 1994 private farmers received less than 2 percent of rural investment funds even though a series of laws adopted in 1990 had stipulated that private farmers would receive no less than 15 percent of rural investment funds.

The implications of the differences between state policy and rural liberal policy preferences are important. First, they mean that the regime confronted not only the rural conservatives who were ideologically opposed to reform policies but also the growing resistance of rural liberal organizations. Successful reform would require rural liberal support, but state financial policies undermined that support. Second, the weakness of rural liberals is shown in that Russian governmental policy did not side with or implement liberal rural demands on a single financial policy issue.[158] Despite ideological agreement on reform policies, on financial issues rural liberals and state interests clashed to the detriment of private farming.

Conclusion

State financial levers have been an important variable affecting the development, operation, productivity, and ultimately the survival of the private farms. State financial levers had a significant impact on the environment in which these farms operated. The manner in which state levers were used did not facilitate the emergence of a strong, commercially viable private farm sector. Instead, small and economically weak subsistence farms predominated, dependent upon state credits. Land privatization in Russia therefore did not necessarily bring less state influence over the rural sector, just different forms of influence. As is clear from this chapter, state actions continued to be of central importance in the countryside.

During the first few years of agrarian reform, private farmers were affected in fundamental ways by the financial strategies employed by the state. Beginning in 1992, the Russian state used financial levers to support its policy of expanding the number of private farms. The policy of extremely rapid private sector expansion continued for about twenty months, distributing subsidized credits regardless of the fact that the private sector had become nearly as dependent on state support as was the collective agricultural sector. The regime's strategy was successful in rapidly increasing the number of farms, but (as the data presented in this chapter show) private farms with good and bad climatic or soil conditions were "profitable" only because of credits from the Russian Farmer program. Over time it became clear that subsidized credits had led to an excessive demand for credit. The state could not fulfill this demand and still comply with the necessary macroeconomic guidelines to receive Western financial assistance. As a result, the state cut off subsidized credits and forced farmers to pay drastically higher interest rates from either the Central Bank or commercial banks—when credits were available at all. In effect, economic policy clashed with political goals.

As a consequence, the financial condition of private farms deteriorated. Because of their small size and their limited income potentials, significantly higher interest payments forced many farms to fold. Even those who remained afloat faced extreme difficulty in receiving credit. The state Rossel'khozbank often did not receive sufficient allocations of credit from the Ministry of Finance and the Central Bank, and commercial banks were unwilling to lend to clients who were considered an extreme credit risk. In the wake of resolution 975 in September 1993, which ended interest rate subsidies to private farmers, the rate of private farm creation plummeted and the number of farm bankruptcies skyrocketed. As the minister of agriculture Viktor Khlystun stated in 1993, "practically everybody—farmers, joint stock enterprises, and collective farms—are in such a position where they do not have the means to acquire necessary material resources."[159] Production subsidies benefited mostly the larger producers, those with established production capabilities. The problem for private farmers was to accumulate sufficient means to produce.

The political environment also was affected by the manner in which financial levers were used. Just as the private farm movement found its genesis in state initiatives and legislation, so too did AKKOR, conceived as a social organization that would represent and promote private farm interests. AKKOR evolved from state structures and never attained independence politically or economically. In the political realm, the critical shortcomings of AKKOR were twofold. First, because it was linked to the state and because land reform policies contradicted mainstream rural opinion, AKKOR was perceived as a state agent, and therefore its base of rural support was small and tenuous. Second, the resources that AKKOR had for building a rural constituency were limited and also dependent on the state. That is to say, in terms of institution-building in the countryside, AKKOR was at a severe disadvantage. Because it was not a political organization, AKKOR had no legislative representation in oblasts where land reform was implemented (although we should note that in its corporatist relationship with the state its views were reflected in the legislation adopted).

From a political standpoint, the ending of subsidized credits removed much of the incentive for rural liberals to support the Yel'tsin regime. The erosion of liberal support in the countryside was significant because it reduced rural political support for the regime, which was already weak among the rural population. Furthermore, the ending of subsidized credits led to the unity of liberal and conservative rural interests in opposition to rural financial policies pursued by the state, as both rural conservatives and liberals lobbied the government for more financial support for agriculture.

7

The State and Agrarian Reform
An Assessment

The Soviet state intervened in the rural economy as part of a larger strategy to control the organization and operation of food producers, to prevent the formation of class divisions based on private property, and to regulate rural social relationships. The nature of these interventions brought political stability but led to farm inefficiencies, high production costs, and a dysfunctional incentive structure that framed farm decisions.

Russian agrarian reform, on the other hand, was undertaken in order to address systemic flaws from the Soviet past that plagued the agricultural sector. The "answer" that Russian agrarian reform posed to the problems inherited from the Soviet era was privatization of land and agricultural enterprises. Russian reform privatized collective and state farms and created a private farming sector. This strategy of land privatization and the creation of a private farming stratum did not signify the end of state influence in the countryside. Instead, privatization brought about new methods and means of state influence. Using legislative, policy, and financial levers, the Russian state has continued to exert a strong influence in agriculture.

From a political standpoint, the nature of state interventions may be understood as an attempt to undermine conservative opposition in the countryside and to maintain political support among urban dwellers. The Yel'tsin regime and reformers in general have depended upon urban support, as shown by surveys and voting results. Thus, the economics of agrarian reform have a political and ideological underpinning.

How are we to understand the results of agrarian reform in Russia after more than six years? A common position of Western analysts is that agrarian reform was not very successful because reform measures were never implemented. This position is tautological. According to this argument, we would know reform had been implemented if it were successful. Second, it is necessary to distinguish between implementation and the performance of reform policies. The score on reform implementation was in fact quite impressive during the first six years of agrarian reform. The real question is not whether reform was implemented, but why the agricultural sector performed so poorly and what relevant factors affected its performance.

The answer is that transformation of the agricultural sector, reversal of Soviet-era legacies, and agricultural performance suffered because rural social policy did not change significantly from its Soviet past. Why did rural social policy not change significantly? There is a direct link between state actions and rural social policy. Patterns of rural social policy and the incentives that derived from it were the result of state-sponsored reform legislation and policies. Reform policies, which were created from above, failed to address the worst aspects of the Soviet legacy, not because the state was weak but because of the way these policies were defined. The results of this study, therefore, reject the notion of a "weak state" in Russian agriculture and instead emphasize that the nature of state-defined reform fundamentally affected reform outcomes.

The State and Agrarian Reform

Continuities in Soviet rural social policy must be understood as the outcome flowing from state-defined reform and the manner in which the postcommunist Russian state intervened in the rural sector. Agrarian reform policies were introduced from above, not because of urban pressures for better food supply or because of rural dwellers' demanding social equality. Conversations with agricultural and land officials in Russia consistently yielded the view that the center was driving reform and there was little local initiative or possibility for reform "from below." For some local officials this was a comfortable condition and provided ready-made excuses as to why reform was not working; but for others who wanted more economic freedom, state reform policies were a hindrance. The Russian state—by defining reform legislation, influencing farm operations, and affecting the larger economic environment—lies at the center of reform results.

If the state was central to the outcome of agrarian reform, and if the state was central to rural social policy, what explains the continuities of Soviet rural social policy? Several reasons may be cited to explain this. First, reform legislation and strategies were badly designed and conceptually flawed.

There were both economic and political reasons for poor reform design. Economically, rural reform from above was combined with a fundamental misunderstanding of rural conditions and a sense of what the rural sector needed in order to be reformed successfully. Politically, part of the reason reform from above was badly designed was the need to maintain urban support. Even as the countryside was transformed, the cities still needed to eat. Since large urban centers provided most of the political support for market and democratic reforms in general, the needs of urban consumers were emphasized.

Reform legislation and strategies were badly designed for another reason—ideological influences. In particular, the advice given by Western policy advisors and organizations attempted to create reform policies that were inappropriate for Russian conditions. Often this advice was based on ideological correctness rather than on economic utility. That is to say, reform legislation and strategies were conceptually flawed because agrarian reform was intended to solve economic problems, but in fact reform was pursued for political goals.

A final consideration concerned continuities in political culture. The degree to which Russian reform institutions reflected continuities from the past should not be underestimated. Emphasis on egalitarianism and collectivism in the rural sector predate the Soviet period. Political culture influences belief systems and values, which in part are reflected in reform institutions and policies. Political culture continuities, for instance, help to explain why AKKOR and the Agrarian Party, which on several fundamental issues were ideologically divided, agreed that reform should preclude the rise of large private estates. Thus, in a broader context reform behaviors must be understood as resulting from continuities in political culture on the part of those who govern and those who are governed.[1]

To conclude, state-defined reform measures were designed and implemented in both the private and the collective rural sectors. However, reform legislation and strategies performed poorly because they did not create new incentives that would lead to behavioral changes by individuals and collectives. We can demonstrate the successes and failures of agrarian reform by returning to the areas in which state interventions influence rural social policy: incentives, economic environment, and rural socioeconomic conditions.

Incentives

At the beginning of this study, we postulated that Russian agrarian reform would need to change the legacy of Soviet rural egalitarianism by moving away from individual, farm, and regional egalitarianism and by revising state subsidy, zonal pricing, and price supplement systems. To what degree have these goals been achieved? In the Soviet period, rural egalitarianism penalized the productive and the efficient and undermined incentives to work hard—aspects that affected food production and ultimately food supply, prices, and variety. Land privatization was the strategy adopted by the state to rectify the problem of excessive egalitarianism. The goal of land privatization was to create incentives that would facilitate improved agricultural performance. Thus reversing rural egalitarianism was the means to achieve the goal of higher cost effectiveness.

Despite reform goals, state-defined reform legislation governing intrarural relations continued to be egalitarian. Rural egalitarianism was important because Russian society as a whole became notably less egalitarian after 1991, with wage differentials on average reaching an estimated 15:1 between highest- and lowest-paid workers during 1995. Thus the countryside remained more egalitarian than society as a whole, and this egalitarianism was inherent even to the new rural private sector.

Egalitarianism in the Private Sector. The definition of reform legislation and the incentives that flowed from reform fundamentally affected the development of private farms and their income potential. Property rights within farms were established, but this was done in an egalitarian manner that distributed land equally without regard to ability for production potential or efficiency. Even the radical Nizhniy model of farm privatization served only to reduce the size of production units. It did nothing to create market relations, introduce competition, or facilitate the creation of a larger market environment in which the newly created farms were to work.

Reform legislation also mandated small land plots as part of a deliberate policy to prevent large landed estates. Small plots of land translated into limited income potential, making it difficult to afford farm equipment, hire labor, or expand operations, all of which affected farm production. Limited income potential exacerbated private farm debt, making it harder to pay off past debt, obtain bank credits, purchase machinery and other inputs, or produce enough to satisfy collateral demands by commercial banks. Limited income also meant that many farmers could not purchase additional land; and farmers argued that small plots made it impossible to conduct effective commercial operations. Thus, small land plots were at the heart of the strategy to prevent rural "latifundia," and state-defined legislation fundamentally affected prospects for economic success.

Rural Egalitarianism in the Collective Sector. Despite the reorganization of state and collective farms, many continuities from the Soviet era were evident in the collective sector. Reform legislation and attempts to reorganize farms were unsuccessful in significantly changing farm operations from their Soviet past. This meant that incentive structures affecting individuals were inadequate, as wage relationships were not significantly changed and wage rates were simply indexed to inflation. In general, price supplements continued to benefit the least efficient and highest-cost farms, although there was constant talk about changing those policies. The allocation of subsidies continued to divert resources to weak farms. Production subsidies failed to motivate efficient use of inputs. Furthermore, the continued program to invest in non–black earth areas signified a continued concern for the poorest farming areas.[2] Some progress was made in privatizing the weakest and most unprofitable farms—as in Nizhniy Novgorod. In Nizhniy and elsewhere, those farms that privatized were among the weakest; even after privatization they required enormous subsidies and special advantages in order to prove that the privatization model "worked."

Economic Environment

The second aspect of rural social policy that needed to change from its Soviet past concerned the economic environment in which farms worked. We postulated that successful reform would remove restrictions on the private rural sector, including both personal plots and private farms; and land privatization would be institutionalized. Furthermore, an economic environment would be created in which all food producers would prosper, and in particular economic conditions would facilitate a new class of prosperous private farmers.

Restrictions on the Private Sector. Clearly, during Russian agrarian reform success was achieved in removing restrictions on the rural private sector, as millions of land plots were privatized and 280,000 private farms were established. Overall, on the institutional level a great deal of transformation occurred. For example, since 1992 private property was legalized, land shares were distributed to rural dwellers, a private farm stratum was created, state and collective farms were reorganized, a land market was created and began to function, and a farm privatization program was implemented in several oblasts. Moreover, success was achieved in the institutionalization of privatization, as groups along the entire political spectrum accepted the principle of private farms and private ownership of land, although disputes continued over rights of land disposal. By early 1997 more than 90 percent of agricultural land had been privatized.

Financial Environment. Despite success in deregulating the private sector and institutionalizing its acceptance, state financial levers had a significant impact on the economic environment in which private farms operated. The manner in which state financial levers were used failed to create an economically strong private sector that was productive and efficient. Instead, small private farms—economically weak and dependent upon state credits—were the norm not the exception in the Russian countryside. Reform legislation that granted small plots of land went hand in hand with credit that was limited. Credits were distributed so that many farms received small amounts of credit, sums that decreased in value (in constant rubles) over time. As a result, the state credit system did not facilitate the emergence of a strong, commercially viable private farm sector, and the newly emergent commercial banking and credit system was unwilling to assume the financial risk of issuing credits to private farmers. Over time the private sector became impoverished, measured by rising farm bankruptcies, and private farmers were differentiated primarily by their degree of debt.

As a result of state financial policies, during 1994 more than 57 percent of agricultural enterprises (state and collective farms and their successors) were unprofitable; this increased to 70 percent during 1995 and 75 percent in 1996. Private farms suffered as well, as less than 20 percent were profitable by mid-1995. The repercussions of a hostile economic environment led to increasing private closures, fourteen thousand in 1993, twenty-six thousand in 1994, and over twenty-five thousand in 1995 and 1996. Meanwhile, the rate of new farm creation plummeted.

Creation of a New Rural Class. Regarding the creation of a new rural class, two aspects warrant mention. First, it may be argued that the distribution of land and farm equipment gave rise to a new class of private food producers in the Russian countryside. However, during the land distribution process not much land was distributed to persons and enterprises who might use it more efficiently, and therefore the economic base for new social classes was weak and tenuous. During the farm reorganization process, relatively few state and collective farms actually disbanded, while an overwhelming percentage of reorganized farms either retained their collective farm status or became joint stock farms. Thus, in most cases the farm and farm property remained intact. Former state and collective farms remained intact by forming joint stock farms, limited partnerships, or retaining their previous status. For example, about 83 percent of former state and collective farms either retained their previous status or reorganized as a joint stock farm.[3] As a result, a relatively small amount of land was distributed free of charge or made available for purchase to private farmers wishing to increase the size of their operations. On January 1, 1995, former state and collective farms and their successors still held about 87 percent of agricultural land (not including land reserves and forest funds), although technically this land was now private, not state, property. Even if the economic environment were such that private farmers could produce to their potential, nationwide, private farmers had a little over 6 percent of Russia's agricultural land during 1996, or about 11 million hectares.[4] Thus, there are strong indicators that the building of a new rural class was ineffective and that rural social structure did not change appreciably. That is to say, despite farm and land privatization, from a sociological perspective the Russian countryside looked much the same after several years of reform as it did before.

Second, the crucial question is not whether a new class exists, but what is its political and economic strength. Politically, we saw evidence that rural liberal organizations that drew support from private farmers were weak and therefore dependent upon alliances with urban-based political parties. Out of necessity, rural liberal organizations pursued a corporatist relationship with the state, thus linking their fate to state policy. The linkage of proreform rural groups to the state meant that independent (nonstate) groups failed to arise or to become significant political actors. It also meant that as state policies turned antirural, the popularity of rural proreform groups declined even further.

Rural Socioeconomic Conditions

The third area of rural social policy that needed to change from its Soviet past concerned socioeconomic conditions. Here we postulate that rural living standards would need to increase, urban-rural wage differences should be narrowed, economic conditions would facilitate the retention of the rural young and skilled, rural infrastructure would improve, and demographic problems would begin to be rectified.

Rural Standard of Living. The ending of the social contract and the introduction of reform measures decreased rural standards, although one could argue that, with a lower standard of living to begin with, the rural population and farm members had less to lose than urban dwellers. Moreover, with access to production from personal plots there is some validity to the assertion that rural dwellers' standard of living did not decline as much as that of urban residents. It does well to remember, in addition, that farm members had access to food supplies in ways that are often difficult to measure (pilfering, underreporting production, payments in-kind, and consumption of personal plot produce), and therefore it is possible that rural standards of living did not decline as much as is commonly believed. It is also true that parent farms continue to provide subsidized food to their members.

Not only did the standard of living decrease, but rural unemployment increased. Although much of rural unemployment is concealed (as is true for the economy as a whole), estimates range between 1.8 and 2.0 million rural unemployed in 1994,[5] equal to about 33–37 percent of total unemployment at the end of 1994, or about 10 percent of the rural working-age labor force. To put it a different way, the number of rural unemployed equaled 20 percent of persons actually employed in agricultural production in 1994. In contrast, urban unemployment equaled about 5 percent of the urban working-age labor force during 1994.[6]

Between 1992 and 1994 rural unemployment grew by more than 370 percent, and in 1994 alone rural unemployment doubled. Nearly two-thirds of rural unemployment occurred in five economic regions: the Central,[7] the Volga, the Northern Caucasus, the Urals, and the Western Siberian regions. Among the rural unemployed, two main groups dominated: women, who comprised about 38 percent of the rural work force, accounted for nearly two-thirds of the rural unemployed. The second group, the rural young (persons less than thirty years of age), accounted for over one-third of rural unemployment (young women are counted in both categories). Among the rural young, the age group most affected by unemployment were persons between twenty-two and twenty-nine years of age.[8]

Urban-Rural Wage Inequality. Russian agrarian reform intensified urban-rural inequality, evidenced by widening urban-rural wage differences and disadvantageous terms of trade for food producers. At the individual level, since agrarian reform was begun, no other occupational group has suffered a greater erosion in relative wage levels. From an average of 93 percent of an industrial worker's average monthly income in 1990, by the first half of 1996 agricultural workers were earning about 36 percent of the average industrial monthly income, a percentage that actually understates the degree to which farmworkers and production personnel saw their incomes erode. By the first half of 1995 agriculture was the lowest paid profession in the nation, measured in absolute terms (although wage levels do not account for income derived from personal plot sales). When one considers chronic nonpayment of wages and that a higher percentage of agricultural enterprises were in arrears on wages than in any other branch of the economy, it is clear that farmworkers were big losers in terms of personal income. In effect, decades of narrowing the gap between urban and rural wages were wiped out in four short years.

At the enterprise level, the terms of trade turned drastically against food producers. From 1991 through 1995 prices for industrial goods used by food producers increased 2,230 percent, while purchase prices for agricultural products rose only 752 percent. This fact was important because it influenced domestic food production, the ability to reform farm operations, and the need for food imports. Through 1996 the trends were not favorable, as food production and net output continued to decrease; farms were too busy trying to survive and therefore placed little emphasis on transformation; and food imports, particularly of high preference goods, increased.

The intensification of urban-rural inequality had three main consequences. First, it galvanized rural political opposition and gave conservatives political ammunition to use against agrarian reform policies and radical reformers in Moscow. The Agrarian Party portrayed itself as the defender of all rural socioeconomic and political interests.

Second, intensified urban-rural inequality also had economic ramifications as the countryside became significantly poorer through policies that facilitated unequal terms of trade, growing price disparities, and resource extraction. Economic effects were felt by farmers in both the private and the collective spheres. The intensification of urban-rural inequality was important because some 39 percent of the population lived in rural areas. Urban bias theorists hypothesize that governments will protect urban interests in order to ensure political stability, but one wonders how many societies, not to mention transitional societies with weak civil traditions, could withstand the social tensions of an urban-rural divide such as that demonstrated in Russia. Quite simply, how long could post-Soviet "urban bias" endure without a backlash from the rural sector?

Last, intensified urban-rural inequality had sociodemographic consequences. As rural conditions deteriorated, the gains in rural social development made under Brezhnev were lost. Rural life became more difficult and less attractive. As a consequence, the young and the skilled continued to leave farms in large numbers, which raised doubts about the efficacy of dividing farmland and property among those who remained.

Rural Out-migration. Agrarian reform had the dual purpose of driving surplus farm personnel off the farm, at the same time attracting workers with higher education and skill levels to become or remain employed in agricultural production. Before reform was begun, as well as to the present, Russia has suffered from the odd combination of having a high percentage of the work force employed in agriculture (15 percent of the labor force worked in agriculture in 1994) while also experiencing annual labor shortages necessitating the mobilization of military personnel, students, and even faculty to help with the fall harvest.

Since 1991, the number of farm personnel—workers involved in agricultural production—dropped considerably. For example, in Kostroma Oblast, whose agricultural situation was broadly representative of other Central region oblasts, the number of workers engaged in agricultural production on agricultural enterprises declined from 46,283 to 34,300 between 1991 and the end of 1995.[9] Nationwide, the number of persons employed in the agroindustrial complex (APK) declined from 16.97 to 16.41 million from 1990 to 1994.[10] Among professions in the APK the largest decline occurred in agricultural production, with a decrease of about 200,000 persons. The decline in farmworkers was of course a long-term trend, but since agrarian reform began the exodus has accelerated. The socioeconomic reasons for the exodus from farms are well known: low levels of rural housing amenities, poor educational opportunities, poor quality medical care and primitive facilities, the lack of adequate recreational facilities, poorly stocked stores, lower levels of trade turnover, and relatively low purchasing power.

Rural Investments. During agrarian reform, much-needed social development of the countryside went unfulfilled, and by the beginning of 1995 capital investments devoted to the rural sector were one-third of their 1991 level when calculated in constant rubles. Rural amenities, services, and facilities deteriorated. Health and education suffered, as evidenced by a skyrocketing rural death rate, widespread school closings, and a large exodus from the teaching profession. Rural nurseries, kindergartens, schools, hospitals, clubs and services for the elderly, and service centers closed because of the lack of funds to operate them.

More than one-half of children's preschools were in need of major structural repair. In 1994 the construction of rural homes and schools declined by 250 percent in comparison with 1990; the construction of rural kindergartens declined 770 percent, rural recreational clubs declined 430 percent, and the construction of rural roads declined 1,200 percent. Some 300,000 rural population points did not have hard-paved roads, and about 7 percent of former state and collective farms still lacked hard-paved roads connecting them to a raion center.[11] Even in mid-1995 it was reported that only 22 percent of rural dwellings were equipped with gas lines, only 20 percent of rural housing had indoor plumbing, and only 15 percent had running water inside their house.[12]

Rural Demography. Since the early 1990s there has occurred (1) a deterioration in the rural birth-death rate; (2) net rural in-migration from the near abroad and from urban centers; and (3) net growth of the rural population. Most notable was a surge in the rural death rate and decrease in the birth rate, particularly in the non–black earth zones of central Russia. For the rural sector throughout Russia, from 1990 through 1994, the birth rate declined from 15.5 to 11.4 per thousand persons. During the same time period, the death rate increased from 13.3 to 17.5 per thousand persons. Overall, these trends meant that the "natural increase" coefficient moved from a +2.2 in 1990 to –6.1 in 1994 (the urban coefficient moved from +2.3 to –6.1 during the same time period).[13] This overall measure masked the severe deterioration in the rural death rate in several oblasts, particularly in the Russian non–black earth zone where the negative coefficient was in double digits for nearly every oblast. This in turn meant that the rural death rate was two or three times as high as the birth rate in the Russian non–black earth zone.

At first glance it would appear that migratory inflows into the countryside since 1991 would begin to address the age-structure problems caused by decades of rural out-migration. As streams of Russians moved into Russia from the near abroad, a majority settled in rural areas. Based on local examples we know that a significant percentage of rural arrivals were young and educated. However, about 80 percent of arrivals were former urbanites. It remained unclear whether these people would settle permanently in the countryside or were simply awaiting an improvement in urban conditions. Furthermore, despite net growth in the rural population as a whole, the farm population continued to decline, in fact even more rapidly than before. This meant that farm labor would remain in short supply, that the farm labor pool was not being replenished, and that the age structure of farm labor remained quite old. Particularly in the Russian non–black earth, it was not uncommon for one-half or more of a farm's membership to be of pension age.

Therefore, rural in-migration from abroad and within Russia have merely masked the sharp deterioration in rural births and the increase in rural deaths, a result of decades of age-specific out-migration. Although the population of the Russian countryside began to increase in 1992 for the first time since World War II, this trend masked other processes, namely that the population in the countryside continued to age. In 1995, persons aged sixty and above accounted for 23 percent of the rural population, up from 20 percent in 1979 and 22 percent in 1989.[14]

The State and Agrarian Reform

I have argued that the post-Soviet state has exerted a profound impact on the nature and development of agrarian reform in Russia by influencing the legislative, economic, and social environment. The Russian state was unbalanced, with a strong state in agriculture while, in urban policy areas, its weakness was apparent as urban civil society grew and strengthened. Thus, in agriculture the task of "getting the state out" would be greater than mere privatization or denationalization. Genuine reform would need to modify state capacities. Changing state capacities would require the emergence of other actors who could fill the void of a receding state. Given state strength in agriculture and its effect on reform, it would be useful to consider how the role of the state might be reduced in agriculture. Two questions need asking. What are the sources for changing the role of the state in agriculture? Can other actors assume a greater role in influencing rural social policy?

The most obvious source of potential change in state influence is privatization. The privatization of land and enterprises was intended to lessen the role of the state by creating independent rural organizations and actors throughout the food system. Independent rural organizations and actors would, ideally, mobilize and generate policy alternatives to state legislative initiatives. Privatization, however, did not lead inexorably to decreased state power, and in a comparative sense Russia was not entirely unique. Privatization in Russia, as elsewhere, provided the state new means of control over the economic environment through the use of budgetary, tax, and credit instruments.[15] Furthermore, privatization was used to build political support. Last, through land distribution to millions, the cost of collective action rose. Rather than a large unified rural group able to act independently of state wishes, instead, rural interests became so fractured that the state was able to dominate rural groups.

A second source of potential change was through the rise of effective rural interest groups who could peruse and implement policies independent of state initiatives. However, rural liberals were not well represented in the legislative branch of government; they had weak popular support among the rural population and were therefore vulnerable to the dominant interests of their urban coalition partners, and they faced significant organizational and financial constraints. Rural conservatives, although better represented in the legislature and government, had their impact muted because of ideological differences with state policies. Rural conservatives overall were ineffective in pursuing their party program, policies, and goals. Part of the reason was that political parties remained weak and unconsolidated.[16] In general, however, the main reason was structural; the Russian presidential system gave enormous state powers to the executive and much less power to other political actors.[17]

A third source of potential change would be the rise of independent commercial banks that could influence the overall economic environment by fulfilling the functions of state funding. To be sure, the banking system did undergo significant reform, with the Central Bank and its regional representatives gaining new independence. Commercial banks arose that exerted an influence on the economic environment, though not always in positive ways. In general the reformed banking system was characterized by a number of shortcomings and limitations, including few financial levers, a rudimentary payment and clearing system, high percentages of credits from the federal budget, and continued bureaucratization, secretness, and centralization.[18]

In the agricultural sphere, starting in 1991 peasant cooperative banks were created, backed by AKKOR. In general, however, these banks were small and poorly capitalized and provided loans mainly to their depositors. These banks did not have enough resources to exert a significant economic influence on the rural economic environment. Therefore, the predominant agricultural banking institution is Rossel'khozbank, the former state Agroprombank. Although formally privatized, Rossel'khozbank is a state bank in the sense that it remained dependent on the receipt of state credits, without which it would surely have collapsed. Rossel'khozbank continues to disperse a majority of the seasonal credit provided to food producers.[19]

A final source of potential change would be a shift in popular attitudes toward the state, its role, and its policies. A well-established trend in pubic attitudes showed a generational effect in which older persons tended to be less supportive of reform and more supportive of communist economic policies. Richard Rose demonstrated this link between generational effects and attitudes toward reform for a number of post-Soviet nations, including Russia.[20] The link between age and attitudes is important, for the Russian rural population was older in general than the urban population and had a higher percentage of pension-aged individuals. The link was further confirmed as rural conservatism was clearly demonstrated in voting behaviors in 1993, 1995, and again in 1996, resulting in an urban-rural divide.

Moreover, not only did the rural age structure exert generational effects on voting results, but also it affected specific aspects of economic reform. In particular, an older, less educated population displayed more supportive attitudes toward state-regulated egalitarianism.[21] Post-Soviet public opinion continued to support state intervention to pursue welfare state policies. For example, those with lower incomes expressed strong desire for the state to provide economic well-being.[22] Older persons also supported economic reform, specifically a market economy, to a lesser extent than younger persons with more education.[23] Furthermore, public opinion continued to support public ownership of basic industries, as well as state control over wages, prices, and profits.[24] The point is that popular opinion in general supported specific aspects of a strong state in the economy, and in particular the sociodemographic characteristics of the rural population were supportive of welfare state policies that would require a strong and active state in the rural economy. The characteristics enumerated above—generational effects, lower levels of education, lower average incomes—typify the rural population. Rural public opinion would not be a driving force in changing the role of the state in the rural economy because it favored state intervention in order to shield the rural sector from the effects of market reform.

Recent inflows of migrants to rural areas, tending to be former urbanites and more highly educated, were not sufficient to change rural public opinion in the short term because the flows were relatively modest compared to decades of rural out-migration, and because it was not clear that those arrivals would remain in the countryside over the longer term. The main questions regarding state strength in agriculture are, therefore, Where would sources of change come from? How would they exert an influence? Who would exert pressure for the state to recede? Who would assume the roles of a receding state? The insufficiency of sources of change—combined with continued state capacities, levers, and advantages—suggests that the Russian state will remain the dominant actor in the agricultural sector for the foreseeable future.

Notes

Index

Notes

All translations, unless otherwise noted, are by the author. All interviews were conducted by the author.

Chapter 1. Agriculture and the State

1. The United Nations, *Statistical Yearbook, 1993* (New York: United Nations, 1995), pp. 236–44. For Russia, see *Rossiyskiy statisticheskiy ezhegodnik, 1995* (Moscow: Goskomstat, 1995), p. 245; for the United States, see *Statistical Abstract of the United States, 1995* (Washington, D.C.: Department of Commerce, 1995), p. 452.

2. See Keith Griffin, *Land Concentration and Rural Poverty* (New York: Holmes and Meier, 1976); and John Sheanan, *Patterns of Development in Latin America: Poverty, Repression, and Economic Strategy* (Princeton: Princeton University Press, 1987).

For a brief summary of the history of land reform and its results in Latin America, see Peter Dorner, *Latin American Land Reforms in Theory and Practice: A Retrospective Analysis* (Madison: University of Wisconsin Press, 1992), esp. chap. 3; Solon Barraclough, ed., *Agrarian Structure in Latin America* (Lexington, Mass.: Lexington Books, 1973), chaps. 5–11; and William C. Thiesenhusen, ed., *Searching for Agrarian Reform in Latin America* (Boston: Unwin Hyman, 1989).

3. For revolutions, see Eric R. Wolf, *Peasant Wars of the Twentieth Century* (New York: Harper Torchbooks, 1969). For governments supporting the peasant movement, see Thiesenhusen, *Searching for Agrarian Reform in Latin America*, chap. 10; Wolf, *Peasant Wars of the Twentieth Century*, chap. 1; and John Dunn, *Modern Revolutions: An Introduction to the Analysis of a Political Phenomenon*, 2d ed. (Cambridge, England: Cambridge University Press, 1989), chap. 2.

It is ironic but land reform governments in Latin America do not remain in power very long. With "the exception of Castro in Cuba, no reform-oriented government has remained in power for more than a few years." Dorner, *Latin American Land Reforms in Theory and Practice*, p. 34.

4. The classic work on urban bias is Michael Lipton, *Why Poor People Stay Poor: Urban Bias in World Development* (Cambridge, Mass.: Harvard University Press, 1976). Urban bias is often found in agrarian nations where the rural population is at least 60 percent of the population, where agriculture generates more than 50 percent of gross national product, but where agriculture receives less than 30 percent of investment. Politically, urban bias characterizes states with one-party systems, and authoritarian and communist systems. The urban political elite act as a "class": not as the bourgeoisie as in Marxist writings but as "a committee for managing the common affairs" of the state toward the rural sector. Lipton argues that urban bias is a universal phenomenon in the Third World and he analyzes its effects in a number of countries. Ibid., pp. 27, 74, 77, 340.

5. Ashutosh Varshney, "Urban Bias in Perspective," *Journal of Development Studies* 29, no. 4 (July 1993): 4.
6. Robert H. Bates, *Markets and States in Tropical Africa: The Political Basis of Agricultural Policies* (Berkeley and Los Angeles: University of California Press, 1981), p. 33.
7. Ibid., chaps. 2, 3.
8. See, for example, Robert H. Bates, *Essays on the Political Economy of Rural Africa* (Berkeley and Los Angeles: University of California Press, 1983); Yahya M. Sadowski, *Political Vegetables?* (Washington, D.C.: Brookings, 1991); "Beyond Urban Bias," special issue, *Journal of Development Studies* 29, no. 4 (July 1993); and William C. Thiesenhusen, *Broken Promises: Agrarian Reform and the Latin American Campesino* (Boulder, Colo.: Westview Press, 1995). For a critique of Lipton and the urban bias approach, see T. J. Byres, "Of Neo-Populist Pipe-Dreams: Daedalus in the Third World and the Myth of Urban Bias," *Journal of Peasant Studies* 6, no. 2 (January 1979): 210–44.
9. Theda Skocpol, "Bringing the State Back In: Strategies of Analysis in Current Research," in Peer B. Evans, Dietrich Rueschemeyer, and Theda Skocpol, eds., *Bringing the State Back In* (Cambridge: Cambridge University Press, 1985), p. 28.
10. This is not to deny that political conflict over agricultural policies existed. See, for example, Werner G. Hahn, *The Politics of Soviet Agriculture, 1960–1970* (Baltimore: Johns Hopkins University Press, 1972).
11. This definition is borrowed from Eric Nordlinger, cited in Gabriel A. Almond, "The Return to the State," *American Political Science Review* 82, no. 3 (September 1988): 858.
12. Carl J. Friedrich and Zbigniew K. Brzezinski, *Totalitarian Dictatorship and Autocracy* (New York: Praeger, 1956).
13. Barrington Moore Jr., *Terror and Progress USSR: Some Sources of Change and Stability in the Soviet Dictatorship* (Cambridge, Mass.: Harvard University Press, 1954). Jerry F. Hough, *Soviet Leadership in Transition* (Washington, D.C.: Brookings, 1980). H. Gordon Skilling and Franklyn Griffiths, eds., *Interest Groups in Soviet Politics* (Princeton: Princeton University Press, 1971). Jerry F. Hough, *The Soviet Prefects: The Local Party Organs in Industrial Decision-Making* (Cambridge, Mass.: Harvard University Press, 1969); and for a reexamination of Hough's theses, see Peter Rutland, *The Politics of Economic Stagnation in the Soviet Union: The Role of Local Party Organs in Economic Management* (Cambridge, England: Cambridge University Press, 1993).
14. For an excellent analysis of the Nineteenth Party Conference, see Seweryn Bialer, "The Changing Soviet Political System: The Nineteenth Party Conference and After," in Seweryn Bialer, ed., *Politics, Society, and Nationality Inside Gorbachev's Russia* (Boulder, Colo.: Westview Press, 1989), pp. 193–241.
15. See, for example, Ronald J. Hill, "The CPSU: From Monolith to Pluralist?" *Soviet Studies* 43, no. 2 (1991): 217–36; and Archie Brown, "Political Change in the Soviet Union," in Alexander Dallin and Gail W. Lapidus, eds., *The Soviet System in Crisis* (Boulder, Colo.: Westview Press, 1991), pp. 116–29.
16. There are too many sources to list all here, but for representative examples, see Vladimir Brovkin, "Revolution from Below: Informal Political Associations in Russia," *Soviet Studies* 42, no. 2 (1990): 233–58; Victoria E. Bonnell, "Voluntary Associations in Gorbachev's Reform Program," in Dallin and Lapidus, *The Soviet System in Crisis*, pp. 151–60; Peter Rutland, "Labor Unrest and Movements in 1989 and 1990," in Ed A. Hewett and Victor H. Winston, eds., *Milestones in Glasnost and Perestroyka: Politics and People* (Washington, D.C.: Brookings, 1991), pp. 287–325; Judith B. Sedaitis and Jim Butterfield, eds., *Perestroika from Below: Social Movements in the Soviet Union* (Boulder, Colo.: Westview Press, 1991); and William Moskoff, *Hard Times: Impoverishment and Protest in the Perestroika Years* (New York: M. E. Sharpe, 1993), chap. 5.
17. See Lubomyr Hajda and Mark Beissinger, eds., *The Nationalities Factor in Soviet Politics and Society* (Boulder, Colo.: Westview Press, 1990).
18. Gail Lapidus, "State and Society: Toward the Emergence of a Civil Society," in Dallin and Lapidus, *The Soviet System in Crisis*, pp. 130–50; and S. Frederick Starr, "Soviet Union: A Civil Society," *Foreign Policy*, no. 70 (spring 1988): 26–41.

19. Institutions refer to the set norms and laws that govern the behavior of actors. Policies concern the flow of resources and the terms of trade confronted by actors within the food system. Social relationships refer to rural social policies.

20. Stephen D. Krasner, *Defending the National Interest* (Princeton: Princeton University Press, 1978), p. 56. Joel S. Migdal, *Strong Societies and Weak States: State-Society Relations and State Capabilities in the Third World* (Princeton: Princeton University Press, 1988), p. 8.

21. Krasner, *Defending the National Interest*, p. 56.

22. See, for example, Stephen Kotkin, *Magnetic Mountain: Stalinism as Civilization* (Berkeley and Los Angeles: University of California Press, 1995); William G. Rosenberg and Lewis H. Siegelbaum, eds., *Social Dimensions of Soviet Industrialization* (Bloomington: Indiana University Press, 1993); and Sheila Fitzpatrick, *Stalin's Peasants: Resistance and Survival in the Russian Village After Collectivization* (New York: Oxford University Press, 1994).

23. See J. Arch Getty, *Origins of the Great Purges: The Soviet Communist Party Reconsidered, 1933–1938* (Cambridge, England: Cambridge University Press, 1985).

24. Richard Sakwa, *Russian Politics and Society* (London: Routledge, 1993), p. 166.

25. Nonimplementation is explained by the need to differentiate between weak states and weak governments (taking into account strong societies). An alternative measure of state strength shifts attention away from policy implementation and concentrates on state capacities.

26. Samuel P. Huntington, *Political Order in Changing Societies* (New Haven: Yale University Press, 1968), esp. chap. 1.

27. Here it is necessary to look at state power not only in absolute but in relative terms, comparing state power to the power of other power rivals in society.

28. See, for example, Cynthia S. Kaplan, *The Party and Agricultural Crisis Management in the USSR* (Ithaca: Cornell University Press, 1987).

29. See Don Van Atta, ed., *The Farmer Threat: The Political Economy of Agrarian Reform in Post-Soviet Russia* (Boulder, Colo.: Westview Press, 1993), pp. 197–204.

30. See Almond, "The Return to the State," pp. 853–74.

31. See Alexander Yanov, *The Drama of the Soviet 1960s: A Lost Reform* (Berkeley and Los Angeles: University of California, Institute of International Studies, 1984).

32. *Izvestiya*, October 28, 1991, p. 2.

33. V. Khlystun, "Napravleniya razvitiya agrarnoy reformy," *APK: ekonomika, upravleniye*, no. 11 (November 1992): 22–25.

34. An analysis and comparison of rural and liberal conservative views may be found in Stephen K. Wegren, "Rural Politics and Agrarian Reform in Russia," *Problems of Post-Communism* 43, no. 1 (January–February 1996): 23–34.

35. In the industrial realm, see Michael McFaul, "State Power, Institutional Change, and the Politics of Privatization in Russia," *World Politics* 47, no. 2 (January 1995): 210–43.

36. Robert H. Bates, "'Urban Bias': A Fresh Look," *Journal of Development Studies* 29, no. 4 (July 1993): 225.

37. See, for example, Douglas A. Chalmers, "Corporatism and Comparative Politics," in Howard J. Wiarda, ed., *New Directions in Comparative Politics* (Boulder, Colo.: Westview Press, 1985), pp. 56–79; and Philippe Schmitter, "Still the Century of Corporatism?" in Frederick B. Pike and Thomas Stritch, eds., *The New Corporatism: Social and Political Structures in the Iberian World* (Notre Dame, Ind.: Notre Dame University Press, 1974), pp. 85–131. For applications of corporatism to the Soviet Union, see Valerie Bunce and John M. Echols, "Soviet Politics in the Breznev Era: 'Pluralism' or 'Corporatism'?" in Donald R. Kelley, ed., *Soviet Politics in the Brezhnev Era* (New York: Praeger, 1980), pp. 1–26; and Valerie Bunce, "The Political Economy of the Brezhnev Era: The Rise and Fall of Corporatism," *British Journal of Political Science* 13 (April 1983): 129–58.

38. See Daniel Kelliher, *Peasant Power in China: The Era of Rural Reform, 1979–1989* (New Haven: Yale University Press, 1992), pp. 36–37.

39. See Mancur Olson, *The Logic of Collective Action: Public Goods and the Theory of Groups* (Cambridge, Mass.: Harvard University Press, 1965). In a later work Olson examines signifi-

cant exceptions for occupational organizations by arguing that fragmented groups are inefficient whereas large groups are able to bargain more effectively if umbrella groups encompassing a range of views within a given movement are formed, something akin to peak groups in German politics. See Mancur Olson, *The Rise and Decline of Nations: Economic Growth, Stagflation, and Social Rigidities* (New Haven: Yale University Press, 1982).

Chapter 2. State Interventions in Rural Social Policy During the Soviet Period

1. There were different kinds of brigades. For example, an "integrated" brigade worked in several branches of production and had the most members and worked the most land. There were also "branch" and "specialized" brigades that worked in one sphere of agricultural production such as fieldwork, vegetable growing, animal husbandry, and so on. Finally, there were "mechanized" brigades—and, in collective farms, "tractor" brigades—who serviced a number of other productive brigades. *Ekonomika sel'skokhozyaystvennykh predpriyatiy*, 3d ed. (Moscow: Politizdat, 1964), pp. 237–38.

2. M. A. Malygin, ed., *Azbuka ekonomika kolkhoza*, 3d ed. (Moscow: Politizdat, 1974), pp. 256–58.

3. T. I. Zaslavskaya, *Raspredeleniye po trudu v kolkhozakh* (Moscow: Ekonomika, 1966), pp. 73, 74–102. Other examples suggest even smaller differences. In 1962, a survey of 116 collective farms in Stavropol' kray found a wage differential of 1:1.9 existed between categories one and six. "O progressivnykh formakh oplaty truda v kolkhozakh," *Ekonomika sel'skogo khozyaystva*, no. 5 (July–August 1959): 82.

Mervyn Matthews cites a number of Soviet sources showing that, although clear differentials existed between fieldworkers and mechanized labor, these differentials were approximately 1:2.5. He notes that the range between the highest and lowest paid categories of skilled labor often was less than 20 percent. Mervyn Matthews, *Class and Society in Soviet Russia* (London: Penguin, 1972), pp. 161–62.

4. Kostroma Oblast is located in the Central Region, northeast of Moscow Oblast, in the Russian non–black earth zone.

5. "O kolichestve kolkhozov i ikh preobrazovaniyakh po Kostromskoy oblasti," Kostromskoy gosudarstvennyy arkhiv [hereafter KGA], f. R-2755, o. 6, d. 625, l. 10.

6. State farmworkers were divided into six wage categories, depending on the difficulty and importance of work, with category one the simplest work, not requiring any special qualification, and category six the most difficult and highest paid. There were six categories of wage rates for both manual and mechanized labor. The wage differential between the highest and the lowest wage categories for manual work was 1:1.8, whereas for mechanized labor the wage differential was 1:2.3. *Ekonomika sel'skokhozyaystvennykh predpriyatiy*, pp. 269–70. Seasonal workers were also divided into six wage categories.

7. N. S. Khrushchev, *O merakh dal'neishego razvitiya sel'skogo khozyaystva SSSR. Doklad na Plenume TsK KPSS 3 Sentyabrya 1953g* (Moscow: Gospolitizdat, 1953), pp. 64–65.

8. "Kolkhoz 'Ukraina': vchera, segodnya, zavtra," *Ekonomika sel'skogo khozyaystva*, no. 12 (December 1987): 22.

9. Alec Nove, "Incentives for Peasants and Administrators," in Roy D. Laird, ed., *Soviet Agricultural and Peasant Affairs* (Lawrence: University of Kansas Press, 1963), pp. 55, 60.

10. V. F. Mayer, *Uroven' zhizni naseleniya SSSR* (Moscow: Mysl', 1977), p. 213.

11. Karl-Eugen Wadekin, "Income Distribution in Soviet Agriculture," *Soviet Studies* 27, no. 1 (1975): 17.

12. Toward this end, the May 1966 plenum adopted a guaranteed minimum wage for kolkhoz workers, which removed much of the financial insecurity due to bad harvests. The resolution adopted at the plenum "On Increasing the Material Interest of Collective Farmers in the Development of Social Production" recommended that collective farms guarantee a minimum wage to their workers beginning July 1, 1966, based on wage rates (tariffs) for corresponding categories in state farms, taking into account the classification of the worker, length of service, and work norms. Initially, in 1966 the minimum wage for collective farmers was established at 40 rubles a month, but this was increased to 60 rubles in September 1967 (effec-

tive January 1, 1968), to 70 rubles in 1971, and was increased a number of times thereafter. By 1980 the average wage of a kolkhoznik had reached about 118 rubles a month in direct wages.

13. For example, a resolution adopted in June 1965 allowed tractorists on state farms to purchase up to ten centners of grain per worker in the Far East, Siberia, Urals, Volga, and Kazakhstan, and up to six centners in other regions, at advantageous prices. See K. Garipov, "Oplata truda mekhanizatorov na uborke urozhaya," *Ekonomika sel'skogo khozyaystva*, no. 7 (July 1967): 66–72 (68). There were also special wage bonuses to attract and retain workers in unpopular jobs that were often lowly paid. For example, in 1967 a bonus of 25 percent was introduced for drivers of large automobiles who delivered grain and other agricultural products from the 1967 harvest. Ibid., p. 70.

14. Mervyn Matthews, *Class and Society in Soviet Russia* (London: Penguin, 1972), p. 167.

15. Among collective farm chairmen there was also relative egalitarianism in monthly wages. Basic wages for farm chairmen, as well as specialists, were linked to the gross income of the farm, with ten different farm income categories ranging from 75,000 to 1.5 million rubles. See M. Ivan'kov, "Oplata truda rukoviditeley i spetsialistov kolkhozov i povysheniye effektivnosti proizvodstva," *Voprosy ekonomiki*, no. 2 (February 1969): 71–83.

In Kostroma Oblast in the mid-1970s a collective farm chairman whose farm grossed less than 100,000 rubles annually in sales was paid 140 rubles a month, whereas a farm chairman whose farm earned more than 1.5 million rubles annually was paid 300 rubles a month. In other words, the value of output differed by a factor of 15, while wage differentials were 1:2.1. "Pokazeteli dlya otneseniya kolkhozov Kostromskoy oblasti k gruppam po oplate truda rukovodyashchikh rabotnikov i spetsialistov," KGA, f. R-2755, o. 8, d. 703, l. 85.

16. *Narodnoye khozyaystvo SSSR za 70 let: yubileynyy statisticheskiy ezhegodnik* (Moscow: Finansy i statistika, 1987), pp. 434–35 [hereafter *Narkhoz za 70 let*]; Mayer, *Uroven' zhizni naseleniya SSSR*, p. 212.

17. For an analysis of wage reform, see Janet G. Chapman, "Gorbachev's Wage Reform," *Soviet Economy* 4, no. 4 (October–December 1988): 338–65; L. A. Sholomitskaya, *Zarabotnaya plata: sovremennyye tendentsii razvitiya* (Minsk: Nauka i Tekhnika, 1989); and V. V. Kozlovskiy, *Zarabotnaya plata i rezul'tativnosti truda* (Minsk: Nauka i Tekhnika, 1990).

18. V. F. Mitrofanov, "O vvedenii novukh usloviy oplaty truda v pererabatyvayushchey promyshlennosti," *Ekonomika sel'skokhozyaystvennykh i pererabatyvayushchikh predpriyatiy*, no. 3 (March 1988): 32–33.

19. S. M. Semenov and Yu. P. Shatyrenko, *Reformy zarabotnoy platy: fakty, problemy, kommentarii* (Moscow: Politizdat, 1989), pp. 44–46 (44).

20. Mitrofanov, "O vvedenii novukh usloviy oplaty truda v pererabatyvayushchey promyshlennosti," p. 34.

21. The 1988 draft model charter allowed collective farms to decide "the form and level" of payment. See section 6, point 27, "Primernyy Ustav kolkhoza," *Ekonomicheskaya gazeta*, no. 3 (January 1988): 16.

22. Data from *Narodnoye khozyaystvo Rossiyskoy Federatsii 1992* (Moscow: Goskomstat Rossii, 1992), pp. 405, 408 [hereafter *Narkhoz Rossii 1992*].

23. Stenographic report of the Oblast Council of Collective Farms, December 8, 1971, KGA, f. 2455, o. 8, d. 50, ll. 24–25.

24. Obkom KPSS ispolkom Oblsoveta, "Informatsiya," KGA, f. R-2755, o. 8, d. 707, ll. 3–4.

25. Report of Finansovoye upravleniye, ispolkom oblsoveta, KGA, f. R-1538, o. 11, d. 747, ll. 68–70.

26. One collective farm director in Ostrovskiy raion, Kostroma Oblast, who noted that his farm was "one of the very economically weakest farms in the raion," wrote the chairman of the Oblast Executive Committee of a soviet (nonparty governmental body, *oblispolkom*), A. I. Dontsov, to request both materials to construct sheds for livestock *(shchitovyye doma)* and also farm personnel who could work in animal husbandry and machine operators. The chairmen noted he had only 44 and 67 percent, respectively, of the number of such personnel the farm needed. When the reply was given, the request for building materials was denied (the request

came too late and materials had already been assigned), but the personnel request was granted, indicating the willingness to divert resources into poor areas and farms. Memo to Otdel po trudu, from kolkhoz imeni A. N. Ostrovskogo, KGA, f. R-1538, o. 11, d. 747, l. 7, and reply, l. 8.

27. Memo from KPSS Kostromskoy raikom (party comittee at raion level) to comrade V. I. Toropov, KGA, f. R-1538, o. 11, d. 747, l. 72.

28. Requests were not always granted. In one case in Kostroma Oblast, the chairman of the ispolkom from Buyskiy raion appealed to the oblast agricultural management to increase the price supplement for agricultural products for two collective farms that, besides being heavily in debt, were "economically weak, having large shortcomings in their work force, are located far from the city and procurement organizations, have exceptionally bad roads, and as a consequence they incur large expenses in food production." Communication from A. M. Dubinichev to V. I. Toropov, KGA, f. R-1538, o. 11, d. 752, l. 43. In this particular case the request was denied by the oblast agricultural administration *(upravleniye sel'skogo khozyaystva)*. Communication between A. I. Moshkin and V. I. Toropov, KGA, f. R-1538, o. 11, d. 752, l. 42.

29. Sovet Ministrov RSFSR, "O pogashenii v 1984 godu kolkhozami i sovkhozami zadolzhennosti po suddam zerna," KGA, f. R-1538, o. 11, d. 752, l. 158. Even productive oblasts such as Ul'yanovsk and Rostov Oblasts "poorly repaid their loans."

30. See David S. Mason and Svetlana Sydorenko, "Perestroyka, Social Justice, and Soviet Public Opinion," *Problems of Communism* 34, no. 6 (November–December 1990): 34–43.

31. These data were cited by V. S. Murakhovskiy, head of Gosagroprom, in an interview in *Sovetskaya Rossiya*, August 21, 1987, translated in Foreign Broadcast Information Service, *Daily Report: Soviet Union*, September 4, 1987, p. 36 [hereafter *FBIS*].

32. Ibid. (emphasis in original translation).

33. *Izvestiya*, December 15, 1989, p. 10.

34. *FBIS*, February 8, 1990, p. 110.

35. Ibid., April 2, 1990, p. 43.

36. *Sel'skaya zhizn'*, April 8, 1990, p. 1.

37. N. Borkhunov, "Logika perestroyki zakupochnykh tsen," *APK: ekonomika, upravleniye*, no. 1 (January 1989): 58–59.

38. Data from USSR Ministry of Finance, cited in USDA, Centrally Planned Economics Branch, *CPE Agricultural Report* 3, no. 6 (November–December 1990): 4–5.

39. *Sel'skaya zhizn'*, January 22, 1991, p. 1.

40. Yu. I. Biryukov et al., "Zonal'nym zakupochnym tsenam—nauchnuyu obosnovannost'," in V. A. Kufakov, ed., *Zakpochnyye tseny i ukrepleniye khozraschetnykh otnosheniy v sel'skom khozyaystve* (Moscow: Uchno-issledovatel'skiy institut po tsenoobrazovaniyu, 1982), p. 5.

41. L. A. Ayvazov, "Zakupochnyye tseny—vazhnyy rychag pod'ema sel'skokhzyaystvennogo proizvodstva," in *Sel'skoye khozyaystvo SSSR na sovremennom etape: dostizheniya i perespektivy* (Moscow: Politizdat, 1972), p. 327.

42. To partially correct for this occurrence, as part of the Food Program adopted in May 1982, the number of zones for certain products were reduced by merging contiguous zones. For example, the number of zones for wheat in the RSFSR declined from 73 to 53. During this zone consolidation and the move to more uniform prices, many farms received a twofold increase in prices as they were merged into higher price zones. Karen M. Brooks, "Soviet Agricultural Policy and Pricing Under Gorbachev," in Kenneth R. Gray, ed., *Soviet Agriculture: Comparative Perspectives* (Ames: Iowa State University Press, 1990), p. 121.

43. N. Bokareva, "O differentsiatsii zakupochnykh tsen po zonam," *Voprosy ekonomiki*, no. 12 (December 1969): 30.

44. L. Kassirov, "O metodike differentsiatsii zakupochnykh i sdatochnykh tsen," *Voprosy ekonomiki*, no. 6 (June 1969): 53.

45. Zonal prices were calculated by determining the average cost of production during a three-, five-, or nine-year period, and then "planning" a rate of "profitability" for farms in a given zone based on those prices. Ibid., p. 55.

46. Strong farms were not entirely hurt, however. Stronger farms had higher revenue due

to greater above-plan sales to the state, and this revenue in turn permitted the farm to pay better wages and to acquire needed fertilizers, machinery, and other productive inputs.

47. V. S. Pavlov, *Radikal'naya reforma tsenoobrazovaniya* (Moscow: Finansy i statistika, 1988), p. 73.

48. Ayvazov, "Zakupochnyye tseny," p. 327.

49. V. P. Nikonov, "Polneye zadeystvovat' potentsial agropromyshlennogo kompleksa," *Kommunist*, no. 5 (March 1987): 20.

50. A. Emel'yanov, "Agrarnaya reforma i sotsial'no-ekonomicheskiye uklady sovremennoy derevni," *Voprosy ekonomiki*, no. 5 (May 1990): 48.

51. *Itogi proizvodstvenno-finansovoy deyatel'nosti sovkhozov Kostromskoy oblasti za 1990 god* (Kostroma: Goskomstat, 1991), p. 4; and *Itogi proizvodstvenno-finansovoy deyatel'nosti kolkhozov Kostromskoy oblasti za 1990 god* (Kostroma: Goskomstat, 1991), p. 4.

52. Unpublished information; and interviews with economists from the Karavaevo Agricultural Institute, Kostroma Oblast, March 1990.

53. For example, see L. Vashchukov and V. Nefedov, "Arendnyye otnosheniya i drugiye napravleniya razvitiya APK," *Planovoye khozyaystvo*, no. 1 (January 1989): 101, who claimed that the decrease in unprofitable farms in the RSFSR from 4,600 in 1987 to around 900 in 1988 was due to full khozraschet and self-financing.

54. The Russian term for "private" was *lichnyi* (personal) not *chastnyi* (private). The Russian term for private plot was *lichnoye podsobnoye khozyaystvo*, which translates as "personal auxiliary agriculture," but which was commonly translated simply as "private plot." We will use the term "personal plot" in order to avoid confusion with private farms and private land plots in the post-Soviet period.

55. See Stephen K. Wegren, "Regional Differences in Private Plot Production and Marketing: Central Asia and the Baltics," *Journal of Soviet Nationalities* 2, no. 1 (spring 1991): 118–38.

56. A. F. Kalinkin, *Lichnoye podsobnoye khozyaystvo: kollektivnoye sadovodstvo i ogorodnichestvo* (Moscow: Kolos, 1981), p. 11. By 1988, it was reported that nearly all collective farm families (98 percent) had a personal plot, while 79 percent of state farm families had a plot. Among urban worker and employee families, only about 25 percent had a personal plot. *Lichnoye podsobnoye khozyaystvo naseleniya v 1988 godu* (Moscow: Goskomstat, 1989), p. 3.

57. See, for example, V. Boyev and V. Shlyakhtin, "Podsobnyye khozyaystva predpriyatiy—vazhnoye zveno prodovol'stvennogo kompleksa," *APK: ekonomika, upravleniye*, no. 7 (July 1988): 32–37.

58. Interviews with several flax factory managers in Kostroma Oblast in 1990 revealed that often the output from these endeavors was significant. In 1982 there were 190 subsidiary farms, encompassing over 14.7 thousand hectares in Kostroma Oblast, including 37 subsidiary farms in the city of Kostroma with 2,856 hectares. "Spravka: o razvitii podsobnykh khozyaystv v Kostromskoy oblasti," KGA, f. R-1538, o. 10 (dop.), d. 222, l. 29.

59. Interview with General V. Arkhipov, "Agrarnomu sektoru—dinamizm perestroyki," *Kommunist vooruzhennykh sil*, no. 8 (April 1989): 9–16; and see Laure Despres and Ksenya Khinchuk, "The Hidden Sector in Soviet Agriculture: A Study of the Military Sovkhozy and Auxiliary Farms," *Soviet Studies* 42, no. 2 (April 1990): 269–94.

60. In the 1980s the most common size for a collective orchard or garden was between 301 and 600 square meters. "O razvitii kollektivnogo sadovodstva i ogorodinichestva," *Vestnik statistiki*, no. 12 (December 1987): 54.

61. For a brief review of limitations, see Stefan Hedlund, *Private Agriculture in the Soviet Union* (London: Routledge, 1989), pp. 50–53.

62. G. Shmelev, *Personal Subsidiary Farming Under Socialism* (Moscow: Progress Publishers, 1986), p. 51.

63. For example, by the late 1980s more than 36 million Soviet families had private plots, while almost 7 million were involved in collective orchards, and 6 million in collective vegetable gardens. "O razvitii kollektivnogo sadovodstva i ogorodinichestva," *Vestnik statistiki*, no. 12 (December 1987): 54.

64. Naum Jasny, *The Socialized Agriculture of the USSR: Plans and Performance* (Stanford: Stanford University Press, 1949), p. 384.

65. *Sel'skoye khozyaystvo SSSR: statisticheskiy sbornik* (Moscow: Statistika, 1960), p. 264.

66. Despite the 1939 crackdown, personal plots accounted for a very high percentage of national food production of certain products. For example, in 1940 private plots produced 71 percent of the nation's meat, 77 percent of its milk, and 94 percent of its eggs. Kalinkin, *Lichnoye podsobnoye khozyaystvo: kollektivnoye sadovodstvo i ogorodnichestvo*, p. 13.

67. See William Moskoff, *The Bread of Affliction: The Food Supply in the USSR During World War II* (Cambridge: Cambridge University Press, 1990), p. 94.

68. The percentage of investments in agriculture was 18.9 percent of all investment during 1929–1932, and only 15.2 percent during 1946–1950. Lazar Volin, *A Century of Russian Agriculture: From Alexander II to Khrushchev* (Cambridge, Mass.: Harvard University Press, 1970), pp. 307–08, 348.

69. O. M. Verbitskaya, *Rossiyskoye krest'yanstvo: ot Stalina k Khrushchevu* (Moscow: Nauka, 1992), p. 142.

70. The crackdown against liberal policies extended to other aspects of production as well. Until 1948 consumer cooperatives, which operated primarily in rural areas, had been allowed to buy food from collective farms and resell it at or near market prices in stalls set up in towns, but in August 1949 urban food sales by rural cooperatives were discontinued, and commercial trade outlets were forbidden to pay higher than state procurement prices for above-quota output.

In February 1950 *Pravda* criticized small-scale work units, and any deviation from the collectivist organization of labor was rejected. By 1950 the "link" (*zveno*) system of production and its most pronounced advocate, A. Andreev, had been publicly attacked in the pages of *Pravda*, spelling the end of the experiment until it resurfaced in the early 1960s. See Dimitry Pospielovsky, "The 'Link System' in Soviet Agriculture," *Soviet Studies* 21, no. 2 (1970): 411–35. In addition, Werner G. Hahn describes the fall of N. A. Voznesenskiy and the ideological battles over economic policy in the late Stalin period. See his *Postwar Soviet Politics: The Fall of Zhdanov and the Defeat of Moderation, 1946–1953* (Ithaca, N.Y.: Cornell University Press, 1982), esp. pp. 87–88, 129–34. See also Sidney Ploss, who discusses the development of the link system during the early postwar years in *Conflict and Decision-Making in Soviet Russia: A Case Study of Agricultural Policy, 1953–1963* (Princeton: Princeton University Press, 1965), chap. 1.

For the story behind the link experiment and the man who led it during the Khrushchev period, see Alexander Yanov, *The Drama of the Soviet 1960s: A Lost Reform* (Berkeley and Los Angeles: University of California, Institute of International Studies, 1984).

71. See Ploss, *Conflict and Decision-Making in Soviet Russia*, chap. 1.

72. By 1953 personal plots still accounted for 74 percent of potato production, 49 percent of vegetable production, 52 percent of meat production, 67 percent of milk production, and 84 percent of egg production. Verbitskaya, *Rossiyskoye krest'yanstvo*, p. 75.

73. See J. V. Stalin, *Economic Problems of Socialism in the USSR* (New York: International Publishers, 1952), pp. 63–71.

74. For an overview of agricultural policy under Khrushchev, see Erich Strauss, *Soviet Agriculture in Perspective: A Study of Its Successes and Failures* (New York: Praeger, 1969), chap. 7; David M. Schoonover, "Soviet Agricultural Policies," in Joint Economic Committee, *The Soviet Economy in a Time of Change*, vol. 2, Joint Committee Print (Washington, D.C.: GPO, 1979), pp. 87–115. See also the recollections of the Khrushchev years by the former Politburo member G. I. Voronov who was chairman of the RSFSR Council of Ministers. See "Nemnogo vospominaniy," *Druzhba narodov*, no. 1 (January 1989): 192–201.

75. Volin, *A Century of Russian Agriculture*, p. 323.

76. Khrushchev, *O merakh dal'neishego razvitiya sel'skogo khozyaystva SSSR*, p. 10.

77. The urban population by 1959 constituted 48 percent of the total population, up from 32 percent in 1939. *Itogi vsesoyuznoy perepisi naseleniya 1959 goda* (Moscow: Gosstatizdat, 1962), pp. 20–21.

78. Calculated from *Naseleniye SSSR 1987: statisticheskiy sbornik* (Moscow: Goskomstat SSSR, 1988), p. 8.

79. Khrushchev, *O merakh dal'neishego razvitiya sel'skogo khozyaystva SSSR*, pp. 7–8; and "Ob otmene obyazatel'nykh postavok i naturoplata za raboty MTS, o novom poryadke, tsenakh i usloviyakh zagorovok sel'skokhozyaystvennykh produktov. Doklad na plenume TSK KPSS 17 Iyunya 1958 goda," in N. S. Khrushchev, *Stroitel'stvo kommunizma v SSSR i razvitiye sel'skogo khozyaystva*, vol. 3 (Moscow: Politizdat, 1962), p. 233.

80. Khrushchev, *O merakh dal'neishego razvitiya sel'skogo khozyaystva SSSR*, p. 13.

81. The 1962 resolution also introduced increases in retail prices for livestock products, "for the purpose of reducing losses that the state incurs in the sale of meat to the population." Effective June 1, 1962, retail prices increased an average of 31 percent for beef, 34 percent for lamb, 19 percent for pork, 31 percent for sausage, and 25 percent for butter. There was also a price range based on the fat content of the animal, up to 20 percent above the purchase price for above-fatty cattle, and up to 25 percent below the purchase price for below-fatty cattle. "O povyshenii zakupochnykh (sdatochnykh) tsen na krupnyy rogatyy skot, sviney, obets, ptitsu, maslo zhivotnoye i slivki i roznichnykh tsen na myaso, myasnyye produkty i maslo zhivotnoye. Postanovelniye Soveta Ministrov SSSR ot 17 Maya 1962 g.," in *Spravochnik bukhgaltera kolkhoza*, 2d ed. (Moscow: Kolos, 1964), pp. 620–23.

82. See Martin McCauley, *Khrushchev and the Development of Soviet Agriculture* (London: Macmillan, 1976), chap. 4.

83. Frank A. Durgin, "Russia's Private Farm Movement: Background and Perspectives," *Soviet and Post-Soviet Review* 21, nos. 2–3 (1994): 217.

84. See Karl-Eugen Wadekin, *The Private Sector in Soviet Agriculture* (Berkeley and Los Angeles: University of California Press, 1973), chap. 8, for an analysis of Khrushchev's plot policy from 1953–1958 (quotation is from p. 250).

85. This category includes those held by kolkhozniki, workers and employees, and individual peasants.

86. These data were derived from *Narodnoye khozyaystvo SSSR v 1958: statisticheskiy ezhegodnik* (Moscow: Gosstatizdat, 1959), pp. 447–48 [hereafter *Narkhoz* with relevant date]; and *Sel'skoye khozyaystvo SSSR: statisticheskiy sbornik* (Moscow: Gosstatizdat, 1960), pp. 266–69.

87. V. A. Belyanov, *Lichnoye podsobnoye khozyaystvo pri sotsializme* (Moscow: Ekonomika, 1970), p. 159.

88. Khrushchev, *Stroitel'stvo kommunizma v SSSR i razvitiye sel'skogo khozyaystva*, vol. 3, pp. 391–92.

89. I. A. Benediktov, "O Staline i Khrushcheve," *Molodaya Gvardiya*, no. 4 (April 1989): 56.

90. Which is what happened. See *Narkhoz 1958*, pp. 447–48; and *Sel'skoye khozyaystvo SSSR* (1960), pp. 266–69.

91. V. I. Novikov, "V gody rukovodstva N. S. Khrushcheva," *Voprosy istorii*, no. 2 (February 1989): 111.

92. "Chislennost' naseleniya Rossii," *Voprosy statistiki*, no. 1 (January 1994): 43.

93. During these years, the republic with the largest loss of rural dwellers was the RSFSR, which experienced a net reduction of 12 percent of its rural population. Within the RSFSR, the Central, Volga-Vyatska, Northwestern, Western Siberian, and Central Black Earth economic regions experienced the sharpest declines in rural population. V. Perevedentsev, "Migratsiya naseleniya i ispol'zovaniye trudovykh resursov," *Voprosy ekonomiki*, no. 9 (September 1970): 34–35.

94. *Itogi Vsesoyuznoy perepisi naseleniya 1979 goda*, vol. 1 (Moscow: Goskomstat, 1989), p. 27.

95. There were several trends to note. First, production from personal plots (the main food goods from plots comprised potatoes, vegetables, meat, milk, eggs, and wool) continued to decline as a percentage of total output in the nation. Second, between 1958 and 1965 the number of head of livestock on personal plots declined. Third, production of all but potatoes and milk increased during this period. Fourth, while production increased, the volume of production sold through collective farm markets decreased. Fifth, collective farm trade declined as a

percentage of total trade turnover. See Belyanov, *Lichnoye podsobnoye khozyaystvo pri sotsializme,* pp. 173–82.

96. Ibid., pp. 160–61.

97. *Resheniya partii i pravitel'stva po khozyaystvennym voprosam,* vol. 5, *1962–1965* (Moscow: Politizdat, 1968), p. 517 [hereafter *Resheniya*].

98. Ibid.

99. L. I. Brezhnev, *Leninskim kursom,* vol. 1 (Moscow: Politizdat, 1970), p. 21 [hereafter *L.K.*].

100. By contrast, the 1935 model charter "recommended" a plot size of between one-quarter and one-half a hectare. For a discussion of collective farm legislation, see A. K. R. Kiralfy, "The History of Soviet Collective Farm Legislation," in William E. Butler, ed., *Russian Law: Historical and Political Perspectives* (Leyden: A. W. Sijthoff, 1977), pp. 193–214.

101. *Pravda,* December 14, 1969 in *Current Digest of the Soviet Press (CDSP)* 21, no. 1 (1969): 14–15.

102. Belyanov, *Lichnoye podsobnoye khozyaystvo pri sotsializme,* p. 53.

103. *Narkhoz 1965,* p. 277; and *Narkhoz 1970,* p. 290.

104. *Narkhoz 1970,* p. 357.

105. I. Buzbalov, "Perevod sel'skogo khozyaystva na industrial'nuyu bazu," *Voprosy ekonomiki,* no. 12 (December 1976): 85.

106. *L.K.* 6, p. 140. The increasingly expensive commitment to agriculture was reflected in the fact that the percentage of total state capital investments to agriculture grew from 20 percent during 1961–1965 to 24 percent during 1966–1970, and to 27 percent during 1971–1975. *Narkhoz za 70 let,* p. 275.

107. A complement to Brezhnev's unfolding plot policy was the adoption of a joint resolution by the CPSU Central Committee and USSR Council of Ministers in July 1977 entitled "On Measures for the Further Development of Trade," which contained provisions to strengthen kolkhoz trade. "O merakh po dal'neyshemu razvitiyu torgovli," *Resheniya,* vol. 12 (July 1977–March 1979), pp. 44–50.

The July 1977 decree on further developing trade contained provisions for developing a network of "bureaus of trade services," which were specially designed to help peasants sell their plot output. By 1978 there were more than seven hundred such bureaus—one for about every ten kolkhoz markets. The establishment of these bureaus were expected to cut down on the estimated annual loss of 200 million man-days, which rural dwellers spent selling their produce on collective markets. V. Voronin, "Lichnyye podsobnyye khozyaystva i torgovlya," *Voprosy ekonomiki,* no. 6 (June 1980): 119, 123.

108. *Resheniya,* vol. 12 (July 1977–March 1979), pp. 104–11.

109. The quotes are taken from the resolution as published in *Sel'skaya zhizn',* January 18, 1981. The resolution is also published in *Resheniya,* vol. 13 (April 1979–March 1981), pp. 533–44, which is an expanded and more detailed version than that in the newspaper.

110. A comprehensive overview of legislation on the personal plot, including an analysis of the January 1981 decree, may be found in M. I. Kozyr', "Sovershenstvovaniye pravovogo regulirovaniya lichnogo podsobnogo khozyaystva grazhdan SSSR," *Sovetskoye gosudarstvo i pravo,* no. 8 (August 1981): 37–45.

111. Hungary had been doing this since 1974. In addition, another experiment was tried in Latvia during 1980 in Bauska and Liepaja districts. The experiment provided state and collective farms to be given a single plan for meat procurements, with no distinction between meat procured from communal and from personal sectors. The change in planning was to give districts a stronger incentive to help people raise livestock on personal farming operations.

112. *Resheniya,* vol. 13 (April 1979–March 1981), p. 300.

113. Although personal plots were given legal standing in the Land Principles and 1977 Constitution, the actual granting of land came from the state or collective farm. There were no inherent "rights" to land, land use was regulated, plot size was limited and quite small, and of course land could not be bought or sold because it was state land (although we should note

that the house, farm animals, and farm tools were considered private property). The January 1981 law did allow farmers to lease additional land from the farm in order to grow fodder for their cattle. State land was also to be made available for leasing under the February 1981 law, but was of poor, usually unusable quality. Interview with Fedor Burlatskiy, September 22, 1988, Duke University.

114. At the same time that personal plots were being promoted, an effort was made to increase production from collective gardens and collective orchards. In addition, at least since early 1979, a concerted effort was made to increase production from subsidiary operations of enterprises. More than a dozen archival documents from Kostroma Oblast showed that campaigns, similar to that for personal plots, were waged for both of these sets of operations. Plans were elaborated, responsibility assigned, and an oblast commission formed, production and sales planned, and the amount of land per raion to be assigned for such uses was determined. Documents indicate that the campaign extended at least until 1984 for collective gardens of individuals. Between 1978 and 1984, 675 hectares of land were allocated for collective fruit gardens, which were operated by more than 22,000 families, and almost 30,500 families had a land plot within a collective vegetable garden. Memo of RSFSR Ministry of Agriculture, Upravleniye Sel'skogo Khozyaystva, "Informatsiya," KGA, f. R-1538, o. 10 (dop.) d. 238, ll. 19–20.

115. Ispolitel'nyy komitet Kostromskoy oblastnogo Soveta narodnykh deputatov, Otdel po delam stroitel'stva i arkhitektury, "O narushenii razmerov sadovodcheskikh domikov," KGA, f. R-1538, o. 10 (dop.), d. 238, ll. 50–51. Because available archive records stopped after 1984, it is unknown if this crackdown was part of a wider effort, if it signified a reversal of previous policy, or if it was focused solely on these abuses.

116. See Zhores A. Medvedev, *Soviet Agriculture* (New York: Norton, 1987), pp. 380–81.

117. Kalinkin, *Lichnoye podsobnoye khozyaystvo*, p. 13.

118. Ligachev supported the expansion of private plots because they did not require the dissolution of state and collective farms and, in fact, would remain wedded to them for supplies and material assistance. In 1990 he indicated that personal plots could be increased by three hectares if the family could "assimilate" that much land, and he indicated that in Central Asia "hundreds of thousands" of families had received supplementary plots of irrigated land. *FBIS,* April 3, 1990, p. 61.

119. "Sovremennaya sel'skokhozyaystvennaya rabota i eyo problemy," *Kommunist,* no. 14 (September 1985): 74–75.

120. Moscow Television, January 10, 1986.

121. *Pravda,* July 24, 1987, p. 1.

122. The draft Model Collective Farm Charter was published in *Ekonomisheskaya gazeta,* no. 3 (January 1988). Section 9 on p. 17 covered personal plots. The final version of the farm charter was adopted in August 1988. See *Sel'skaya zhizn',* August 4, 1988, p. 1.

123. See Stephen K. Wegren, "Private Agriculture in the Soviet Union Under Gorbachev," *Soviet Union* 16, nos. 2–3 (1989): 111–15; and see *Lichnoye podsobnoye khozyaystvo naseleniya v 1988 godu,* p. 108.

124. *Narodnoye khozyaystvo Kostromskoy oblasti v 1991 godu: statisticheskiy sbornik* (Kostroma: Goskomstat RSFSR, 1992), p. 32 [hereafter *Narkhoz Kostromskoy 1991*].

125. *Proizvodstvo i zakupki sel'skokhozyaystvennoy produktsii v lichnykh podsobnykh khozyaystvakh naseleniya* (Kostroma: Goskomstat RSFSR, 1991), pp. 3–4; and *Lichnyye podsobnoyye khozyaystva naseleniya Kostromskoy oblasti* (Kostroma: Goskomstat Rossiyskoy Federatsii, 1992), p. 6.

126. *Narkhoz Rossii 1992,* p. 196.

127. Ibid., p. 472.

128. *Nechernozemnaya zona Rossiyskoy Federatsii v tsifrakh 1992* (Moscow: Goskomstat Rossii, 1992), p. 115.

129. *Narkhoz 1990,* p. 502.

130. See *Ekonomika i zhizn',* no. 6 (February 1991): 10, and no. 40 (October 1991): 15.

131. On the Baltics, see, for example, William H. Meyers et al., *Agricultural Transformation*

and Privatization in the Baltics, Report 92-BR-7, Center for Agricultural and Rural Development, Iowa State University, December 1992.

132. For example, the Law on Property in the USSR, adopted in March 1990 by the USSR Supreme Soviet stated that citizens' rights to property (article 7) included residential buildings, dachas, garden-plot buildings, the crops from personal plots, the instruments used to operate private plots, means of transportation, money and stocks, personal items, and so on. No mention was made of the right to own land. *Izvestiya*, March 10, 1990, p. 2.

133. Citizens who were not members of collective or state farms could also obtain land for peasant farming, but only through a land reserve fund.

134. There were less than 21,000 peasant farms throughout the USSR as of April 1990. Of that number, over 12,000 were located in Georgia and another 7,620 in the Baltic states. *Ekonomika i zhizn'*, no. 30 (July 1990): 5.

135. Section 2, article 8, point 3, "O krest'yanskom (fermerskom) khozyaystve," in *Sbornik zakonodatel'nykh aktov, postanovleniy, ukazaniy i rekomendatsiy po zemel'noy reforme i krest'yanskim (fermerskim) khozyaystvam*, chast' 1 (Moscow: Gosudarstvennyy komitet RSFSR po zemel'noy reforme, 1991), p. 28.

136. *Izvestiya*, January 7, 1991, p. 1.

137. Robert Conquest, *The Harvest of Sorrow: Soviet Collectivization and the Terror-Famine* (New York: Oxford University Press, 1986).

138. See Sheila Fitzpatrick, *Stalin's Peasants: Resistance and Survival in the Russian Village After Collectivization* (New York: Oxford University Press, 1994).

139. James Millar and Alec Nove, "A Debate on Collectivization: Was Stalin Really Necessary?" *Problems of Communism* 25, no. 4 (July–August 1976): 49–62.

140. Jerzy F. Karcz, "From Stalin to Brezhnev: Soviet Agricultural Policy in Historical Perspective," in James R. Millar, ed., *The Soviet Rural Community* (Urbana: University of Illinois Press, 1971), pp. 36–72.

141. Holland Hunter, "Soviet Agriculture With and Without Collectivization, 1928–1940," *Slavic Review* 47, no. 2 (summer 1988): 215.

142. Stephen Merl, "Did the Kolkhoz System Really Fulfill the Initial Aims of the Party in the 1930s?" in Josef C. Brada and Karl-Eugen Wadekin, eds., *Socialist Agriculture in Transition: Organizational Response to Failing Performance* (Boulder, Colo.: Westview Press, 1988), p. 90.

143. The percentage of investments to agriculture was 18.9 percent of all investment during 1929–1932, but only 15.2 percent during 1946–1950. Volin, *A Century of Russian Agriculture*, pp. 307–08, 348.

144. In 1950, potatoes were 75 percent of their 1937 price and meat was 73 percent. See Volin, *A Century of Russian Agriculture*, p. 309. On the other hand, purchase prices for "industrial" crops such as sugar beets, cotton, and flax were increased.

145. See Alec Nove, *An Economic History of the USSR* (Middlesex, England: Penguin Books, 1982), p. 300.

146. Strobe Talbott, trans. and ed., *Khrushchev Remembers: The Last Testament* (Boston, Mass.: Little, Brown, 1974), p. 112.

147. David W. Bronson and Constance B. Kreuger, "The Revolution in Soviet Farm Household Income, 1953–1967," in Millar, ed. *The Soviet Rural Community*, p. 229. These percentages do not include income from personal plots.

148. *Narkhoz za 70 let*, pp. 275–76; *SSSR v tsifrakh v 1982 godu: kratkiy statisticheskiy sbornik* (Moscow: Finansy i statistika, 1983), p. 154.

149. Mayer, *Uroven' zhizni naseleniya SSSR*, p. 212. These data do not include income from personal plots.

150. M. Sidorova, "Sblizheniye urovnya i usloviy zhizni gorodskogo i sel'skogo naseleniya," *Ekonomika sel'skogo khozyaystva*, no. 11 (November 1987): 48.

151. Doklad General'nogo sekretarya TsK KPSS tovarishcha Gorbacheva M. S., "O khode realizatsii resheniy XXVII s"ezda KPSS i zadachakh po urublenniyu perestroki," *XIX vsesoyuznaya konferentsiya kommunisticheskoy partii Sovetskogo Soyuza: stenograficheskiy otchet* (Moscow: Politizdat, 1988), pp. 25–26.

152. *Narkhoz 1990*, pp. 38–39. These data do not include income from personal plots.
153. *Narkhoz Rossii 1992*, pp. 372, 405, 408. These data do not include income from personal plots.
154. *Narodnoye khozyaystvo Kostromskoy oblasti v 1990 godu: statisticheskiy sbornik* (Kostroma: Goskomstat RSFSR, 1991), pp. 234, 258, 262; *Narkhoz Kostromskoy 1991*, pp. 213, 236, 240.
155. *Narkhoz Rossii 1992*, pp. 141, 405. A large part of the increase in urban incomes was due to compensation payments in light of the April 2, 1991, price increase for food products sold in state stores.
156. For early trends, see Stephen K. Wegren, "Rural Reform in Russia," *RFE/RL Research Report* 2, no. 43 (October 29, 1993): 52.
157. See Alfred Evans Jr., "Equalization of Urban and Rural Living Levels in Soviet Society," *Soviet Union* 8, no. 1 (1981): 38–61; and Gertrude E. Schroeder, "Rural Living Standards in the Soviet Union," in Robert C. Stuart, ed., *The Soviet Rural Economy* (Totowa, N.J.: Rowman and Allanheld, 1983), pp. 241–57.
158. "O torgovom i bytovom obsluzhivanii naseleniya v sel'skoymestnosti," *APK: ekonomika, upravleniye*, no. 3 (March 1990): 20. During the tenth FYP the average rate of growth in trade turnover per capita was 4.2 percent in rural areas and 2.9 percent in urban centers; during the eleventh FYP the corresponding rates were 2.2 and 1.3 percent respectively; and in the twelfth FYP the rate of growth per capita was only 0.7 percent in rural areas and 2.8 percent in urban centers. As a consequence, total trade turnover for food products in Russia was not only much lower in rural villages than in cities but also declining over time. *Rossiyskaya Federatsiya v tsifrakh v 1993 godu* (Moscow: Goskomstat Rossii, 1994), p. 256.
159. *Torgovlya SSSR: statisticheskiy sbornik* (Moscow: Finansy i statistika, 1989), p. 109.
160. See E. Pavlovskaya and L. Pisareva, "Material'nyye i kul'turno-bytovyye usloviya zhizni kolkhoznikov," *APK: ekonomika, upravleniye*, no. 3 (March 1989): 78; *Narkhoz za 70 let*, pp. 444–45; and *Narkhoz 1990*, pp. 113–14.
161. *Agropromyshlennyy kompleks SSSR: statisticheskiy sbornik* (Moscow: Goskomstat SSSR, 1990), p. 153; "Blagoustroystvo zhilishchnogo fonda kolkhozov na nachalo 1988 g.," *APK: ekonomika, upravleniye*, no. 5 (May 1989): 114; and "Blagoustroystvo zhilishchnogo fonda sovkhozov na nachalo 1988 g.," ibid, no. 6 (June 1989): 20. These were national averages and there was considerable differentiation among the republics with Central Asia republics even less developed.
162. "Gasifikakatsiya kvartir v 1989 godu," *APK: ekonomika, upravleniye*, no. 8 (August 1990): 47.
163. A. Buzlayeva, "Usloviya zhizni na sele i sotsial'naya spravedlivost'," *APK: ekonomika, upravleniye*, no. 9 (September 1990): 69.
164. L. Vashchukov and V. Nefedov, "Uskorit' razvitiye agropromyshlennogo kompleksa," *Vestnik statistiki*, no. 1 (January 1989): 15.
165. In the Ukraine, for example, 8,400 rural population points (29 percent) did not have any public health enterprises, including 300 villages with 500 or more people. The social problems of the countryside had to be solved by the farms themselves. In the first half of the 1980s kolkhozy in the Ukraine built nearly all the rural hospitals, 80 percent of the clubs, and 60 percent of the schools. O. Vasil'yev and V. Kozlov, "Optimizatsiya sotsial'nogo ravnovesiya mezhdu gorodom i derevney," *APK: ekonomika, upravleniye*, no. 9 (September 1990): 85.
166. Buzlayeva, "Usloviya zhizni na sele i sotsial'naya spravedlivost'," p. 69.
167. *Sotsial'noye razvitiye SSSR: statisticheskiy sbornik* (Moscow: Goskomstat SSSR, 1990), p. 214.
168. Buzlayeva, "Usloviya zhizni na sele i sotsial'naya spravedlivost'," p. 69.
169. "Sel'skoye khozyaystvo," *Politicheskoye obrazovaniye*, no. 6 (April 1989): 14.
170. *Narkhoz Rossii 1992*, p. 508.
171. G. V. Ioffe, *Sel'skoye khozyaystvo nechernozem'ya: territorial'nyye problemy* (Moscow: Nauka, 1990), p. 87.
172. T. Levina, "Demograficheskaya situatsiya v sel'skoy mestnosti," *Vestnik statistiki*, no. 1 (January 1992): 11.

173. A net decrease of 101,000 persons occurred per year after factoring in births. Ibid., p. 10.

174. See Matthews, *Class and Society in Soviet Russia*, pp. 202–03, 110.

175. "Svedeniya o rodivshikhsya po ocherdnosti ikh rozhdeniya i vozrastu materi za 1980 god (sel'skiye mestnosti)," KGA, f. R-1951, o. 15, d. 141. However, we know that the average Russian rural family included just over three persons, so on average relatively few women had a second child. *Kolkhozy SSSR: kratkiy statisticheskiy sbornik* (Moscow: Finansy i statistika, 1988), p. 181. This is important because even though there was a net gain (+158) of rural women aged from twenty to twenty-two in 1982, statistically speaking those women did not offset the net loss (–415) in a younger-aged cohort (aged from seventeen to nineteen). The number of children born to rural women was below family replacement levels. (Net gain was calculated from archival data indicating age and sex of persons leaving rural areas and of those arriving in rural areas.)

Chapter 3. Reform of the Collective Agricultural Sector

1. Jaclyn Y. Shend, *Agricultural Statistics of the Former USSR Republics and the Baltic States*, statistical bulletin no. 863 (Washington, D.C.: USDA, Economic Research Service, 1993), pp. 92–95, 184.

2. See Frank A. Durgin, "Appendix: Trends in Production and Consumption and the Problems of Increasing Shortages and Imports, 1965–1990," *Soviet and Post-Soviet Review* 21, nos. 2–3 (1994): 247.

3. Shend, *Agricultural Statistics of the Former USSR Republics and the Baltic States*, p. 231.

4. See data in Durgin, "Appendix," pp. 247–52.

5. This is not to deny the enormous and debilitating deficiencies in storage, transportation, and distribution of food that resulted in huge losses and long lines for consumers.

6. Karl-Eugen Wadekin, "Potentials and Deadlocks in the Soviet Food Economy," in Michael P. Claudon and Tamar L. Gutner, eds., *Putting Food on What Was the Soviet Table* (New York: New York University Press, 1992), p. 28.

7. Production brigades did not work on a crop through its full production cycle, that is, from planting to harvest. Since a brigade's work was not directly tied to the final results, when plans were underfulfilled it was difficult to identify the culprit. Brigade workers were not directly responsible for final output, which led to a constant search for higher productivity and a way to tie remuneration to output.

8. When these words were spoken in 1992, Vershinin was the chairman of the Moscow Peasant Union Concil, and co-chairman of the Peasant Democratic Party of Russia.

9. D. Gale Johnson, "Agricultural Organization and Management," in *The Soviet Economy: Toward the Year 2000* (London: George Allen and Unwin, 1983), pp. 135, 137, emphasis added.

10. One of the liberal reforms in labor organization was the "link" *(zveno)* system, which occurred under Khrushchev and was designed to address the problem of responsibility for final results. This was an attempt at reform from below. Links were distinct from the usual brigade in two key ways. First, the link was smaller: it usually consisted of less than twenty-five members and could number as few as eight or ten. Second, rather than work on only a part of the overall production cycle, a link was responsible for the entire production cycle. Thus, the link was directly responsible for its harvest and was paid according to its output.

The link form of labor organization and remuneration represented an attempt to cut the cost of production and to increase both output production and incentives to perform quality work. The best known of the link experiments occurred in Kazakhstan (the Akchi model) and was led by I. Khudenko. The results of this experiment led to a threefold increase in output while cutting the average number of workers by 85 percent. The link system also allowed the state to lower procurement prices but did not, however, allow the free marketing of produce.

11. This quote, from G. Shmelev, was cited in *Radio Liberty Research Report*, RL 61/88 (February 15, 1988): 3.

12. Vladimir Tikhonov was a primary "liberal" in the agrarian reform movement. For an example of his views, see V. Tikhonov, "Kontseptsiya radikal'noy perestroiki khozyaystvennogo mekhanizma APK," *Planovoye khozyaystvo,* no. 4 (April 1987): 12–22, in which he criticized the administrative and pricing system, while advocating greater farm autonomy.

13. Lecture by V. Tikhonov, Kennan Institute, Woodrow Wilson Center, Washington, D.C., February 15, 1989.

14. S. S. Shatalin, *Perekhod k rynku: kontseptsiya i programma* (Moscow: Arkhangel'skoye, 1990), p. 174.

15. The plan revealed that as of January 1, 1990, one in six state and collective farms was considered economically weak (meaning that farm profitability was less than 10 percent). In 1990 there were 23,500 state farms and 29,100 collective farms in the USSR (not including Lithuania and Estonia), meaning that about 8,400 farms were economically weak. These farms lacked the means to repay 38.4 billion rubles in credits; it was estimated that 70 percent of farm financial resources came from credits.

16. Shatalin, *Perekhod k rynku,* pp. 175–76.

17. M. S. Gorbachev, *Potentsial kooperatsii-delu perestroyki* (Moscow: Politizdat, 1988), p. 31.

18. Speech at Twenty-Eighth Party Congress, July 1990, in *Pravda,* July 3, 1990, pp. 2–4.

19. Gorbachev, *Potentsial Kooperatsii-delu perestroyki,* pp. 31–32.

20. "Primernyy Ustav kolkhoza," *Ekonomicheskaya gazeta,* no. 3 (January 1988): 15–18.

21. Ibid., section 2, point 7. This was a stipulation that was not included in the original 1935 kolkhoz statute and one that was treated ambiguously in the 1969 statute. Although farm members had certainly left farms in the past, the 1988 statute was the first document to make explicit the right to resign.

22. In February 1992 a Russian Congress of Collective Farms of Russia was held, and it adopted a new set of statutes. According to this document a person was to receive his work record and any wages due him on the day he ceased to be a farm member. See section 4, point 12, "Primernyy Ustav kolkhoza (sel'skokhozyaystvennogo pooizvodstvennogo kooperativa)," *APK: ekonomika, upravleniye,* no. 5 (May 1992): 15.

23. See *Sel'skaya zhizn',* August 4, 1988, p. 1, and August 21, 1988, p. 2.

24. *Pravda,* June 26, 1987, pp. 1–4.

25. Ibid., May 15, 1988, p. 2.

26. *Sel'skaya zhizn',* August 27, 1988, p. 1.

27. After the March 1989 plenum the USSR Council of Ministers passed a resolution on leasing. "Ob ekonomicheskikh i organizatsionnykh osnovakh arendnykh otnosheniy v SSSR," *Ekonomicheskaya gazeta,* no. 19 (May 1989): 7–8. This resolution was followed by the Supreme Soviet Law "On Leasing," which specified terms and conditions of land leasing. See the law passed by the USSR Supreme Soviet, in *Pravda,* December 1, 1989, pp. 3–4. The law took effect in January 1990. Land leasing was also codified in the Law on Land, adopted at the end of February 1990 and published in March 1990, which provided for the leasing of land for the creation of independent peasant farms. *Izvestiya,* March 7, 1990, pp. 1–2. None of the laws cited above, however, allowed land to be sold, purchased, or mortgaged.

28. Rental payments were linked to production costs, land location, and land quality and were differentiated by region. Rent was charged by the hectare leased or per head of livestock. Rent could be paid in-kind or in cash. See A. Unukovich and A. Golos, "Rentyye platezhi i arendnaya plata za zemlyu," *APK: ekonomika, upravleniye,* no. 6 (June 1990): 56–60; and A. Molodtsov and A. Nikoforov, "Arenda i oplata truda," ibid., no. 7 (July 1990): 82–87.

29. A. Nikoforov and A. Molodtsov, "Arenda na sele: oplata i stimulirovaniye proizvodstva konechnogo produkta," *Planovoye khozyaystvo,* no. 10 (October 1989): 76.

30. *Ekonomika i zhizn',* no. 17 (April 1991): 8.

31. Don Van Atta, "Full-Scale, like Collectivization, but Without Collectivization's Excesses: The Campaign to Introduce the Family and Lease Contract in Soviet Agriculture," *Comparative Economic Studies* 32, no. 2 (summer 1990): 109–43.

32. See Karen Brooks, "Lease Contracting in Soviet Agriculture in 1989," *Comparative Economic Studies* 32, no. 2 (summer 1990): 85–108.

33. "Arenda i fermerstvo: po dannym Goskomstata SSSR," *Voprosy ekonomiki*, no. 5 (May 1990): 81.

34. In May 1991, for example, a set of recommendations advocated the privatization of state and collective farms, and there was at least one report that a state farm had been privatized in Siberia. *Izvestiya*, May 23, 1991, p. 1.

35. "O poryadke reorganizatsii kolkhozov i sovkhozov," *Zemlya i lyudi*, no. 2, January 10, 1992, pp. 1, 3.

36. Point 3, "O poryadke reorganizatsii kolkhozov i sovkhozov." Ibid., p. 1.

37. Whatever remained after the auction was to be sold at an open auction for "all those who wish to undertake agricultural activities on the territory of the former farm." Point 14, "O poryadke reorganizatsii kolkhozov i sovkhozov." Ibid., p. 3.

38. See "O programme vozrozhdedniya Rossiyskoy derevni i razvitiya agropromyshlennogo kompleksa" from December 3, 1990; and "O sotsial'nom razvitii sela" from December 21, 1991, in *Zemel'naya reforma v Rossii* (Moscow: Yuntus, 1991), pp. 5–17.

39. This argument was made at the Congress of Collective Farms of Russia in February 1992. See *Sel'skaya zhizn'*, February 15, 1992, p. 1.

40. *Kostromskoy kray*, January 9, 1992, p. 1.

41. Moscow television, translated in Foreign Broadcast Information Service, *Daily Report: Central Eurasia* (hereafter *FBIS:CE*), January 24, 1992, p. 50.

42. "O khode i razvitii agrarnoy reformy v Rossiyskoy Federatsii," *Krest'yanskiye vedomosti*, no. 14, March 31–April 6, 1992, p. 6.

43. Interview with N. Kuchenko, chairman of Rus joint stock farm in Salsk raion, Rostov Oblast, July 1993; and see *Kak preobrazovat' kolkhov ili sovkhoz v aktsionernoye obshchestvo zakrytogo tipa* (Moscow: Rossiyskaya akademiya predprinimatel'stva, 1992), pp. 12–15.

44. Interview with Sergei Zanozin, Kostroma Gorkomzem, July 1995; and see *Krest'yanskiye vedomosti*, no. 23, June 6–12, 1994, p. 13. However, a government resolution adopted in February 1995 allowed a farm member to lease his land share or to invest the use right for a term not exceeding three years, during which time he retains ownership of the landshare. After three years the farm member may request to withdraw his landshare from the parent farm in order to start a private farm. According to the resolution, the request has to be approved unanimously by all other landshare owners in the parent farm. See "O poryadke osshchestvleniya prav sobstvennikov zemel'nykh doley i imushchestvennykh paev," *Sobraniye zakonodatel'stva Rossiyskoy Federatsii*, no. 7 (February 13, 1995): 1112–40.

45. Persons were entitled to withdraw from a farm with land in order to undertake private farming, cultivate private plots, or for any other number of small-scale agricultural purposes. See article 7 of the 1991 RSFSR land code, contained in *Krest'yanskiye vedomosti*, no. 16, May 21, 1991.

46. Interview with Sergei Zanozin, Kostroma Gorkomzem, July 1995.

47. Stephen K. Wegren, "New Perspectives on Spatial Patterns of Agrarian Reform: A Comparison of Two Russian Oblasts," *Post-Soviet Geography* 35, no. 8 (October 1994): 455–81; and Stephen K. Wegren, "Agrarian Reform in Moscow Oblast," paper presented at the Twenty-Sixth National Convention of the American Association for the Advancement of Slavic Studies, Philadelphia, Pennsylvania, November 17–20, 1994.

48. *Interfax Food and Agriculture Report* 5, no. 6 (February 2–9, 1996): 3.

49. See Stephen K. Wegren, "The Development of Market Relations in Agricultural Land: The Case of Kostroma Oblast," *Post-Soviet Geography* 36, no. 8 (October 1995): 496–512; Institute of Sociology, Kaluga Oblast, *Formirovanie rynka zemli v Rossii: regional'nyye aspekty* (Kaluga: Institut sotsiiologii, 1997); and E. N. Krylatykh, "Stanovleniye i razvitiye sistemy ekonomicheskogo regulirovaniya zemel'nykh otnosheniy," *Problemy pronozirovaniya*, no. 1 (1997): 31–39.

50. Interview with M. Boytsov, People's Deputy, Kostroma Oblast, and head of the Committee on Agriculture and Land in the Oblast Soviet, July 1993.

51. See the definitions of open and closed types in *Krest'yanskaya Rossiya*, March 19, 1992, p. 3.

52. Interview with N. Kuchenko, chairman of Rus joint stock farm in Salsk raion, Rostov

Oblast, July 1993; and A. Yugay, "Privatsizatsiya i stanovleniye aktsionernykh predpriyatiy na sele," *APK: ekonomika, upravleniye,* no. 7 (July 1991): 52–53.

53. "Polozheniye o reorganizatsii kolkhozov, sovkhozov i privatizatsii gosudarstvennykh sel'skokhozyaystvennykh predpriyatiy," *Ekonomika i zhizn',* no. 41 (October 1992), insert *Vash Partner* (October 1992), p. 6.

54. When the right to land sales was granted, if a person had remained on the farm it was these shares of land that were sold. Unless a person left the farm his land remained in the farm and was not distributed.

55. See Point 14, "Polozheniye o reorganizatsii kolkhozov, sovkhozov i privatizatsii gosudarstvennykh sel'skokhozyaystvennykh predpriyatiy," p. 6; *Kak preobrazovat' kolkhov ili sovkhoz v aktsionernoye obshchestvo zakrytogo tipa,* pp. 20–26; and G. Grigor'yev, "Aktsionernoye obshchestvo v kolkhoze: opyt sozdaniya i pervyye rezul'taty," *APK: ekonomika, upravleniye,* no. 6 (June 1991): 87.

56. Grigor'yev, "Aktsionernoye obshchestvo v kolkhoze: opyt sozdaniya i pervyye rezul'taty," p. 87.

57. "O vyplate dividendov," *Ekonomika sel'skokhozyaystvennykh i pererabatyvayushchikh predpriyatiy,* no. 6 (June 1994): 46.

58. Ibid., pp. 45–46.

59. Grigor'yev, "Aktsionernoye obshchestvo v kolkhoze: opyt sozdaniya i pervyye rezul'taty," pp. 87–88.

60. See *Kak preobrazovat' kolkhoz ili sovkhoz v assotsiatsiyu fermerskikh khozyaystv, malykh kooperativov i predpriyatiy* (Moscow: Rossiyskaya akademiya predprinimatel'stva, 1992), p. 48.

61. *Reorganizatsiya kolkhozov i sovkhozov Rossiyskoy Federatsii (po sostoyaniyu na 1.1.1994g.)* (Moscow: Goskomstat, 1994), p. 1.

62. Rutskoy's views on agrarian reform are best represented by a long article in *Sel'skaya zhizn',* April 15, 1992; and his book entitled *Agrarnaya reforma v Rossii* (Moscow: RAU korporatsiya, 1993).

63. *Zemlya i lyudi,* no. 13, March 27, 1992, p. 1.

64. *Krest'yanskaya Rossiya,* April 3, 1992, p. 1.

65. This position was similar to the one held by former Politburo member Yegor Ligachev.

66. For a fascinating view of personalities and the power structure among agrarian interests during 1992–1993, see Craig L. Infanger, "An Inside View of Russian Agrarian Reform," *Soviet and Post-Soviet Review* 21, nos. 2–3 (1994): 189–209. Craig Infanger served as a policy advisor to the Ministry of Agriculture from the U.S. Department of Agriculture.

67. See, for example, interviews with two of Khlystun's first deputy ministers, V. N. Shcherbak in *Zemlya i lyudi,* nos. 2–3, January 15, 1993, pp. 1–2, and A. A. Chernyshev, in ibid., no. 1, January 1, 1994, pp. 1–2.

68. *Krest'yanskiye vedomosti,* no. 40, October 3–9, 1994, pp. 1–2 (2).

69. *Krest'yanskaya Rossiya,* no. 44, November 7–13, 1994, p. 4. However, in May 1996 Khlystun was reappointed minister of agriculture, purportedly because he felt he could make more of an impact on policy.

70. Interview with Eugeniya Serova, August 1996. Ms. Serova served as Khlystun's assistant during his first term as minister of agriculture.

71. Rural opinion has been measured by surveys and through voting results. For results from opinion polls, see Stephen K. Wegren, "Rural Reform and Political Culture in Russia," *Europe-Asia Studies* 46, no. 2 (1994): 215–41; Stephen Whitefield, "Social Responses to Reform in Russia." in David Lane, ed., *Russia in Transition: Politics, Privatization and Inequality* (London and New York: Longman, 1995), chap. 6. See also the survey results published by the All-Russian Center for Public Opinion in its journal *Ekonomicheskiye i sotsial'nyye peremeny: monitoring obshchestvennogo mneniya.*

On voting results, with specific attention to the urban-rural electoral divide, see Jerry F. Hough, "The Russian Election of 1993: Public Attitudes Toward Economic Reform and Democratization." *Post-Soviet Affairs* 10, no. 1 (1994): 1–37; Jerry F. Hough, Evelyn Davidheiser, and Susan Goodrich Lehmann, *The 1996 Russian Presidential Election* (Washington, D.C.: Brookings,

1996); Ralph S. Clem and Peter R. Craumer, "The Politics of Russia's Regions: A Geographical Analysis of the Russian Election and Constitutional Plebiscite of December 1993," *Post-Soviet Geography* 36, no. 2 (February 1995): 67–86; Ralph S. Clem and Peter R. Craumer, "A Rayon-Level Analysis of the Russian Election and Constitutional Plebiscite of December 1993," *Post-Soviet Geography* 36, no. 8 (October 1995): 459–75; Ralph S. Clem and Peter R. Craumer, "The Geography of the Russian 1995 Parliamentary Election: Continuity, Change, and Correlates," *Post-Soviet Geography* 36, no. 10 (December 1995): 587–616; and Robert W. Orttung and Anna Paretskaya, "Presidential Election Demonstrates Urban-Rural Divide" *Transition* 2, no. 19 (September 20, 1996): 33–38.

72. The survey was conducted among eighty-three agricultural enterprises and thirteen private peasant farms. Before reorganization about 90 percent had worked in a state or collective farm, which meant that most of them "were witnesses and direct participants in agrarian reform." Some 84 percent of the respondents were aged between twenty-five and forty-nine, and more than 50 percent had a specialized secondary agricultural education. N. Proka, "Agrarnaya reforma i motivatsiya truda rabotnikov sel'skogo khozyaystva," *APK: ekonomika, upravleniye,* no. 3 (March 1995): 30–31. Similar results were found by Western researchers. A World Bank–sponsored survey found in 1994 that "many farm employees surveyed in the winter of 1992–1993 were not aware that their farms had changed organizational form." Zvi Lerman and Karen Brooks, "Russia's Legal Framework for Land Reform and Farm Restructuring," *Problems of Post-Communism* 43, no. 6 (November–December 1996): 54.

73. Compiled from *Agropromyshlennyi kompleks Rossiyskoy Federatsii, 1992* (Moscow: Goskomstat, 1992), p. 62; *Razvitiye agropromyshlennogo kompleksa i fermerstva v Rossiyskoy Federatsii,* 2d ed. (Moscow: Goskomstat, 1992), p. 106; *Razvitiye agropromyshlennogo kompleksa i fermerstva v Rossiyskoy Federatsii,* 1st ed. (Moscow: Goskomstat, 1993), p. 41; and *Reorganizatsiya kolkhozov i sovkhozov Rossiyskoy Federatsii (po sostoyaniyu na 1.10.1993g.)* (Moscow: Goskomstat, 1993), p. 1.

74. The percentage of farms using collective labor organization could be as high as 91 percent, depending on the definition of "other forms," which were undefined in statistical compilations.

75. *Reorganizatsiya kolkhozov i sovkhozov Rossiyskoy Federatsii (po sostoyaniyu na 1.01.1994g.),* p. 2. In addition, out of the reorganization process some 81,628 private peasant farms were created as of January 1994.

76. There is disagreement as to how many people work as private farmers. Official government statistics have maintained that for the Russian Federation as a whole, there are three members per private peasant farm, although there are regional differences. This estimate coincides with rural demographic statistics showing that the average rural Russian family was 3.2 persons according to the 1989 census. If this figure is correct it would mean that less than 1 million people were private farmers as of early 1994. On the other hand, in September 1993, the president of AKKOR, V. Bashmachnikov, claimed that more than 1.5 million Russians "live and work" on private farms. *Rossiyskiy fermer,* no. 34, September 28–October 4, 1993, p. 1. At AKKOR's fifth congress in February 1994, he mentioned this figure once again, this time stating that "on such farms 1.5 million persons work." *Zemlya i lyudi,* no. 7, February 18, 1994, p. 3.

77. Karen Brooks et al., *Agricultural Reform in Russia: A View from the Farm Level,* World Bank Discussion Papers, no. 327 (Washington, D.C.: World Bank, 1996), p. 33.

78. See article by Evgeniya Serova in *Rossiyskiy fermer,* no. 40, November 9–16, 1993, p. 3; and for Serova's views and role in agrarian reform, see Craig L. Infanger, "An Inside View of Russian Agrarian Reform," pp. 192–97.

79. The city of Kostroma is located in the southwest corner of the oblast where climate and land quality are relatively better than in other parts of the oblast, even though overall Kostroma is a poor agricultural region, specializing in dairy products and cattle. The four surrounding raions that had zero private farms created from farm reorganization included Krasnol'skiy raion, Nerekhtskiy raion, Sudislavskiy raion, and Susaninskiy raion. For a map of the oblast and a general description of the oblast and its climate, see Stephen K. Wegren, "Agricultural

Reform in the Nonchernozem Zone: The Case of Kostroma Oblast," *Post-Soviet Geography* 32, no. 10 (December 1992): esp. 645–54.

80. Calculated from "Nalichiye zemel' sel'skokhozyaystvennykh predpriyatiy i grazhdan (po sostoyaniyu 1 Yanvarya 1994 goda)," unpublished data from Committee on Land Reserves, Kostroma Oblast, 1994; and *Krest'yanskiye khozyaystva oblasti na 1 Yanvarya 1994 goda (po rezul'tatam obsledovaniya)* (Kostroma: Goskomstat, 1994), p. 14.

Patterns of reorganization within raions further illustrate what occurred during the reorganization process. As of January 1, 1994, fourteen state and collective farms in Kostromskoy raion had reregistered. Of farms that reregistered, one farm formed a joint stock farm of the open type, six formed joint stock farms of the closed type, three formed agricultural cooperatives, and four retained their previous status. No peasant farms were created out of reformed state and collective farms in that most important raion. In Nerekhtskiy raion, directly to the south of Kostromskoy raion, of the fourteen state and collective farms, all had reorganized into joint stock farms of the closed type, and again no peasant farms were created out of reformed farms. Farm reorganization in Krasnosel'skiy raion (located to the southeast of Kostromskoy raion), in Susaninskiy raion to the east, and in Sudislavskiy raion also to the east did not result in the creation of any private farms, and in each of these raions most—if not all—farms reorganized into joint stock farms of the closed type. "Reorganizatsiya kolkhozov i sovkhozov oblasti na 1 Yanvarya 1994 goda," unpublished data from Goskomstat, Kostroma branch, 1994.

81. Only nine of the oblast's twenty-four raions had peasant farms created out of the reorganization process, and only three of those raions could be considered to lie in a semifavorable geographical and climatic location (Buyskiy, Antropovskiy, and Ostrovskiy).

82. "Khod agrarnoy reformy v Kaluzhskoy oblasti na 1 Yanvarya 1994 goda," unpublished information from Goskomstat, Kaluga Oblast branch, 1994. Both raions are relatively unproductive agricultural regions when measured in gross ruble value.

83. *Kaluzhskaya oblast' v tsifrakh v 1993 godu* (Kaluga: Goskomstat Rossii, 1994), p. 5; *Pokazetli ekonomicheskogo i sotsial'nogo razvitiya gorodov i raionov Kaluzhskoy oblasti* (Kaluga: Goskomstat Rossiyskoy Federatsii, 1993), pp. 4–7; and *Chislennost' naseleniya Rossiyskoy Federatsii po gorodam, rabochim poselkami raionam na 1 Yanvarya 1994 g.* (Moscow: Goskomstat Rossii, 1994), pp. 108–09.

84. *Sel'skaya zhizn'*, July 17, 1993, p. 2.

85. Interview with farm director, Chernopenskiy state farm, Kostroma Oblast, June 1992.

86. Price liberalization brought with it a drastic worsening of the terms of trade for the agricultural sector. For example, during 1992 prices for mineral fertilizers increased by a factor of 21, agricultural machinery and equipment by a factor of 19, tractors by a factor of 25, and fuels by a factor of 35. In comparison, purchase prices for agricultural products increased by a factor of 10. Overall, during 1992 prices for basic industrial inputs used by farms and farmers increased by a factor of 26. *Ekonomika i zhizn'*, no. 5 (February 1993): 4; and *Krest'yanskiye vedomosti*, no. 6, February 15–21, 1993, p. 2.

87. "O merakh po stabilizatsii ekonomiki agropromyshlenogo kompleksa," *APK: ekonomika, upravleniye*, no. 7 (July 1992): 5. This decree was signed in early April 1992.

88. "O vvedenii v 1992 godu dotatsiy na zhivotnodcheskuyu produktsiyu," ibid., no. 8 (August 1992): 6–7.

89. I. Terent'yev, "Itogi raboty agropromyshlennogo kompleksa v 1994 g.," *Ekonomist*, no. 4 (April 1995): 56.

90. Economic Research Service, *Former USSR: Situation and Outlook Series*, WRS-95-1 (Washington, D.C.: USDA, 1995), p. 28.

91. *Sel'skoye khozyaystvo Rossii* (Moscow: Goskomstat, 1995), pp. 335–38.

92. *Krest'yanskiye vedomosti*, no. 26, June 23–29, 1992, p. 2.

93. Ibid, no. 29, July 14–20, 1992, p. 4.

94. For Khlystun, *Izvestiya*, December 12, 1992, p. 1; for promises, ibid., January 29, 1993, p. 1, and February 12, 1993, p. 1; for policy revealed, *Rossiyskaya gazeta*, February 2, 1993, p. 5.

95. For Chernomyrdin, see *Krest'yanskiye vedomosti*, no. 5, February 8–14, 1993, p. 5; for subsidies, *Krest'yanskaya Rossiya*, no. 11, March 29–April 4, 1993, p. 3.

96. *Zemlya i lyudi*, no. 16, April 23, 1993, p. 3, and no. 17, April 30, 1993, p. 3.

97. *Krest'yanskaya Rossiya*, no. 11, March 29–April 4, 1993, p. 3.

98. *Rossiyskiy fermer*, no. 43, November 30–December 6, 1993, p. 2; *Sel'skaya zhizn'*, February 26, 1994, pp. 1–2, and March 16, 1995, pp. 1–2; and *Rossiyskiy fermer*, nos. 27–28, October 10–16, 1995, p. 2. *Sobraniye zakonodatel'stva Rossiyskoy Federatsii*, no. 7 (February 12, 1996): 1753–65; *Zemlya i lyudi*, no. 16, April 1996, p. 4; *Sobraniye zakonodatel'stva Rossiyskoy Federatsii*, no. 27, (July 1, 1996): 6777–81; and *Sobraniye zakonodatel'stva Rossiyskoy Federatsii*, no. 10 (March 10, 1997): 2123–30. In addition, a special federal program on the stabilization and development of agroindustrial production from 1996 through 2000 was adopted in mid-1996. See *Sobraniye zakonodatel'stva Rossiyskoy Federatsii*, no. 26 (June 24, 1996): 6337–78.

99. Interview, Agricultural Administration, Kostroma Oblast, July 1993. Raw data on agricultural subsidies by raion during the first half of 1993 are available from author.

100. *Krest'yanskaya Rossiya*, no. 26, July 10–16, 1995, p. 1.

101. I. Bukhtoyatov, "Problemy agrarnoy reformy," *APK: ekonomika, upravleniye*, no. 9 (September 1993): 29.

102. For representative examples of World Bank views, see Karen Brooks and Zvi Lerman, *Land Reform and Restructuring in Russia*, World Bank Discussion Papers, no. 233 (Washington, D.C.: World Bank, 1994).

103. See *Zemlya i lyudi*, no. 27, July 8, 1994, p. 1.

104. *Krest'yanskaya Rossiya*, no. 12, March 28–April 3, 1994, p. 2.

105. *Krest'yanskiye vedomosti*, no. 12, March 28–April 3, 1994, p. 2.

106. *Sel'skaya zhizn'*, April 8, 1994, p. 1.

107. *Krest'yanskaya Rossiya*, no. 21, May 30–June 5, 1994, p. 1.

108. *Rossiyskiy fermer*, no. 18, May 10–16, 1994, p. 1.

109. See M. I. Palladina, "O dal'neyshem razvitii agrarnoy reformy i nekotorykh aspektakh pravovogo statusa sel'skokhozyaystvennykh predpriyatiy i organizatsiy," *Gosudarstvo i pravo*, no. 1 (1997): 45–46.

110. *Izvestiya*, February 7, 1991, p. 2.

111. Ibid., September 1, 1994, p. 1.

112. *Sel'skaya zhizn'*, February 17, 1994, p. 2.

113. Cited in Bolus Ignovich Poshkus, "Economic Reforms and the Agricultural Situation in Russia," Working Paper 93-WP 104 (Center for Agricultural and Rural Development, Iowa State University, February 1993), pp. 2–3.

114. Daniel W. Bromley, "Revitalizing the Russian Food System: Markets in Theory and Practice," *Choices* 8, no. 4 (fourth quarter 1993): 8.

115. Robert J. McIntyre, "Phantom of the Transition: Privatization of Agriculture in the Former Soviet Union and Eastern Europe," *Comparative Economic Studies* 34, nos. 3–4 (fall–winter 1992): 91.

116. Some of the material describing farm privatization in Nizhniy Novgorod draws from Stephen K. Wegren, "Farm Privatization in Nizhnii Novgorod: A Model for Russia?" *RFE/RL Research Report* 3, no. 21, May 27, 1994, pp. 16–27. See also V. Uzun, "Nizhegorodskaya model' reformirovaniya sel'skokhozyaystvennykh predpriyatiy," *Voprosy ekonomiki*, no. 1 (January 1995): 57–65.

117. *Finansovyye izvestiya*, no. 30, July 7–13, 1994, p. 5.

118. International Finance Corporation, *Land Privatization and Agricultural Enterprise Reorganization in Russia: The Nizhny Novgorod Model*, unpublished draft document, February 1994, pp. 7–9.

119. *Izvestiya*, March 12, 1994, p. 1.

120. *Rossiyskaya gazeta*, April 26, 1994, p. 4.

121. IFC, "Russian Federation Nizhny Novgorod Land Privatization Project: Annex I" (unpublished document, March 1994), p. 1.

122. "Reformirovanii sel'skokhozyaystvennykh predpriyatiy s uchetom praktiki Nizhegorodskoy oblasti," *Krest'yanskaya Rossiya*, no. 37, September 19–25, 1994, pp. 12–17.

123. *Rossiyskaya gazeta*, February 1, 1995, insert *Biznes v Rossii*, pp. 4–5.
124. IFC, "Land Reform in Russia," Operational Report, November 1996; and V. A. Uzun, ed., *Sotsial'no-ekonomicheskiye posledstviya privatizatsii zemli i reorganizatsii sel'skozyaystvennykh predpriyatiy (1994–1996 gg.)* (Moscow: Entsiklopediya rossiyskikh dereven', 1997), p. 8.
125. The words "entitlements" and "shares" are used interchangeably. "Entitlements" is the word used in IFC documents, whereas the word "share" (*dolya*, "pay") is used in Russian sources.
126. IFC, *Land Privatization and Agricultural Enterprise Reorganization in Russia*, p. 60.
127. It should be emphasized that the ruble value of land used for this measurement has no relation to the monetary value of land. Farm management was to discourage land shareholders from associating the ruble value with actual value of land. It has been recommended that Russian law be changed to remove ruble values from land certificates in order to preclude confusion.
128. *Krest'yanskiye vedomosti*, no. 1, January 3–9, 1994, p. 2.
129. "Reformirovanii sel'skokhozyaystvennykh predpriyatiy s uchetom praktiki Nizhegorodskoy oblasti," p. 14.
130. IFC, *Land Privatization and Agricultural Enterprise Reorganization in Russia*, p. 70.
131. Ibid., pp. 67–68.
132. *Krest'yanskiye vedomosti*, no. 10, March 14–20, 1994, p. 2. The same article noted that "many peasants, especially pensioners, leased their hectares to the joint stock farm or to farmers."
133. IFC, *Land Privatization and Agricultural Enterprise Reorganization in Russia*, p. 71.
134. Ibid., p. 73.
135. *Zemlya i lyudi*, no. 12, March 25, 1994, p. 3.
136. Land auctions are held first, followed by property auctions. "Reformirovanii sel'skokhozyaystvennykh predpriyatiy s uchetom praktiki Nizhegorodskoy oblasti," p. 15.
137. At the auction, only entitlement holders are eligible to participate and only share rubles and ballohectares are used as means of payment.
138. The application lists the land and property lots that the group or individual wants to obtain, the intended form of the enterprise, a list of the land and property shares and the value of each, and the signature of each of the members in the new enterprise. The farm commission is entrusted with checking that all parts of the application are completed, that the applicants have documentation for all shares listed, and that the lots listed in the applications do not exceed the applicants' purchasing power. IFC, *Land Privatization and Agricultural Enterprise Reorganization in Russia*, pp. 97–100. However, the same document stipulates that on an "exceptional basis" an applicant can apply for lots whose value exceeds his purchasing power, but this can be done only once per applicant.
139. IFC, *Land Privatization and Agricultural Enterprise Reorganization in Russia*, pp. 102–03. The terms or conditions of "exceptional" were not defined.
140. Ibid.
141. *Zemlya i lyudi*, no. 12, March 25, 1994, p. 3; IFC, *Land Privatization. . . . Russia*, pp. 100–11.
142. It is interesting to note that many newly privatized farms opted for a collective form of labor organization by combining land and property shares, specifically closed joint stock farms (though on a smaller scale than state and collective farms turned joint stock) even after "decollectivization." The primary reason seems to be the economic environment, although cultural considerations cannot be dismissed.
143. See, for example, the coverage of the former state farm in Rostov-on-Don in *Krest'yanskaya Rossiya*, no. 50, December 25–30, 1995, p. 2.
144. IFC, *Land Privatization and Agricultural Enterprise Reorganization in Russia*, pp. 7–10.
145. IFC, "Land Reform in Russia," Operational Report, March 1–April 30, 1995, p. 6.
146. Uzun, *Sotsial'no-ekonomicheskiye posledstviya privatizatsii zemli i reorganizatsii sel'skozyaystvennykh predpriyatiy*, pp. 64–93.
147. IFC, "Land Reform in Russia," p. 4.

148. See V. Kuznetsov and A. Shlyakhetskiy, "Problemy formirovaniya mnogoukladnoy ekonomiki na sele," *APK: ekonomika, upravleniye,* no. 6 (June 1994): 5.
149. Information from Gretchen Wilson, an official in the IFC.
150. *Sel'skaya zhizn',* January 20, 1994, p. 2.
151. *Rossiyskiy fermer,* nos. 12–13, April 11–17, 1995, p. 3.
152. *Izvestiya,* November 25, 1994, p. 4; *Rossiyskiy fermer,* no. 19, May 16–22, 1995, p. 3.
153. *Izvestiya,* March 22, 1994, p. 2.
154. *Rossiyskaya gazeta,* May 17, 1995, p. 1.
155. *Sel'skaya zhizn',* January 20, 1994, p. 2.
156. *Rossiyskaya gazeta,* April 26, 1994, p. 4.
157. *Krest'yanskaya Rossiya,* no. 12, April 3–9, 1995, p. 1.
158. *Trud,* October 10, 1995, p. 2; translated in *FBIS:CE,* December 8, 1995, p. 21.
159. *Krest'yanskaya Rossiya,* no. 28, July 4–10, 1994, p. 3.

Chapter 4. Financial Levers and the End of the Social Contract

1. For a discussion of the social contract, see Gail Lapidus, "Social Trends," in Robert F. Byrnes, ed., *After Brezhnev: Sources of Soviet Conduct in the 1980s* (Bloomington: Indiana University Press, 1983); Seweryn Bialer, *Stalin's Successors: Leadership, Stability, and Change in the Soviet Union* (New York: Cambridge University Press, 1980), esp. pp. 158–65; Peter A. Hauslohner, "Gorbachev's Social Contract," *Soviet Economy* 3, no. 1 (January–March 1987): 54–89; Walter D. Connor, *Socialism's Dilemmas: State and Society in the Soviet Bloc* (New York: Columbia University Press, 1988), esp. pp. 67–85; Ed A. Hewett, *Reforming the Soviet Economy: Equality Versus Efficiency* (Washington, D.C.: Brookings, 1988), esp. pp. 39–50; and Linda J. Cook, *The Social Contract and Why It Failed* (Cambridge, Mass.: Harvard University Press, 1993).

2. Stephen K. Wegren, "The Social Contract Reconsidered: Peasant-State Relations in the USSR," *Soviet Geography* 32, no. 10 (December 1991): 653–82. Gorbachev himself in his memoir wrote, "Statements claiming that agriculture was 'unprofitable' were found to be wrong. All data pointed to the fact that much more was siphoned off from agriculture than invested in it. And, of course, the nation's economic development had been achieved largely at the expense of the countryside" (*Memoirs* [New York: Doubleday, 1996], p. 120).

3. On Soviet standards of living in general, see Mervyn Matthews, *Class and Society in Soviet Russia* (London: Penguin, 1972); Mervyn Matthews, *Poverty in the Soviet Union: The Life-Styles of the Underprivileged in Recent Years* (Cambridge: Cambridge University Press, 1986); Mervyn Matthews, *Patterns of Deprivation in the Soviet Union Under Brezhnev and Gorbachev* (Stanford: Hoover Institution Press, 1989); Alastair McAuley, *Economic Welfare in the Soviet Union: Poverty, Living Standards, and Inequality* (Madison: University of Wisconsin Press, 1979); and A. P. Tyurina, *Sotsial'no-ekonomicheskoye razvitiye Sovetskoy derevni 1965–1980* (Moscow: Mysl', 1982).

4. See Sheila Fitzpatrick, *Stalin's Peasants: Resistance and Survival in the Russian Village After Collectivization* (New York: Oxford University Press, 1994).

5. Cook, *The Social Contract and Why It Failed,* p. 3.

6. This December 3, 1990, resolution, "Programme vozrozhdeniya Rossiyskoy derevni i razvitiya agropromyshelnnogo kompleksa," is reprinted in *Zemel'naya reforma v Rossii* (Moscow: Yuntus, 1991), pp. 5–7.

7. "O sotsial'nom razvitii sela," *Ekonomika i zhizn',* no. 4 (January 1991): 19.

8. "O politicheskom i sotsial'no-ekonomicheskom polozhenii v RSFSR i merakh po vykhodu iz krizisa," *Zemlya i lyudi,* nos. 14–15, April 10, 1992, p. 3.

9. The agreement was signed by Yegor Gaydar, along with then minister of agriculture V. Khlystun and A. E. Vorontsov of the Agrarian Union of Russia and the Russian Council of Collective Farms and Other Forms of Farming. *Sel'skaya zhizn',* July 3, 1992, p. 2.

10. See Don Van Atta, "The Second Congress of the Russian Agrarian Union," *RFE/RL Research Report* 2, no. 31 (July 30, 1993): 42–49.

11. The leadership's attitude toward urban dwellers stood in stark contrast. Gorbachev, for instance, revealed that "any noticeable increase in retail food prices was resolutely rejected.

The problem was . . . a purely political issue. The alternative was to increase subsidies from the state budget." Gorbachev, *Memoirs*, p. 121.

12. Michael McFaul, "State Power, Institutional Change, and the Politics of Privatization in Russia," *World Politics* 47, no. 2 (January 1995): 235.

13. Robert H. Bates, *Markets and States in Tropical Africa: The Political Basis of Agricultural Policies* (Berkeley and Los Angeles: University of California Press, 1981), p. 112.

14. *Krest'yanskaya Rossiya*, no. 12, March 28–April 3, 1994, p. 2.

15. See Robert J. McIntyre, "The Phantom of the Transition: Privatization of Agriculture in the Former Soviet Union and Eastern Europe," *Comparative Economic Studies* 34, nos. 3–4 (fall–winter 1992): 81–95.

16. See David A. J. Macey, "Demise of the Moral Imperative: Agricultural Reform in Russia Today," in Michael Kraus and Ronald Liebowitz, eds., *Russia and East Europe in Transition* (Boulder, Colo.: Westview Press, 1995).

17. The International Monetary Fund and its sister institution, the World Bank, extended billions of dollars worth of loans and credits to the Yel'tsin government. For example, in March 1996 an agreement was reached in which the IMF would extend a 10.2 billion dollar loan to Russia over a three year period. Western assistance was linked to the Russians implementing a reform program that had the approval of the IMF. In mid-1996 the IMF suspended payments when Russian reforms were not meeting certain conditions to which assistance was linked. The involvement of the IMF in Russian economic reforms led some Russian analysts and conservative politicians to accuse the IMF of having a political agenda and interfering in Russian domestic politics. See *New York Times*, July 23, 1996, p. A1; ibid., August 22, 1996, p. A5; and ibid., October 25, 1996, p. A6. In mid-1997 it was reported that the World Bank launched an expansion of loans to Russia that would total over 6 billion dollars during 1998–1999. See *Financial Times*, June 12, 1997, p. 2. For a Russian view of the role of the World Bank in agrarian reform, see E. Zhogoleva, "Vsemirnyi bank i ego podderzhka sel'skogo khozyaystva Rossii," *APK: ekonomika, upravleniye*, no. 6 (June 1994): 77–78.

18. In one reported case, the IMF instructed the Russian government not to accept any variant of a land code that did not allow the free purchase and sale of land without restriction. See *Zemlya i lyudi*, no. 50, December 15, 1995, p. 4.

19. Vladimir Popov, "Will Russia Create a Western-style Capitalist Society? Not Likely, There Are Too Many Cultural Chains," *Geonomics* 4, no. 1 (January 1992): 3.

20. This conclusion is based on the numerous USAID-sponsored program proposals I have read since 1993 that were intended to help Russia privatize its agrarian sphere. The language is quite explicit about "adapting American institutions to the Russian situation" and some even spoke of their activities as "interventions."

21. For early interviews with Khlystun, see *Izvestiya*, February 17, 1992, p. 2, and June 10, 1992, p. 2, in which he explicitly supports reform efforts. For Infanger's experiences and impressions, see Craig L. Infanger, "An Inside View of Russian Agrarian Reform," *Soviet and Post-Soviet Review* 21, nos. 2–3 (1994): 189–210.

22. See, for example, *Krest'yanskiye vedomosti*, no. 49, December 13–19, 1993, p. 6, and nos. 51–52, December 27, 1993–January 2, 1994, p. 20.

23. *Krest'yanskaya Rossiya*, no. 15, April 24–30, 1995, p. 3.

24. *Krest'yanskiye vedomosti*, no. 48, December 6–12, 1993, p. 4.

25. A. Zaveryukha, "Problemy agrarnoy reformy v Rossii," *APK: ekonomika, upravleniye*, no. 7 (July 1995): 3–12; M. El'diev, "Agrarnyy krizis preodolim," ibid., no. 12 (December 1996): 13–22; A. Rutskoy and N. Radugin, "Agrarnyy krizis prodolzhaetsya," ibid., no. 1 (January 1997): 3–7.

26. "O katastroficheskom polozhenii v agropromyshlennom komplekse Rossiyskoy Federatsii," *Sobraniye zakonodatel'stva Rossiyskoy Federatsii*, no. 13 (March 31, 1997): 2624–25.

27. See Stephen K. Wegren, "From Farm to Table: The Food System in Post-Communist Russia," *Communist Economies and Economic Transformation*, 8, no. 2 (1996): 149–83.

28. "Agropromyshlennyy kompleks Rossii v 1994g.," *APK: ekonomika, upravleniye*, no. 4 (April 1995): 48.

29. *Zemlya i trud,* no. 6, February 6–12 1996, p. 1.
30. *Sel'skaya zhizn',* November 5, 1994, p. 2.
31. *Finansovyye izvestiya,* no. 53, July 27, 1995, p. 4.
32. *Zemlya i trud,* no. 44, October 31–November 6, 1995, p. 3.
33. *Sel'skaya zhizn',* July 28, 1994, p. 1, and December 3, 1994, p. 1.
34. Ibid., November 5, 1994, p. 2.
35. The conversion of rubles to dollars is complicated by inflation and by the eroding value of the ruble against the dollar. In January 1994, for example, the ruble was traded at 1,255 rubles to the dollar; by December 1994 it had reached 3,249 rubles to the dollar. It is hard, therefore, to establish a precise conversion of the 1994 budget into an annual dollar equivalent because some monies were dispersed early in the year whereas most were dispersed in the third and fourth quarters.
36. I. Terent'yev, "Itogi raboty agropromyshlennogo kompleksa v 1994 g.," *Ekonomist,* no. 4 (April 1995): 52.
37. *Sel'skaya zhizn',* June 8, 1995, p. 1.
38. *Sotsial'no-ekonomicheskoye polozheniye Rossiya,* no. 12 (Moscow: Goskomstat Rossii, 1995), p. 183.
39. *Sotsial'no-ekonomicheskoye polozheniye Rossiya,* no. 12 (Moscow: Goskomstat Rossii, 1996), p. 145.
40. *Sotsial'no-ekonomicheskoye polozheniye Rossiya,* January–May 1997 (Moscow: Goskomstat Rossii, 1997), p. 148.
41. M. Kozlov, "Finansovoye sostoyaniye sel'skokhozyaystvennykh tovaroproizvoditeley v usloviyakh reformy," *APK: ekonomika, upravleniye,* no. 9 (September 1995): 39.
42. When the president of AKKOR, V. Bashmachnikov, claimed in early 1995 that the "private sector" produced 50 percent of the nation's food in 1994, he was including the production from personal plots and other small-scale privatized plots of land whose intent was—and remains—primarily to supplement the family food supply. See *Krest'yanskaya Rossiya,* no. 10, March 20–26, 1995, p. 1; and *Rossiyskiy fermer,* nos. 9–10, March 14–20, 1995, pp. 1–2. Similar assertions were made in subsequent years, as USDA reported that 48 percent of food production in 1996 came from the "private sector," which includes private farms and production from the population. In reality, about 46 percent of private production came from the population's agricultural plots, and only 2 percent from private farms (Economic Research Service, USDA, *Newly Independent States and the Baltics Situation and Outlook Series,* WRS-97-1 [Washington: D.C.: USDA, 1997], pp. 43, 45). As one set of authors argued, however, subsidiary plots of the population should not be considered "private" at all because of their close links and dependence on collective farms for supplies and services (Zvi Lerman et al., "Self-Sustainability of Subsidiary Household Plots: Lessons for Privatization of Agriculture in Former Socialist Countries," *Post-Soviet Geography,* 35, no. 9 [November 1994]: 526–42).
43. A. Borisenko, "Strukturnaya politika i investitsionnaya deyatel'nost' APK," *APK: ekonomika, upravleniye,* no. 7 (July 1997): 24–25.
44. *Razvitiye agropromyshlennogo kompleksa i fermerstva v Rossiyskoy Federatsii,* vypusk 1 (Moscow: Goskomstat Rossii, 1993), pp. 33–38.
45. M. P. Kazakov, "Finansovyye problemy sel'skogo khozyaystva," *Ekonomika sel'skozyaystvennykh i pererabatyvayushchikh predpriyatiy,* no. 11 (November 1995): 7.
46. "Kapital'noye stroitel'stvo v otrasliakh APK Rossii," *APK: ekonomika, upravleniye,* no. 12 (December 1994): 25. Moreover, federal capital investments became more restricted as specific uses became targeted. As a result, about 13 percent of federal investment monies were allocated to 42 specific programs designed to promote production of a particular food product. "Agropromyshlennyy kompleks Rossii v 1993 g.," *APK: ekonomika, upravleniye,* no. 5 (May 1994): 23.
47. *Sel'skaya zhizn',* April 1, 1995, p. 1; *Razvitiye agropromyshlennogo kompleksa i fermerstva v Rossiyskoy Federatsii,* vypusk 1, p. 33; and "O polozhenii v agropromyshlennom komplekse Rossii," *APK: ekonomika, upravleniye,* no. 1 (January 1995): 39; *Sotsial'no-ekonomicheskoye polozheniye Rossiya,* no. 12 (1995), p. 59.
48. *Sotsial'no-ekonomicheskoye polozheniye Rossiya,* no. 12 (1996), p. 44.

49. *Sel'skaya zhizn'*, April 13, 1993, p. 1; and A. Ogarkov, "Investitsionnaya politika v sel'skom khozyaystve Rossii," *APK: ekonomika, upravleniye*, nos. 11–12 (November–December 1993): 9.
50. *Sotsial'no-ekonomicheskoye polozheniye Rossiya*, no. 12 (Moscow: Goskomstat Rossii, 1994), p. 47.
51. *Sotsial'no-ekonomicheskoye polozheniye Rossiya*, no. 12 (1996), p. 44.
52. V. Mashenkov and Ye. Lysenko, "Sotsialnoye razvitiye sela," *APK: ekonomika, upravleniye*, no. 9 (September 1994): 3.
53. Terent'yev, "Itogi raboty agropromyshlennogo kompleksa v 1994 g.," p. 61.
54. *Izvestiya*, January 12, 1995, p. 2.
55. *Ekonomika i zhizn'*, no. 5 (February 1993): 4. These 1992 aggregated indexes included grain, which rose by a factor of 23, vegetables and potatoes by a factor of 8–9, cattle and poultry by a factor of 6, milk by a factor of 7, and eggs by a factor of 8. At the same time, feed increased in price by a factor of 16, mineral fertilizer by a factor of 21, agricultural machinery and equipment by a factor of 19, tractors by a factor of 25, and fuel and lubricating oils by a factor of 35. *Razvitiye agropromyshlennogo kompleksa i fermerstva v Rossiyskoy Federatsii*, vypusk 1, p. 29.
56. *Sel'skaya zhizn'*, November 3, 1994, p. 2.
57. N. Kuznetsova, "O regulirovanii tsen na produktsiyu agropromyshlennogo kompleksa," *Ekonomist*, no. 4 (April 1995): 66.
58. *Zemlya i lyudi*, nos. 11–12, March 24, 1995, p. 2.
59. *Sel'skaya zhizn'*, June 20, 1995, p. 1. See *Zemlya i lyudi*, no. 43, October 28, 1994, p. 2.
60. *Sel'skaya zhizn'*, August 17, 1993, p. 1.
61. This point is taken from interviews with agricultural personnel at the Novosel'skoye sovkhoz in Kaluga Oblast, July 1994, and economists at the Kostroma Agricultural Institute during 1992–1993.
62. *Finansovyye izvestiya*, no. 39, September 1–7, 1994, p. 1.
63. *Sel'skoye khozyaystvo Rossii* (Moscow: Goskomstat, 1995), p. 76; and *Osnovnyye pokazateli agropromyshlennogo kompleksa Rossiyskoy Federatsii (Yanvar'–Sentyabr' 1996 goda)* (Moscow: Goskomstat Rossii, 1996), p. 34.
64. Tax rates on farm profits were increased to 39 percent. A. Nazarchuk, "APK na sovremennom etape ekonomicheskoy reformy," *Ekonomist*, no. 3 (March 1995): 15. According to the chief accountant at the Kostroma Agricultural Administration, collective agricultural enterprises paid several different types of taxes. In 1994 farms paid the following taxes: land taxes; 35 percent profit tax (similar to the U.S. income tax); 2 percent transport tax; 1 percent wage fund tax; value added tax; import and export taxes (sometimes referred to as tariffs); and then several taxes to special funds, such as taxes to state pension funds, medical insurance, unemployment taxes, road fund taxes, and taxes to support agricultural sciences. Interview, Kostroma Agricultural Administration, August 1994.
65. *Kostromskoy kray*, July 18, 1994, p. 1.
66. See Wegren, "From Farm to Table," pp. 149–83.
67. V. Timofeyev, "Sotsial'noye obustroystvo sela—zadacha obshchenarodnaya," *APK: ekonomika, upravleniye*, no. 9 (September 1993): 15.
68. *Zemlya i lyudi*, no. 13, March 27, 1992, p. 3.
69. Timofeyev, "Sotsial'noye obustroystvo sela—zadacha obshchenarodnaya," pp. 15–16.
70. L. Prozorina, "O realitzatsii programm po sotsial'nomy razvitiyu sela," *Ekonomika sel'skogo khozyaystva Rossii*, no. 3 (March 1994): 8.
71. *Zemlya i trud*, no. 49, December 7–13, 1993, p. 2.
72. *Razvitiye agropromyshlennogo kompleksa i fermerstva v Rossiyskoy Federatsii*, vypusk 2 (Moscow: Goskomstat, 1993), pp. 43–48.
73. *Sel'skaya zhizn'*, December 16, 1993, p. 1.
74. "Ob osnovnykh itogakh raboty APK za 1994 god," unpublished document P-2-24/20, dated January 5, 1995, p. 7. I thank Christian Foster of USDA for making this document available to me.

75. Terent'yev, "Itogi raboty agropromyshlennogo kompleksa v 1994 g.," p. 61.
76. *Rossiyskaya gazeta*, February 20, 1993, p. 5.
77. For a survey of rural living standards, see *Zemlya i trud*, no. 45, November 7–13, 1995, p. 5.
78. This conclusion is based upon personal information the author has of Dallas-based companies contracting to construct suburban housing using Western architectural plans and standards. The author aided in the translation of technical documents and contracts.
79. For an early report on this occurrence, see Stephen K. Wegren, "Rural Reform in Russia," *RFE/RL Research Report* 2, no. 43 (October 29, 1993): 52.
80. *Razvitiye agropromyshlennogo kompleksa i fermerstva v Rossiyskoy Federatsii*, vypusk 1, p. 30.
81. *Zemlya i trud*, no. 1, January 2–8, 1996, p. 3.
82. *Sotsial'no-ekonomicheskoye polozheniye Rossiya*, no. 12 (1996), p. 163.
83. *Rossiya v tsifrakh 1997* (Moscow: Goskomstat Rossii, 1997), p. 46.
84. *Itogi proivodstvenno-finansovoy deyatel'nosti sel'skokhozyaystvennykh predpriyatiy Kostromskoy oblasti za 1995 god* (Kostroma: Goskomstat, 1996), p. 3.
85. *Zemlya i lyudi*, no. 49, December 8, 1995, p. 1.
86. The amount of land cultivated for grain was less in 1995 than in 1950. Grain yields in 1995 were similar to those of the early 1960s.
87. A. Selezenev, "Nekotoryye itogi raboty APK v 1995 g.," *Ekonomist*, no. 6 (June 1996): 81; *Sotsial'no-ekonomicheskoye polozheniye Rossiya*, no. 12 (1996), p. 49.
88. *Sotsial'no-ekonomicheskoye polozheniye Rossii*, no. 12 (1995), p. 74.
89. These data actually understate the magnitude of the production decline in the collective agricultural sector because they include production by private farms and by personal plots, both of which have increased their production significantly since 1991. From 1989 to 1996 the percentage of gross volume of output from state and collective farms (and their successors) declined from almost 78 percent of production to 52 percent; meanwhile the percentage of output from personal plots increased from about 22 to over 35 percent in 1993 and was estimated at 38 percent in 1994, nearly 44 percent in 1995, and 46 percent in 1996. Private farm output remained at about 2 percent during 1992–1996.
90. A. Nazarchuk, "Sovremennoye sostoyaniye sel'skokhozyaystvennogo proizvodstva v Rossiyskoy Federatsii," *APK: ekonomika, upravleniye*, no. 1 (January 1996): 12; Economic Research Service, USDA, *Newly Independent States and the Baltics: Situation and Outlook Series*, WRS-97-1 (Washington, D.C.: USDA, 1997), p. 21.
91. ERS, USDA, *Newly Independent States and the Baltics*, p. 51.
92. *Agrokhleb byulleten'*, no. 5 (March 1995): 2–3.
93. For the situation in southern black earth regions, see *Zemlya i lyudi*, no. 42, October 21, 1994, pp. 1, 3; V. Kuznetsov, "Agropromyshlennyy kompleks Severnogo Kavkaza," *APK: ekonomika, upravleniye*, no. 7 (July 1997): 3–10.
94. *Krest'yanskiye vedomosti*, no. 5, February 5–11, 1996, p. 2.
95. L. Belokopytova, "Vliyaniye agrarnoy reformy na razvitiye sel'skogo khozyaystva v Povolzh'e," *APK: ekonomika, upravleniye*, no. 9 (September 1995): 52.
96. B. K. Markin, "Stanovleniye krest'yanskiye (fermerskikh) khozyaystv v Povolzh'e," *Ekonomika sel'skokhyzystvennykh i pererabatyvayushchikh predpriyatiy*, no. 11 (November 1995): 27.
97. Jaclyn Y. Shend, *Agricultural Statistics of the Former USSR Republics and the Baltic States*, statistical bulletin no. 863 (Washington, D.C.: USDA, Economic Research Service, 1993), p. 231.
98. Ann M. Lane, Ruth M. Marston, and Susan O. Welsh, "The Nutrient Content of the Soviet Food Supply and Comparisons with the U.S. Food Supply," in Joint Economic Committee, *Gorbachev's Economic Plans*, vol. 2, Joint Committee Print, 100th Cong., 1st sess. (Washington, D.C.: GPO, 1987), pp. 79–100 (quote p. 79, data p. 83).
99. During the ten years between 1981 and 1990, state subsidies to agriculture increased from 25 to over 100 billion rubles annually, three-quarters of which went to animal husbandry products. See E. Serova, "Predposylki i sushchnost' sovremennoy agrarnoy reformy v

Rossii," *Voprosy ekonomiki,* no. 1 (January 1995): 32–46 (subsidy statistic p. 33).

100. "Agropromyshlennyy kompleks Rossii v 1993 g.," *APK: ekonomika, upravleniye,* no. 5 (May 1994): 22.

101. For evidence on hunger, see Stephen K. Wegren and Frank A. Durgin, "Why Agrarian Reform Is Failing," *Transition* 1, no. 19 (October 20, 1995): 52.

102. For example, one article cited the use of ration coupons in Krasnoyarsk and Khabarovsk krays, and in Amur Oblast. *Zemlya i lyudi,* no. 48, December 1, 1995, p. 1.

103. *Zemlya i lyudi,* no. 2, January 20, 1995, p. 1; these estimates were supported by studies conducted at the Institute of Socioeconomic Problems in the Development of the Agroindustrial Complex, the Russian Academy of Sciences, located in Saratov. See *Zemlya i lyudi,* no. 3, February 2, 1996, p. 2.

104. Because it was an experimental farm its reform options were only to disband or to remain a state farm. No one left the farm to become a private farmer, while the average in the oblast was for two or three families to leave per farm.

105. Kozlov, "Finansovoye sostoyaniye sel'skokhozyaystvennykh tovaroproizvoditeley v usloviyakh reformy," pp. 38–39.

106. The usual stipulations about the inapplicability of Western concepts of profitability and unprofitability continue to apply.

107. The farm reorganization decree was directed at unprofitable and low profitable, a sum that equaled about 10 percent of all farms.

108. During 1993, about 57 percent of state and collective farms obtained subsidized credits, and the predominant lender was Rossel'khozbank.

109. A. Nazarchuk, "APK na sovremennom etape ekonomicheskoy reformy," *Ekonomist,* no. 3 (March 1995): 15; and "Ob osnovnykh itogakh raboty APK za 1994 god," unpublished document P-2-24/20, dated January 5, 1995, p. 5.

110. Selezenev, "Nekotoryye itogi raboty APK v 1995 g.," p. 81; and *Krest'yanskiye vedomosti,* no. 2, January 20–26, 1997, p. 2.

111. The Russian food market was not fully competitive and functioned somewhat differently than in the West because of deficiencies in bank transfers, lack of access to time-urgent price information, and the absence of a culture that recognized the primacy of contractual obligations. See Stephen K. Wegren, "Building Market Institutions: Agricultural Commodity Exchanges in Post-Communist Russia," *Communist and Post-Communist Studies* 27, no. 3 (1994): 195–224. While the "food market" had a number of different outlets, it ws characterized by increased trade freedom and an ability to withdraw from ventures that were economically unattractive, a common feature found among food producers around the world. See Bates, *Markets and States in Tropical Africa,* pp. 82–95.

112. I. Gridasov, "Sostoyaniye zernogo khozyaystva," *Ekonomist,* no. 1 (January 1996): 92–93.

113. "O formirovanii gosudarstvennykh prodovol'stvennykh fondov na 1992 godu," *APK: ekonomika, upravleniye,* no. 4 (April 1992): 3–4. Excluded from the list were personal plot operators, collective gardens and orchards, suburban cooperatives, nonagricultural enterprises, and subsidiary agricultural operations of enterprises and organizations.

114. Moscow Television, August 15, 1992, in *FBIS: CE* August 17, 1992, p. 11.

115. In August 1992 state procurement prices rose from an average of 10,000 rubles to 12,000 rubles per ton for hard wheat (class 3), about $28 per ton at exchange rates in effect at the time. In November 1992 grain purchase prices were increased to an average of 25,000 rubles per ton (still $28 a ton). Economic Research Service, *Former USSR: Situation and Outlook Series,* WRS-94-1 (Washington, D.C.: USDA, 1994), p. 52.

116. The year 1991 generally was considered a disastrous year, with food rationing and threats of hunger in cities. For 1992 output, see "Sel'skoye khozyaystvo Rossii v 1992 godu," *APK: ekonomika, upravleniye,* no. 4 (April 1993): 56–57; and Christian Foster, "The Former USSR: 1992 Agricultural Performance Down," *Newsletter for Research on Soviet and East European Agriculture* 14, no. 4 (December 1992): 1, 7.

117. *Izvestiya*, January 28, 1993, p. 2. The decision to create food funds was first revealed in late December 1992.
118. "O formirovanii federal'nykh i regional'nykh prodovol'stvennykh fondov v 1993 godu," *APK: ekonomika upravleniye*, no. 5 (May 1993): 7–8.
119. *Rossiyskaya gazeta*, December 25, 1992, p. 1.
120. "O formirovanii federal'nykh i regional'nykh prodovol'stvennykh fondov v 1993 godu," p. 8.
121. "Zakon o zerne," *Sel'skaya zhizn'*, June 1, 1993, p. 2.
122. Roskhleboprodukt was privatized in 1994, although 51 percent of the shares were to remain in state hands for a period of three years. *Izvestiya*, December 28, 1993, p. 1.
123. "O liberalilzatsii zernovogo rynka v Rossii," *Vash partner*, no. 4 (January 1994): 5.
124. Roskhleboprodukt was the organization in charge of purchasing domestic grain; Exportkhleb was the organization in charge of the import and export of grain.
125. In mid-September 1993 the government announced that, beginning in October 1993, the indexation of purchase prices would end. *Rossiyskiy fermer*, no. 32, September 14–20, 1993, p. 1.
126. See the interview with the head of the Federal Food Corporation in *Zemlya i lyudi*, no. 1, January 1996, p. 5.
127. "O zakupkakh i postavkakh," *Rossiyskiy fermer*, no. 1, January 3–9 1995, article 6, point 4.
128. *Krest'yanskaya Rossiya*, no. 11, March 29–April 4, 1993, p. 3; *Sel'skaya zhizn'*, February 26, 1994, pp. 1–2, and March 16, 1995, pp. 1–2.
129. *Finansovyye izvestiya*, no. 11, February 16, 1995, p. 4.
130. *Krest'yanskiye vedomosti*, no. 5, February 5–11, 1996, p. 2.
131. "Sel'skoye khozyaystvo Rossii v 1996 godu," *APK: ekonomika, upravleniye*, no. 3 (March 1997): 10.
132. Robert H. Bates and William P. Rogerson, "Agriculture in Development: A Coalitional Analysis," *Public Choice* 35, no. 5 (1980): 523.
133. See Don Van Atta, "Political Mobilization in the Russian Countryside: Creating Social Movements from Above," in Judith B. Sedaitis and Jim Butterfield, eds., *Perstroika from Below: Social Movements in the Soviet Union* (Boulder, Colo.: Westview Press, 1991), pp. 53–56.
134. *Sel'skaya zhizn'*, June 2, 1990, p. 2; and *Izvestiya*, June 15, 1990, p. 3.
135. *Sel'skaya zhizn'*, March 15, 1990, p. 2.
136. Ibid., March 27, 1990, p. 2.
137. *Izvestiya*, April 28, 1990, p. 2.
138. *Sel'skaya zhizn'*, April 27, 1990, p. 3.
139. Van Atta, "Political Mobilization in the Russian Countryside," p. 53.
140. The party is headed by Mikhail Lapshin, who previously headed the Russian Agrarian Union, after V. Starodubtsev. Lapshin was born in 1934 to a peasant family in Altai kray. He graduated as an agronomist from the Timiryazev Agricultural Academy in Moscow and later graduated from a foreign language institute in Moscow. In 1961 he became the director of a state farm in Moscow Oblast and remained its director until 1992 when the farm transformed into a joint stock company, of which he was elected president. In February 1993 he was elected chairman of the Agrarian Party of Russia (APR). *Zemlya i trud*, no. 48, November 30–December 6, 1993, p. 2.
141. Ibid., no. 10, March 9–15, 1993, p. 1, and no. 51, December 20–26, 1994, pp. 3–4.
142. See, for example, peasant demands to the government in *Sel'skaya zhizn'*, April 13, 1995, p. 1; and coverage of the annual (since 1994) All Russian Agricultural Meeting *(skhod)* organized by the APR, in *Zemlya i trud*, no. 46, November 14–20, 1995, p. 2.
143. *Zemlya i trud*, no. 48, November 30–December 6, 1993, p. 3; no. 36, September 6–12, 1994, p. 2; no. 16, April 18–24, 1995, p. 2; and no. 24, June 13–19, 1995, p. 1.
144. *Krest'yanskaya Rossiya*, no. 28, July 24–30, 1995, p. 2.
145. *Zemlya i trud*, no. 50, December 13–19, 1994, p. 4.

146. Ibid., no. 40, October 4–10, 1994, p. 3.
147. In the Federation Council, Evengii Savchenko was selected to head the Committee for Agrarian Policy. He was appointed by presidential decree in 1993 to head Belgorad Oblast. In the December 1995 election he was reported to support Chernomyrdin's Russia Is Our Home Party but ran as an independent candidate.
148. ERS, *Former USSR: Situation and Outlook Series*, WRS-94-1 (1994), p. 52.
149. *Rossiyskaya gazeta*, November 11, 1993, p. 2.
150. *Sel'skaya zhizn'*, November 11, 1993, p. 2, and November 20, 1993, p. 1; *Vash partner*, no. 7 (February 1994): 3.
151. *Zemlya i lyudi*, no. 14, April 8, 1994, p. 1.
152. *Krest'yanskiye vedomosti*, no. 11, March 21–27, 1994, pp. 1, 12; *Izvestiya*, July 2, 1994, p. 2.
153. For example, the mayor of Moscow, Yu. Luzhkov, denounced the import tariffs and demanded their immediate elimination. *Krest'yanskiye vedomosti*, no. 33, August 15–21, 1994, p. 12.
154. Although total meat imports in 1994 constituted only about 68 percent of the 1991 level, meat imports in 1994 were over 500 percent of the 1993 level. Economic Research Service, *Former USSR: Situation and Outlook Series*, WRS-95-1 (Washington, D.C.: USDA, 1995), p. 31. Meat imports continued to rise during 1995 and 1996, so that by the end of 1996 total meat imports (beef, pork, and poultry) were nearly 3 times their 1993 level. Poultry comprised the single largest component of meat imports. Poultry imports also accounted for more than one-half of Russian consumption of poultry products. ERS, *Newly Independent States and the Baltics*, p. 15.
155. *Sel'skaya zhizn'*, June 3, 1995, p. 2.
156. *Sel'skoye khozyaystvo Rossii*, p. 125; *Rossiyskiy statisticheskiy ezhegodnik 1996*, p. 349.
157. V. V. Buryukov, "Ob importno-eksportnykh i vnutrennikh tsenakh na sel's-kokhozyaystvennuyu produktsiyu," *Ekonomika sel'skokhozyaystvennykh i pererabatyvayushchikh predpriyatiy*, no. 8 (August 1996): 10–11. In 1994, Russian domestic producers were paid 21 trillion rubles, but the nation spent 26 trillion on imports; in 1995 Russian domestic producers were paid 50 trillion rubles but the nation spent 60 trillion in imports. The primary reason for this is that Russia purchased more meat and "high preference" items that were more expensive.
158. See Stephen K. Wegren, "Rural Politics and Agrarian Reform in Russia," *Problems of Post-Communism* 43, no. 1 (January 1996): 23–34.
159. *Zemlya i trud*, no. 2, January 10–16, 1995, p. 3.

Chapter 5. Land Reform and the Development of Private Farming

1. Privatization in the agricultural sector was multifaceted and included land, food-producing enterprises, food-processing plants, procurement and storage enterprises, and food-marketing channels. Among those aspects the focus here will be on land privatization and food producers because land is the basic unit of production and the fundamental source from which rural stratification would occur. For a description of privatization in other spheres of the agricultural sector, see A. E. Chernomorets, *Pravo sobstvennosti v sel'skom khozyaystve Rossiyskoy Federatsii* (Moscow: Institut gosudarstva i prava, 1993).
2. Barrington Moore, *Social Origins of Dictatorship and Democracy: Lord and Peasant in the Making of the Modern World* (Boston: Beacon Press, 1966).
3. Among other factors was the fact that many regions in the non–black earth zone of Russia suffered a dearth of rural dwellers who were willing to take a risk. Part of the reason was political (many did not believe in the longevity of agrarian reform); part of the reason was innate conservatism and risk-aversion; and part was the result of decades of rural out-migration, which left an extremely weak demographic base. During December 1991 a state farm director on the outskirts of the city of Kostroma remarked that the main problem was "finding someone who wanted to work." See Stephen K. Wegren, "Agricultural Reform in Kostroma Oblast," *Newsletter for Research on Soviet and East European Agriculture* 14, no. 1 (March 1992):

31–35, for a report on findings from that trip. Economists at the Kostroma branch of Goskomstat during interviews seconded this view, explaining that "few people want to become peasant farmers," and maintained in their official report that "peasant farms are clearly linked to collective and state farms and depend on them for support; however, seldom [do peasant farmers] have negative relations from farm directors." *Razvitiye krest'yanskikh khozyaystv oblasti* (Kostroma: Goskomstat of Kostroma, 1991), p. 2.

Part of the reason was the high cost and high risk associated with private farming. Early data on farm income from Kostroma Oblast showed that peasant farm expenditures were on average about 4.5 times higher than income per farm, meaning that, in order to survive, peasant farms would need subsidies and credits also. *Razvitiye krest'yanskikh khozyaystv oblasti v 1991 godu* (Kostroma: Goskomstat of Kostroma, 1992), p. 16.

4. According to the Yel'tsin decree published in March 1991, the term was shortened to a week. *Sel'skaya zhizn'*, March 20, 1991, p. 2.

5. Section 4, article 26, "O zemle," *Pravda*, March 7, 1990, p. 2.

6. Interview with a peasant farmer from Kostromskoy raion, August 1994.

7. See *Rossiyskaya gazeta*, April 27, 1992, p. 2, on farm refusals to reorganize.

8. *Informatsiya o khode zemelnoy reformy na 1.1.93 g.* (Moscow: Komitet po zemel'noy reforme i zemel'nym resursam pri pravitel'stve Rossiyskoy Federatsii, 1993), pp. 86–88. See Stephen K. Wegren, "Trends in Russian Agrarian Reform," *RFE/RL Research Report* 2, no. 13 (March 26, 1993): 55–56, for patterns of fines.

9. Survey of five oblasts in geographically distinct areas in 1992. See Karen Brooks and Zvi Lerman, *Land Reform and Farm Restructuring in Russia*, World Bank Discussion Papers, no. 233 (Washington, D.C.: World Bank, 1994), pp. 66–67.

10. Karen Brooks et al., *Agricultural Reform in Russia: A View from the Farm Level*, World Bank Discussion Papers, no. 327 (Washington, D.C.: World Bank, 1996), pp. 33, 54.

11. *Krest'yanskikh khozyaystv oblasti na 1 Yanvarya 1994 goda (po rezul'tam obsledovaniya)* (Kostroma: Goskomstat of Kostroma, 1994), p. 5.

12. Unpublished information from the Committee on Land Resources, Kostroma Oblast.

13. *Krest'yanskaya Rossiya*, no. 2, January 16–22, 1995, p. 6.

14. For an analysis of early reform legislation, see Stephen K. Wegren, "Private Farming and Agrarian Reform in Russia," *Problems of Communism* 41, no. 3 (May–June 1992): 107–21; and Stephen K. Wegren, "Political Institutions and Agrarian Reform in Russia," in Don Van Atta, ed., *The Farmer Threat: The Political Economy of Agrarian Reform in Post-Soviet Russia* (Boulder, Colo.: Westview Press, 1993), pp. 121–47. A more recent review of relevant legislation may be found in Zvi Lerman and Kren Brooks, "Russia's Legal Framework for Land Reform and Farm Restructuring," *Problems of Post-Communism* 43, no. 6 (November–December 1996): 48–58.

15. "O krest'yanskom khozyaystve" (draft), *Ekonomika i zhizn'*, no. 28 (July 1990): 15–17; and "O krest'yanskom khozyaystve," *Sbornik zakonodatel'nykh aktov, postanovleniy ukazaniy i rekomendatsiy po zemel'noy reforme i krest'yanskim (fermerskim) khozyaystvam*, chast' 1 (Moscow: Gosudarstvennyy komitet RSFSR po zemel'noy reforme, 1991), pp. 20–47.

16. See article 2 of the law, "O zemel'noy reforme," *Sbornik zakonodatel'nykh aktov, postanovleniy ukazaniy i rekomendatsiy po zemel'noy reforme i krest'yanskim (fermerskim) khozyaystvam*, pp. 13–14. The final form of this law was then adopted with changes on December 27, 1990.

17. "O sobstvenosti v RSFSR," *Ekonomika i zhizn'*, no. 3 (January 1991): 13–14.

18. *Krest'yanskiye (fermerskiye) khozyaystva Rosskiysoy Federatsii (po dannym obsledovaniya na 1 Yanvarya 1992 goda)* (Moscow: Goskomstat Rosskiyskoy Federatsii, 1992), p. 37. Vladeniye was even more popular in the non–black earth zone, with more than 69 percent of private farms registered on this basis.

19. *Krest'yanskie (fermerskiye) khozyaystva Rosskiysoy Federatsii (po dannym obsledovaniya na 1 Yanvarya 1993 goda)* (Moscow: Goskomstat Rosskiyskoy Federatsii, 1993), p. 28.

20. "Zemel'nyy kodeks RSFSR," *Krest'yanskiye vedomosti*, no. 16, May 21, 1991, pp. 3–11.

21. For extracts of the Civil Code that pertain to landownership, see "Grazhdanskiy kodeks Rossiyskoy Federatsii," *Krest'yanskaya Rossiya*, no. 3, January 23–29, 1995, pp. 17–20.
22. See section 2, article 4, of this law.
23. Section 2, article 4, point 1, "O krest'yanskom (fermerskom) khozyaystve." The draft law included language that indicated preference would be given to those families with many children. Section 2, article 1, "O krest'yanskom khozyaystve" (draft), *Ekonomika i zhizn'*, no. 28 (July 1990): 15. This provision was dropped in the final version of December 1990.
24. *Krest'yanskaya Rossiya*, no. 13, September 14, 1991, p. 3.
25. *Izvestiya*, December 29, 1992, p. 1.
26. *Zemlya i lyudi*, no. 20, May 21, 1993, p. 1.
27. "O poryadke organizatsii krest'yanskikh (fermerskikh) khozyaystv v Moskovoy oblasti," *Vestnik Moskovskogo oblastnogo soveta narodnykh deputatov*, no. 18 (September 1993): 10–17.
28. Ibid.; and *Rossiyskiy fermer*, no. 27, August 10–16, 1993, p. 3.
29. Private farms based on leased land were regulated by article 13 of the Land Code, which stipulated that leases were obtained through the local Soviet of People's Deputies and the landowner, which in most cases meant a state or collective farm. Rent was negotiable but could not exceed the land tax, and once-owned land could be converted into leased land only under certain conditions such as the owner's temporary incapacity to work, service in the military, or entrance into an educational facility. The size of the land plot to be leased was limited by the lessee's ability to pay land rent and the land tax, not by any legal restriction. "Zemel'nyy kodeks RSFSR," p. 4.
30. *Izvestiya*, January 7, 1991, p. 1.
31. Point 1, "O dopolnitel'nykh merakh po uskoreniyu provedeniya zemel'noy reformy v RSFSR," *Sel'skaya zhizn'*, March 20, 1991, p. 2.
32. Section 2, article 6, point 1, "O krest'yanskom (fermerskom) khozyaystve."
33. Ibid., section 2, article 5, point 3.
34. It is important to note a rough correlation between quantity of land available, location, and quality. The more remote the location and the lower the land quality, the larger the quantity of land available per raion.
35. The Law on Land stated that non-farm citizens who wanted land could receive land according to the stipulations established in article 38 of the law. Section 3, article 26, "O zemle." Article 38 then defined this land fund as "all land not assigned ownership rights or in constant use." Section 8, article 38, "O zemle."
36. Section 3, chapter 10, article 60, "Zemel'nyy kodeks RSFSR," p. 7.
37. "Otchet o nalichii i ispol'zovaniya fonda pereraspredeleniya zemel'," unpublished information from the Committee on Land Resources, Moscow Oblast.
38. "Otchet o nalichii i ispol'zovaniya fonda pereraspredeleniya zemel'," unpublished information from the Committee on Land Resources, Kostroma Oblast.
39. Calculated from *Itogi proizvodstvenno-finansovoy deyatel'nosti sel'skokhozyaystvennykh predpriyatiy Kostromskoy oblasti za 1994 god* (Kostroma: Goskomstat of Kostroma, 1995), p. 5.
40. See Stephen Wegren, "Early Trends in Peasant Farming in the RSFSR," *Newsletter for Research on Soviet and East European Agriculture* 13, no. 4 (December 1991): 16–17.
41. "O plate za zemlyu," *Krest'yanskiye vedomosti*, no. 39, October 29, 1991, insert pp. 1–8.
42. Land taxes in Russia are regulated by the federal law "On Payment for Land" which entered into force on January 1, 1992. The text of the original law "On Payment for Land" and subsequent government resolutions are reprinted in *Krest'yanskiye vedomosti*, no. 19, May 13–19, 1996, insert pp. 7–10. It was subsequently amended in July 1994 by the Duma and Federation Council, changes that were approved in August 1994 by Yeltsin, who signed the new version of the law. See I. Vyskrebentsev, "Plata za zemlyu," *APK: ekonomika, upravleniye*, no. 8 (August 1995): 38–49, for an analysis of the changes. Normative prices for land and land taxes are regulated by government resolutions no. 1204, issued in November 1994, resolution no. 562 issued in June 1995, and resolution no. 378 issued in April 1996; and by a series of

"instructions" issued by the State Tax Service of the Russian Federation. A list of instructions from the tax service are found in *Krest'yanskiye vedomosti*, no. 33, August 19–25, 1996, insert p. 7. A new law on the payment for land was expected to be drafted by February 1997, but at the time of this revision had not yet been published. For an analysis of the land taxes in urban and rural sectors and for the land market in general, see Stephen K. Wegren, "Land Reform and the Land Market in Russia: Operations, Constraints, and Prospects," *Europe-Asia Studies* 49, no. 6 (1997): 959–87.

43. See Wegren, "Private Farming and Agrarian Reform in Russia," esp. pp. 112–14.

44. "O neotlozhnykh merakh po osushchestvleniyu zemel'noy reformy v RSFSR," *Rossiyskaya gazeta*, December 31, 1991, p. 3.

45. Interview with Sergei Zanozin, Kostroma Gorkomzem, July 1995.

46. The decree was necessary in order to secure property rights for landshare holders. Many persons who had received land did not possess a land title or deed to their land. Land titles would not only ensure ownership but were also necessary if land was to be exchanged in a land market. Thus, one of the notable features of the decree was to order the distribution of land deeds. "O regulirovanii zemel'nykh otnosheniy i razvitiiagrarnoy reformy v Rossii," *Krest'yanskiye vedomosti*, no. 43, November 1–7, 1993, pp. 8–9.

47. For a complete analysis of the decree, see Stephen K. Wegren, "Yel'tsin's Decree on Land Relations: Implications for Agrarian Reform," *Post-Soviet Geography* 35, no. 3 (March 1994): 166–78.

48. The certificate was based upon shares that had been previously established during farm reorganization. That is, property shares were based upon length of service to the farm and amount of work performed, although land shares were equal in size to all members of the farm.

49. Owners of land shares were free to voluntarily combine land and land shares, even among nonrelated people, thus overcoming past restrictions that allowed only the creation of family farms.

50. Unpublished data from Kostroma Committee on Land Resources. The landowner had to initiate action to receive his land certificate by going to the Oblast Committee on Land Resources. In Kostroma, land shareholders totaled over 73,000 persons, and personal plot holders totaled over 153,000 individuals. Persons who held land leases were not eligible for ownership certificates.

51. *Krest'yanskiye vedomosti*, no. 38, September 23–29, 1996, p. 10.

52. "O privedenii zemel'nogo zakonodatel'stva Rossiyskoy Federatsii v sootvetstviye s Konstitutsiey Rossiyskoy Federatsii," *Krest'yanskaya Rossiya*, nos. 2–3, January 24–30, 1994, p. 11; and for an explanation of this decree, see *Krest'yanskiye vedomosti*, nos. 20–21, May 23–29, 1994, p. 8.

53. Point 4, "O neotlozhnykh merakh po osushchestvleniyu zemel'noy reformy v RSFSR," *Rossiyskaya gazeta*, December 31, 1991. Land norms were essentially land-to-labor ratios within a given raion, which meant that the total amount of farmland was divided by the number of participants in distribution. For the sake of simplicity, the sum of rural dwellers was used as the basis for "participants." The government's Resolution "On the Procedure for Reorganizing Collective and State Farms" of December 29, 1991, stipulated that state and collective farm members, as well as those on pension, were eligible to receive a free land plot; and with the approval of the farm collective, so too were workers in the social sphere who worked on the farm. Point 9, "O poryadke reorganizatsii kolkhozov i sovkhozov," *Zemlya i lyudi*, no. 2, January 10, 1992, p. 2. For farm members, these norms remained in place even after Soviet-era political institutions were disbanded as a result of the events of October 1993.

54. *Krest'yanskaya Rossiya*, no. 32, August 23–29, 1993, p. 6.

55. Interview with officials at Kostroma Gorkomzem, August 1994. For Moscow, see *Kak poluchit' zemel'nyy uchastok v podmoskov'ye* (Moscow: Agrodom, 1992).

56. *Krest'yanskaya Rossiya*, no. 26, July 4–10, 1994, p. 6.

57. *Sel'skaya zhizn'*, August 27, 1988, p. 1; and *Pravda*, March 16, 1989, pp. 1–4.

58. *Izvestiya*, March 7, 1990, pp. 1–2.
59. Points 12 and 13, "O neotlozhnykh merakh po osushchestvleniyu zemel'noy reformy v RSFSR."
60. *Rossiyskaya gazeta*, October 15, 1992, p. 5; and for an analysis, see *Sel'skaya zhizn'*, November 24, 1992, p. 1.
61. *Sel'skaya zhizn'*, October 27, 1992, p. 1.
62. Ibid., December 31, 1992, p. 1.
63. In December 1993 a presidential decree was signed that defined the tax obligations for sellers of land. See *Krest'yanskaya Rossiya*, no. 49, December 20–26, 1993, p. 2.
64. *Krest'yanskaya Rossiya*, no. 29, August 2–8, 1993, pp. 1–2. There were other restrictions that did not affect the creation of a land market. For example, a peasant farmer had to resign from his previous job "at the moment" he registered the creation of his personal farm. In addition, free plots of land for personal plots henceforth would be assigned only to agricultural enterprises and rural dwellers; urban dwellers no longer were eligible to receive land for personal plots (which is not to say they could not operate plots around their house or dacha; and in any event urbanites could purchase land for personal plot operation).
65. "O regulirovanii zemel'nykh otnosheniy i razvitiiagrarnoy reformy v Rossii," *Krest'yanskiye vedomosti*, no. 43, November 1–7, 1993, pp. 8–9.
66. *Krest'yanskaya Rossiya*, no 29, July 25–31, 1994, p. 6.
67. Interview with P. Tutun, Committee on Land Reform, Kostroma Oblast, June 1995.
68. See the article by Vasiliy Starodubtsev, former head of the Agrarian Union in *Rossiyskiy fermer*, no. 40, November 9–16, 1993, p. 2.
69. See the article by Evgeniya Serova in ibid., p. 3; and interview with B. Varenov, head of the Committee on Land Resources in Moscow Oblast, August 1994.
70. Yavlinsky's Yabloko Party and Zhirinovsky's Liberal Democratic Party did not vote.
71. *Izvestiya*, July 15, 1995, p. 1. For an overview of some important points in the draft land code, see *Krest'yanskaya Rossiya*, no. 8, February 28–March 6, 1994, pp. 1–2, and *Rossiyskiy fermer*, no. 20, May 24–30, 1994, p. 2. For a draft of the new Law on Peasant Farming, see *Rossiyskiy fermer*, no. 4, February 1–7, 1994, pp. 4–5.
72. *Kommersant*, no. 33, September 12, 1995, p. 4.
73. *Sel'skaya zhizn'*, April 13, 1996, p. 2.
74. *Krest'yanskaya Rossiya*, no. 28, July 22–28, 1996, p. 1.
75. *Sel'skaya zhizn'*, December 15, 1996, p. 2.
76. *Rossiyskiy fermer*, no. 61, October 29, 1996, p. 2.
77. Jamestown Monitor, June 11, 1997; Johnson's Russia List, June 11, 1997. The revised code would allow individuals or legal entities to own or lease land that previously belonged to municipalities. Former members of state or collective farms could own or lease farmland, but its purchase and sale would still be prohibited. The code also would not allow foreigners to buy land, although foreigners could lease land for a maximum of 50 years.
78. Interview with Yuriy Lebedev, head of the APR in Kostroma Oblast, July 1995.
79. In Kostroma, at least, private farmers talked of peasant associations doing little more than giving money to friends and bureaucrats, or of doing "illegal" things with farmer dues. There was evidence of corruption at the national level as well, as questions were raised over the method and uses of AKKOR monies.
80. "Gosudarstvennaya programma podgotovki fermerov v Rossiyskoy Federatsii," *APK: ekonomika, upravleniye*, no. 6 (June 1993): 8–10.
81. For more on agricultural education, see Organization for Economic Cooperation and Development, "Review of the Agricultural Education and Training System of Russia," working document dated January 7, 1994.
82. As of early 1994 there were some fifty-two agricultural institutes in Russia, but only eight universities.
83. Interview with director, Academy of Management and Agrobusiness, August 1994 and July 1995. The courses were free of charge to those who enrolled, paid for by their organization.

84. Of the 188 lecturers hired in 1993, 34 came from the Kostroma Agricultural Institute, 16 from other research institutes, 44 from agricultural enterprises, 66 from the agricultural administration, 24 from other places, and 4 from the academy. Unpublished records from the Academy of Management and Agrobusiness, Kostroma Oblast.

85. Unpublished "Uchebnyy" plan (or curriculum) from the Kostroma Academy of Management and Agrobusiness, Kostroma Oblast.

86. For general information and advice, see, for example, A. M. Kayyali and N. S. Kharitonov, *Posobiye po ekonomike i organizatsii krest'yanskogo (fermerskogo) khozyaystva* chast' 1–2 (Moscow: Nauchno-vnedrencheskiy tsentr "Selo," 1990); *Bibliotechka fermera* (Moscow: Informagrotex, 1991); V. A. Udalov, *Nastol'naya kniga Rossiyskogo fermera (osnovnyye dokumenty po sozdaniyu i deyatel'nosti krest'yanskogo khozyaystva s kommentariyami)* (Moscow: Znaniye, 1993); and for a technical assistance handbook, see *Spravochnik fermera* (Moscow: Informagrotex, 1992).

87. See Wegren, "Private Farming and Agrarian Reform in Russia," pp. 118–20.

88. *Krest'yanskiye vedomosti,* no. 21, May 31–June 5, 1993, p. 10.

89. Ibid., no. 33, September 17, 1991, p. 9.

90. *Krest'yanskaya Rossiya,* no. 13, September 14, 1991, p. 3.

91. Cited in Kenneth Gray, "Farm Privatization in Georgia, the Baltics and the CIS with Focus on Russia: A Review," *Economies in Transition Agriculture Update* 4, no. 5 (September–October 1992): 48.

92. *Razvitiye krest'yanskikh khozyaystv oblasti na 1 Yanvarya 1993 goda (po rezul'tam obsledovaniya)* (Kostroma: Goskomstat of Kostroma, 1993), p. 5.

93. Unpublished information from Kaluga branch of Goskomstat. It is interesting that only 57 farms were created by former urbanites who relocated into the countryside after receiving a plot of land, whereas another 114 farms were begun by urbanites who remained in a city even after receiving a plot of land for farming.

94. *Itogi khozyaystvennoy deyatel'nosti krest'yanskikh (fermerskikh) khozyaystv Rossiyskoy Federatsii v 1993 godu* (Moscow: Goskomstat Rossii, 1994), p. 4.

95. See Peter R. Craumer, "Regional Patterns of Agricultural Reform in Russia," *Post-Soviet Geography* 35, no. 6 (June 1994): 329–51.

96. See, for instance, the case of Samara Oblast, in *Izvestiya,* November 27, 1992, p. 2.

97. Nor was increased output simply a function of more farms producing. Average output per farm increased between 1992 and 1993 for grain, sugar beets, and milk, while declining for potatoes and vegetables and remaining constant for meat and poultry production.

98. In a survey of peasant farms in Kaluga Oblast during 1993 farmers responded that the primary factors hindering the development of private farming were the following, in order of magnitude: (1) high prices for agricultural machinery and construction materials; (2) high interest rates for credit; (3) shortages of specialized machinery, seed, fertilizer, and so on; (4) lack of faith in the long-term character of land laws and agrarian reform; (5) absence of roads, communications, water, electricity, and gas. Data is taken from unpublished information from Kaluga branch of Goskomstat. The lack of faith in the longevity of agrarian reform was also expressed by 51 percent of private farmers surveyed in Kostroma Oblast during 1994, up from 40 percent in 1993.

99. Goskomstat estimates are used and cited more commonly than those of AKKOR. Private farmers in 1995 had about 5 percent of total agricultural land and 6 percent in 1996.

100. *Krest'yanskiye (fermerskiye) khozyaystva Rossiyskoy Federatsii (po dannym obsledovaniya na 1 Yanvarya 1993 goda)* (Moscow: Goskomstat Rossiyskoy Federatsii, 1993), pp. 64–65.

101. *Razvitiye krest'yanskikh khozyaystv oblasti* (Kostroma: Goskomstat of Kostroma, 1991), p. 3.

102. *Krest'yanskiye khozyaystva oblasti na 1 Yanvarya 1995 goda* (Kostroma: Goskomstat of Kostroma, 1995), p. 4.

103. Wegren, "Early Trends in Peasant Farming in the RSFSR," pp. 16–18.

104. *Izvestiya,* November 25, 1992, p. 1.

105. Nancy J. Cochrane, "Farm Restructuring in Central and Eastern Europe," *Soviet and Post-Soviet Review* 21, nos. 2–3 (1994): 319–35.

106. Land plot norms did not change after soviets were abolished in October 1993 but were retained by the "administrations" that followed.

107. Interview with P. Tutun, Committee on Land Resources, June 1994.

108. For example, see *Sel'skaya zhizn'*, April 29, 1994, p. 2, in which a farmer received seventeen hectares spread among five different locations.

109. Brooks and Lerman, *Land Reform and Farm Restructuring in Russia*, p. 56.

110. Brooks and Lerman, *Land Reform and Farm Restructuring in Russia*, p. 48.

111. See Stephen K. Wegren, "Rural Reform and Political Culture in Russia," *Europe-Asia Studies* (formerly *Soviet Studies*) 46, no. 2 (1994): 224, on raion-level norms for land distribution throughout Kostroma Oblast.

112. Additional land could also be leased without restriction as to size. One private farmer's farm in Salsk raion totaled over fifteen hundred hectares in 1993. Land rent (per hectare) could not exceed the land tax, which made land very cheap and thus made leasing quite advantageous. In Kostroma raion during 1992 additional land could be purchased for 5,000 rubles per hectare (about $22 per hectare). Interview with P. Tutun, official in Kostroma Oblast Committee for Land Reform, June 1992.

113. *Interfax Food and Agriculture Report* 5, no. 6 (February 2–9, 1996), chap. 3.

114. R. Kuchukov, "Mnogoukladnost' i razvitiye rynochnoy ekonomiki," *APK: ekonomika, upravleniye*, no. 10 (October 1995): 58.

115. *Rosskyskiye vesti*, December 29, 1992, p. 1.

116. Wegren, "Rural Reform and Political Culture in Russia," pp. 232–33.

117. Article 8, point 2, "O krest'yanskom khozyaystve" (draft), *Ekonomika i zhizn'*, no. 28 (July 1990). There were different methods used to calculate labor input. One method simply totaled all wages received during employment; another method took the average of the highest five years of income; a third method averaged wages during the past three years.

118. See Brooks and Lerman, *Land Reform and Farm Restructuring in Russia*, p. 49.

119. "Polozheniye o reorganizatsii kolkhozov, sovkhozov i privatizatsii gosudarstvennykh sel'skokhozyaystvennykh predpriyatiy," *Ekonomika i zhizn'*, no. 41 (October 1992): insert *Vash partner*, p. 6.

120. *Krest'yanskaya Rossiya*, no. 91, September 26, 1992, pp. 2, 4.

121. After the national parliament and local governments were abolished in the fall of 1993, local administrations replaced the soviets of the communist era. There are four types of local administrations (local government) that have jurisdiction over land within their boundaries: city, rural, oblast, and village. The boundaries of each administration's jurisdiction are contiguous with another's boundaries. In case of disagreement over land use between the landowner and the local administration, the governor of the oblast would adjudicate land use.

122. *Krest'yanskiye vedomosti*, no. 1, January 3–9, 1994, p. 5.

123. Interview with Kostroma Oblast land officials, Committee on Land Resources, June 1995.

124. See Stephen K. Wegren, "The Development of Market Relations in Agricultural Land: The Case of Kostroma Oblast," *Post-Soviet Geography* 36, no. 8 (October 1995): 496–512.

125. For descriptions of early urban land market sales in Moscow, see *Izvestiya*, July 8, 1994, p. 1, and November 5, 1994, p. 5.

126. *Rossiyskaya gazeta*, February 17, 1995, p. 8.

127. According to Boris Varenov, chairman of the Committee on Land Resources in Moscow Oblast whom I interviewed during the summer of 1994, the starting price of a land plot was set by the seller, without any interference from local officials. In addition, Varenov claimed there were no limits on the size of a land plot. But he did point out the existence of a "black market" in land sales. He defined a process whereby the actual negotiated price between buyer and seller was much higher than the officially registered price, from which taxes

and registration fees were calculated. "Profit" from a privatized land plot sale is taxed as income. In addition, the local committee for land resources charges a transaction fee on a sliding scale, depending on the value of the land. In Kostroma the transaction fee averaged 3 percent of the value of the sale. Interview with P. Tutun, Committee on Land Resources, Kostroma Oblast, August 1994.

128. Unpublished data from Committee on Land Resources, Moscow Oblast.

129. Moscow Oblast (without factoring in the city of Moscow) had 144 persons per square kilometer; if the city of Moscow were to be included, the population density would be 333 persons per square kilometer in 1994.

130. Interviews with P. Tutun, Committee on Land Resources, Kostroma Oblast, August 1994, and June, July, and August 1995; and unpublished information from Committee on Land Resources, Kostroma Oblast. Of those transactions, 8 were concluded by "juridical persons" (that is, organizations and enterprises) and the other 1,049 by individuals.

131. There are twenty-four raions in Kostroma Oblast, but six raions had no land transactions from January 1994 to July 1995.

132. Unpublished information from the Kostroma Committee on Land Resources.

133. Registered land prices were in all likelihood understated and more sales took place than were registered. According to the estimates of Mr. Varenov, actual land prices ranged from $200 to $10,000 per hundred square meters *(sotka)* in the summer of 1994, with higher prices being asked for land closer to the city. Prices and demand for land dropped off considerably as the distance from Moscow increased, declining to about $90 per sotka in raions farther out and even less in surrounding oblasts. *Rossiyskiye vesti,* June 29, 1995, p. 3. For a description of how the state valued land transactions for tax purposes and how the state responded to tax evasion, see Wegren, "The Development of Market Relations in Agricultural Land," pp. 502–05.

134. V. Kuznetsov, "Pochemu neobkhodimo gosudarstvennoye regulirovaniye razvitiya APK," *APK: ekonomika, upravleniye,* no. 5 (May 1995): 61.

135. *Zemlya i lyudi,* no. 31, August 5, 1994, pp. 2–3.

136. See Stephen K. Wegren, Gregory Ioffe, and Tatyana Nefedova, "Demographic and Migratory Responses to Agrarian Reform in Russia," *Journal of Communist Studies and Transition Politics* 13, no. 4 (December 1997) pp. 54–78.

137. *Trud,* October 10, 1995, p. 2; translated in Foreign Broadcast Information Service, *Daily Report Supplement: Central Eurasia,* December 8, 1995, p. 23. A survey of reorganized and nonreorganized farms in Rostov, Nizhniy Novgorod, and Orel Oblasts by the Agrarian Institute found that less than 5 percent of farm members desired to sell their land shares. On reorganized farms 80–90 percent of members leased their land shares, but most often to the parent farm. See V. Ya. Uzun, ed., *Sotsial'no-ekonomicheskiye posledstviya privatizatsii zemli i reorganizatsii sel'skokhozyaystvennykh predpriyatiy (1994–1996)* (Moscow: Entsiklopediya rossiyskikh dereven', 1996), pp. 85–90.

138. *New York Times,* March 17, 1996, pp. A1, A8.

139. See, for example, *Krest'yanskaya Rossiya,* no. 22, June 6–12, 1994, p. 1.

140. By early 1995 this problem had been identified as a primary restraint on the private farming sector. See the resolutions passed at the sixth congress of AKKOR in March 1995. *Rossiyskiy fermer,* nos. 9–10, March 14–20, 1995, p. 3.

Chapter 6. Financial Levers and the Impact on Private Farming

1. See Roy L. Prosterman and Timothy Hanstad, *An Update on Individual Peasant Farming in the USSR,* RDI Monographs on Foreign Aid and Development, no. 8 (Rural Development Institute, Seattle, 1991), pp. 15–16.

2. Due to inflation the average monthly salary in 1991 had jumped to about 580 rubles for an industrial worker, up from about 300 in 1990. Rural wages increased from 307 to 483 rubles a month. The point is that, even at these inflated wages, few persons had large bank accounts, after many years of wages less than 200 rubles a month. In 1992 the average bank account was just over 4,000 rubles, not nearly enough to meet start-up costs.

3. *Izvestiya,* October 28, 1991, p. 2.
4. "O merakh gosudarstvennoy podderzhki krest'yanskikh (fermerskikh) khozyaystv v 1992," *Pravitel'stvennyy vestnik,* no. 7 (February 1992): 2.
5. Unpublished data from officials in the Russian Farmer program. "Other" was not defined.
6. *Rossiyskiy fermer,* no. 7, March 23–29, 1993, p. 1.
7. *V pomoshch' nachinayushchemu fermeru* (Moscow: Association "Dukhovnoye dostoyaniye derevni," 1992), pp. 174–75.
8. Loan guarantee forms are reprinted in *Bank i fermer* (Moscow: Informagrotekh, 1994), pp. 38–40.
9. For local examples on how the system worked, see Stephen K. Wegren, "Agricultural Reform in the Nonchernozem Zone: The Case of Kostroma Oblast," *Post-Soviet Geography* 33, no. 10 (December 1992): esp. 669–71.
10. *Rossiyskiy fermer,* no. 23, July 13–19, 1993, p. 2.
11. Ibid., no. 22, July 6–12, 1993, p. 3, described the Moscow peasant union's attempt to assign a higher percentage of credit based on the number of farms, and less based on the quantity of hectares. This proposal was not approved, and the old percentage guidelines remained intact.
12. There were also regional farmers' banks. Only begun in the second half of 1992, twelve such banks were in operation at the end of 1993 and through October 1993 had distributed 16 billion rubles in short-term credits to farmers. *Rossiyskiy fermer,* no. 47, December 28, 1993–January 3, 1994, p. 2. During the latter half of 1993 and throughout 1994, farmers' banks provided loans at about one-half the interest rate established by the Central Bank. The July 27, 1993, presidential decree suggested channeling state credits through farmers' banks, something that private farmers had repeatedly called for. In early 1994, AKKOR opened its own bank and hoped to open regional branches as well. *Sel'skaya zhizn',* February 8, 1994, p. 3.

These banks represented other options for peasant farmers to receive credits. Even so, there was evidence that farmers' banks did not give loans to all farmers, just as AKKOR was accused of backing loans based on contacts and friendships. In one publicized case a farmers' bank would grant loans only to farmers who deposited money in the bank or became shareholders or founders. See *Rossiyskiy fermer,* no. 14, October 19–25, 1993, p. 1.

13. *Ekonomika i zhizn',* no. 25 (June 1993): 16.
14. It is difficult to convert the dollar equivalent because credits were distributed throughout the year, during which exchange rates were constantly changing. For example, at the beginning of 1992 the exchange rate was about 198 rubles to the dollar, but by the end of 1992 the rate had fallen to 415 rubles to the dollar. If we use the average exchange rate for 1992, then 6.5 billion rubles would equal about $29 million.
15. See the government resolution signed by R. Khasbulatov on March 12, 1992, that compensated 50 percent of fuel costs for agricultural producers. *Rossiyskaya gazeta,* March 19, 1992, p. 2; Yel'tsin's decree of April 1992, which ordered subsidies be provided for animal husbandry products, effective May 1, 1992, "O merakh po stabilizatsii ekonomiki agropromyshlennogo kompleksa," *APK: ekonomika, upravleniye,* no. 7 (July 1992), p. 5. The subsequent government resolution on subsidies that implemented Yel'tsin's decree is found in *Krest'yanskiye vedomosti,* no. 21 (May 19–25, 1992), p. 7. In July 1992 subsidies for animal husbandry products were increased significantly. See *Krest'yanskiye vedomosti,* no. 31 (August 4–10, 1992), p. 2.
16. *Rossiyskaya gazeta,* April 9, 1992, p. 2.
17. For tax obligations of private farmers, see *Krest'yanskaya Rossiya,* no. 15, April 26–May 2, 1993, insert on laws p. 11; and *Krest'yanskiye vedomosti,* no. 13, April 5–11, 1993, p. 4.
18. "O merakh gosudarstvennoy podderzhki krest'yanskikh (fermerskikh) khozyaystv v 1992 godu," *Ekonomika sel'skokhozyaystvennykh i pererabatyvayushchikh predpriyatiy,* no. 9 (September 1992): 27–28.
19. "Dogovor," *Krest'yanskiye vedomosti,* no. 8, February 18–24, 1992, p. 3.
20. *Rossiyskiy fermer,* no. 5, March 9–15, 1993, p. 1.
21. "O chrezvychaynykh merakh finansovoy podderzhki agropromyshlennogo kompleksa

Rossiyskoy Federatsii," *APK: ekonomika, upravleniye,* no. 4 (April 1993): 4–5.

22. "O dopolnitel'nykh merakh po stabilizatsii agropromyshlennogo proizvodstva v 1993 godu," *APK: ekonomika, upravleniye,* no. 7 (July 1993): 21–22.

23. *Rossiyskiy fermer,* no. 5, March 9–15, 1993, p. 1.

24. "O nekotorykh merakh po podderzhke krest'yanskikh (fermerskikh) khozyaystv i sel'skokhozyaystvennykh kooperativov," *Krest'yanskiye vedomosti,* no. 30, August 2–8, 1993, p. 10.

25. *Krest'yanskiye (fermerskiye) khozyaystva Rossiyskoy Federatsii (po dannym obsledovaniay na 1 Yanvarya 1992 goda)* (Moscow: Goskomstat Rossiyskoy Federatsii, 1992), p. 12.

26. *Izvestiya,* July 27, 1992, p. 2.

27. Ibid., November 25, 1992, p. 1.

28. *Razvitiye krest'yanskikh (fermerskikh) khozyaystv Rostovskoy oblasti v 1991–1992 g.g.* (Rostov-on-Don: Rostovskoye oblastnoye upravleniye statistiki, 1993), p. 40.

29. *Krest'yanskiye (fermerskiye) khozyaystva Rossiyskoy Federatsii (po dannym obsledovaniya na 1 Yanvarya 1992 goda),* pp. 86–87.

30. The Law on Peasant Farms allowed private farmers to hire labor. However, low farm incomes constrained the ability to hire labor. According to the law, wages for hired labor were to be paid first and the wages were not dependent on harvest results unless covered by a special agreement. The size of the wage could not be lower than the minimum salary established by law. If a private farm did not make a profit then wages could be covered by bank credits, land mortgage, selling of property, or selling of part of the farm's land. See *Krest'yanskaya Rossiya,* no. 47, December 6–12, 1993, p. 7. On average, private farms used hired laborers only a few days a year because of the severe financial constraints faced by most private farms. In Kostroma Oblast the average in 1994 was ten man-days per year, whereas nationwide in 1993 the average was twenty-one man-days.

31. *Razvitiye krest'yanskikh (fermerskikh) khozyaystv Rostovskoy oblasti v 1991–1992 gg.,* p. 36.

32. *Rossiyskiy fermer,* no. 16, May 25–31, 1993, p. 1.

33. For results from a 1994 World Bank–sponsored survey in five oblasts, see Karen Brooks et al., *Agricultural Reform in Russia: A View from the Farm Level,* World Bank Discussion Papers, no. 327 (Washington, D.C.: World Bank, 1996), p. 54.

34. *Izvestiya,* May 7, 1992, p. 2.

35. *Krest'yanskaya Rossiya,* no. 22, June 6–12, 1994, p. 12.

36. *Rossiyskiy fermer,* no. 3, January 25–31, 1994, p. 3.

37. *Finansovyye izvestiya,* no. 15, April 14–20, 1994, p. 3.

38. *Rossiyskiy fermer,* no. 7, February 22–March 1, 1994, p. 1.

39. *Izvestiya,* October 23, 1992, p. 1.

40. *Krest'yanskiye (fermerskiye) khozyaystva Rossiyskoy Federatsii (po dannym obsledovaniya na 1 Yanvarya 1993 goda)* (Moscow: Goskomstat Rossiyskoy Federatsii, 1993), p. 52.

41. *Krest'yanskaya Rossiya,* no. 42, November 1–7, 1993, p. 2; and *Rossiyskiy fermer,* no. 9, April 6–12, 1993, p. 2.

42. *Krest'yanskaya Rossiya,* no. 34, September 6–12, 1993, p. 3.

43. *Rossiyskaya gazeta,* October 2, 1992, p. 1.

44. *Krest'yanskiye vedomosti,* no. 48, December 6–13, 1993, p. 4.

45. Unpublished data obtained from AKKOR, Kostroma branch, June 1993.

46. Interview with M. Berulin, president of AKKOR in Kostroma Oblast, July 1993. Others told me it had to do with poor leadership and management in AKKOR.

47. *Rossiyskiy fermer,* no. 23, July 13–19, 1993, p. 1; *Krest'yanskaya Rossiya,* no. 16, April 25–May 1, 1994, p. 2.

48. *Rossiyskiy fermer,* no. 2, January 18–24, 1994, p. 2.

49. See complaints about difficulties in receiving credit at the fifth congress of AKKOR in February 1994 in *Rossiyskaya gazeta,* February 16, 1994, p. 2; *Rossiyskiy fermer,* no. 5, February 8–14, 1994, p. 1, and no. 6, February 15–21, 1994.

50. *Izvestiya,* July 1, 1995, p. 1.

51. See the three-part article on conflicts between farmers in Lenigrad Oblast in *Krest'yan-*

skiye vedomosti, no. 35, September 6–12, 1993, p. 4; no. 36, September 13–19, 1993, p. 4; and no. 37, September 20–26, 1993, p. 4.

52. Interview, Council of Federation building, August 1994.

53. *Krest'yanskiye vedomosti,* no. 7, February 21–27, 1994, p. 2.

54. Ken Gray and Yuri Markish, "Russian Land Privatization: Two Decrees Forward, One Decree Backward?" *Newsletter for Research on Soviet and East European Agriculture* 14, no. 1 (March 1992): 12–13.

55. Stephen K. Wegren, "Trends in Russian Agrarian Reform," *RFE/RL Research Report* 2, no. 13 (March 26, 1993): 46–57.

56. Stephen K. Wegren, "Rural Reform in Russia," *RFE/RL Research Report* 2, no. 43 (October 29, 1993): 43–53.

57. *Krest'yanskiye vedomosti,* no. 5, February 8–14, 1993, p. 4.

58. Grain production was chosen as an indicator because it is the most common of private farm activities, undertaken by more than 71 percent of all private farms in Russia during 1992. See *Krest'yanskiye (fermerskiye) khozyaystva Rossiyskoy Federatsii (po dannym obsledovaniya na 1 Yanvarya 1993 goda),* pp. 31–33. Overall, grain production by private farmers was rather insignificant during 1992, accounting for only about 2 percent of all grain harvested and about 3 percent of all state grain purchases. Among regions in European Russia, the Northern Caucasus, Urals, and Volga regions were the most productive grain areas for peasant farmers. In addition, regions with the highest yields per hectare included the Northern Caucasus, Central black earth region, and the Volga-Vyatka region, although the yields even in these areas were below those attained by state and collective farms. During 1992, the three regions with the fewest private farms and lowest grain production received the fewest credits (the Northwestern, Volga-Vyatka, and Northern regions). *Itogi khozyaystvennoy deyatel'nosti krest'yanskikh (fermerskikh) khozyaystv Rossiyskoy Federatsii v 1992 godu* (Moscow: Goskomstat Rossii, 1993), pp. 23, 42.

59. *Krest'yanskaya Rossiya,* no. 10, March 22–28, 1993, p. 2.

60. *Izvestiya,* August 10, 1993, p. 1.

61. *Krest'yanskiye vedomosti,* no. 34, August 30–September 5, 1993, p. 1.

62. *Rossiyskiy fermer,* no. 36, October 12–18, 1993, p. 1.

63. *Izvestiya,* October 26, 1993, p. 1.

64. *Krest'yanskaya Rossiya,* no. 16, April 25–May 1, 1994, p. 2; information on 1994 interest rates from interview with N. Nikitov, Kostroma Agricultural Administration, August 1994.

65. *Krest'yanskiye vedomosti,* no. 26, July 5–11, 1993, p. 2.

66. *Finansovyye izvestiya,* no. 40, August 6–12, 1993, p. 2.

67. *Rossiyskiy fermer,* no. 37, October 19–25, 1993, p. 2.

68. *Krest'yanskaya Rossiya,* no. 10, March 22–28, 1993, p. 2.

69. N. Popov, "Krest'yanskiye (fermerskiye) khozyaystva," *APK: ekonomika, upravleniye,* no. 5 (May 1996), p. 59.

70. Interview at Agricultural Administration, Kostroma Oblast, August 1994.

71. *Kostromskoy kray,* September 30, 1994, p. 1.

72. *Rossiyskiy fermer,* no. 40, November 9–16, 1993, p. 2.

73. "O merakh gosudarstvenoy podderzhki agropromyshlennogo kompleksa v 1993–1994 godakh," *Ekonomika sel'skokhozyaystvennykh i pererabatyvayushchikh predpriyatiy,* no. 2 (February 1994): 30.

74. *Izvestiya,* January 12, 1994, p. 2.

75. *Finansovyye izvestiya,* no. 49, October 8–14, 1993, p. 1.

76. *Rossiyskiy fermer,* no. 3, January 24–30, 1995, p. 1.

77. *Sel'skaya zhizn',* February 26, 1994, p. 2.

78. "Ob organizatsii obespecheniya agropromyshlennogo kompleksa mashinostroitel'noy produktsiey na osnove dolgosrochnoy arendy (lizinga)," *Sobraniye zakonodtel'stva Rossiyskoy Federatsii,* no. 8 (June 20, 1994): 1311–12.

79. "O Federal'noy programme mashinostroeniya dlya agropromyshlennogo kompleksa Rossii," *Sobraniye zakonodtel'stva Rossiyskoy Federatsii,* no. 9 (June 27, 1994): 1459–60.

80. *Sel'skaya zhizn',* March 16, 1995, p. 2.

81. "O gosudarstvennoy podderzhke malogo predprinimatel'stva v Rossiyskoy Federatsii," *Sobraniye zakonodtel'stva Rossiyskoy Federatsii*, no. 21 (May 22, 1995): 3780.

82. *Zemlya i lyudi*, no. 16, April 26, 1996, p. 4.

83. "O federal'noy tselevoy programme razvitiya krest'yanskikh (fermerskikh) khozyaystv i kooperativov na 1996–2000 gody," *Sobraniye zakonodtel'stva Rossiyskoy Federatsii*, no. 1 (January 6, 1997): 207.

84. Ibid., pp. 209–10.

85. Ibid., p. 217.

86. *Sel'skaya zhizn'*, May 31, 1997, p. 1.

87. *Krest'yanskiye vedomosti*, no. 48, December 6–12, 1993, p. 4.

88. *Rossiyskiy fermer*, nos. 9–10, March 14–20, 1995, p. 2.

89. "Predostavleniye zemel' dlya kollektivnogo sadovodstva i ororodnichestva v Rossiyskoy Federatsii," *APK: ekonomika, upravleniye*, nos. 11–12 (November–December 1993): 26.

90. These two oblasts are completely representative of wider national trends.

91. Unpublished information from the Committee on Land Resources, Kostroma Oblast; and unpublished information from Kostroma Oblast branch of Goskomstat.

92. See the two-part article on the revival of personal plots in *Sel'skaya zhizn'*, December 27, 1994, p. 3, and December 19, 1994, p. 3.

93. See Stephen K. Wegren, "Rural Migration and Agrarian Reform in Russia: A Research Note," *Europe-Asia Studies* (formerly *Soviet Studies*) 47, no. 5 (1995): 880–81.

94. A. Puzanovskiy and A. Ivantsov, "Demograficheskiy faktor razvitiya fermerstva," *APK: ekonomika, upravleniye*, no. 2 (February 1994): 65.

95. *Pol i vozrast naseleniya oblasti v rasreze administrativnykh edinits* (Kostroma: Goskomstat RSFSR, 1990), p. 6.

96. *Sel'skaya zhizn'*, February 3, 1994, p. 2.

97. See *Rossiya-1995: Sotsial'no-demograficheskaya situatsiya* (Moscow: Institute of Social-Economic Problems of the Population, 1996), pp. 61–62; and Stephen K. Wegren, Gregory Ioffe and Tatyana Nefedova, "Demographic and Migratory Responses to Agrarian Reform in Russia," *Journal of Communist Studies and Transition Politics*, 13, no. 4 (December 1997).

98. *Itogi khozyaystvennoy deyatel'nosti krest'yanskikh (fermerskikh) khozyaystv Rossiyskoy Federatsii v 1993 godu* (Moscow: Goskomstat Rossii, 1994), p. 5.

99. Early private farmers not only received the best land but also received the best equipment from state and collective farms. "Second generation" private farmers received poorer land and less and poorer equipment.

100. *Krest'yanskiye vedomosti*, no. 21, May 31–June 5, 1993, p. 10.

101. See Stephen K. Wegren, "Agrarian Reform in Moscow Oblast," paper presented at the Twenty-Sixth Annual Convention of the Association for the Advancement of Slavic Studies, Philadelphia, Pennsylvania, November 17–20, 1994.

102. "Novye aktsenty," *Rossiyskaya Federatsiya*, no. 3 (February 1994): 37–39 (39). Unfortunately, Western scholars conducting survey research have neglected to use the urban/rural dimension as an independent variable, although some have looked at the attitudes toward market reform and democracy by occupational groups. For example, see Raymond Duch, "Tolerating Economic Reform: Popular Support for Transition to a Free Market in the Former Soviet Union," *American Political Science Review* 87, no. 3 (September 1993): 590–608; and Stephen Whitefield, "Social Responses to Reform in Russia," in David Lane, ed., *Russia in Transition: Politics, Privatization, and Inequality* (London: Longman, 1995), pp. 91–111.

103. The rate of increase in new private farm creation slowed in Moscow Oblast over time. A total of 2,880 new farms were registered during 1992: from January 1 to July 1, 1992, 2,046 new farms were registered in Moscow Oblast, followed by the registration of another 834 between July 1, 1992, and January 1, 1993. In 1993, a total of 1,585 farms were registered. During the first half of the year, from January 1 to July 1, 1993, 1,227 new farms were created, and then between July 1, 1993, and January 1, 1994, only 358 farms were created. During 1994, the rate decreased even more. From January 1 to July 1, 1994, 533 new peasant farms were registered, and from July 1, 1994, to January 1, 1995, only 183 farms were created.

From January 1, 1995, to January 1, 1997, a net of only 306 farms were created. As we have seen, this slowdown mirrored trends throughout Russia.

104. Despite a larger rural population, the average private farm in Moscow Oblast had three members. Two raions in Moscow Oblast, Balashikhinskiy and Kashirskiy, averaged four persons per farm.

105. Calculated from the population of Moscow Oblast only, not including the population of the city of Moscow.

106. *Kak poluchit'zemel'nyy uchastok v Podmoskov'e* (Moscow: Agrodom, 1992), p. 5.

107. Cited in *Vestnik merii Moskvy*, no. 5 (March 1994): 54.

108. *Moskovskiye novosti*, no. 25, June 19–16, 1994, p. 5.

109. "O programme razvitiya kollektivnogo sadovodstva dlya zhiteliy g. Moskvy do 2002g.," *Vestnik merii Moskvy*, no. 14 (July 1993): 11.

110. Calculated from *Osnovnyye pokazateli khozyaystvennoy deyatel'nosti sovkhozov i kolkhozov za 1991 god* (Moscow: Moskovskiy oblastnoy komitet po statistike, 1992), pp. 5–17. By the end of 1993, agricultural enterprises (state and collective farms and their successors) in these raions had 99,947 hectares, representing 7.7 percent of the oblast's agricultural land. *Osnovnyye pokazateli khozyaystvennoy deyatel'nosti sovkhozov i kolkhozov za 1993 god* (Moscow: Moskovskiy oblastnoy komitet po statistike, 1994), pp. 8–18.

111. *Sel'skaya zhizn'*, May 30, 1995, p. 2.

112. *Itogi proizvodstvenno-finansovoy deyatel'nosti sel'skokhozyaystvennykh predpriyatiy Kostromskoy oblasti za 1994 god* (Kostroma: Goskomstat of Kostroma, 1995), p. 9.

113. *Krest'yanskiye (fermerskiye) khozyaystva Rossiyskoy Federatsii na 1 Yanvarya 1997 goda* (Moscow: Goskomstat, 1997), p. 2.

114. *Krest'yanskaya Rossiya*, March 25, 1992, p. 2.

115. See I. Buzdalov and V. Afanas'yev, "Vozrozhdeniye sel'skokhozyaystvennoy kooperatsii," *APK: ekonomika, upravleniye*, no. 1 (January 1994): 16–23.

116. "Krest'yanskiye (fermerskiye) khozyaystva na 1 Iyulya 1995 goda," unpublished information from Kostroma Goskomstat; and *Krest'yanskiye khozyaystva oblasti na 1 Yanvarya 1995 goda* (Kostroma: Goskomstat of Kostroma Oblast, 1995), p. 5.

117. Private farms received about 90 percent of their revenue from production sales in 1994 (revenues are tabulated from production and from credits, subsidies, and compensations). In this respect in 1994 private farmers earned a higher percentage from production than did collective farms (86 percent) and new agricultural enterprises (87 percent). However, increases in earnings from production were exceeded by increases in input prices. Because private farmers derived more of their revenue from production, production declines took an especially heavy toll on private farming. M. Kozlov, "Finansovoye sostoyaniye sel'skokhozyaystvennykh tovaroproizvoditeley v usloviyakh reformy," *APK: ekonomika, upravleniye*, no. 9 (September 1995): 38–39.

118. These statistics are from *Kostromskoy kray*, March 11, 1995, p. 2, and March 10, 1995, p. 2.

119. Compare *Itogi khozyaystvennoy deyatel'nosti krest'yanskikh (fermerskikh) khozyaystv Rossiyskoy Federatsii v 1992 godu*, p. 6, and *Itogi khozyaystvennoy deyatel'nosti krest'yanskikh (fermerskikh) khozyaystv Rossiyskoy Federatsii v 1994 godu*, p. 10.

120. See *Osnovnyye pokazateli agropromyshlennogo kompleksa Rossiyskoy Federatsii (Yanvar'–Sentyabr' 1996 goda)* (Moscow: Goskomstat Rossii, 1996), p. 34.

121. A. Chernyaev, "Fermerskiye khozyaystva v Povolzh'e," *APK: ekonomika, upravleniye*, no. 4 (April 1992): 28. According to calculations by agricultural economists in the region, a private farm with four persons should have 100–130 hectares of land and 20–25 head of cattle in order to achieve sufficient production to remain viable.

122. Ibid.

123. B. K. Markin, "Stanovleniye krest'yanskikh (fermerskikh) khozyaystv v Povolzh'e," *Ekonomika sel'skokhozyaystvennykh i pererabatyvayushchikh predpriyatiy*, no. 11 (November 1995): 28.

124. Private farmers followed nationwide trends. For the nation as a whole, by mid-1997,

the number of cattle, cows, and pigs on Russian agricultural enterprises had fallen to 1940s USSR levels. Compare *Sostoyaniye zhivotnovodstva na 1 Iyulya 1997 goda* (Moscow: Goskomstat, 1997), p. 4; and *Narodnoye khozyaystvo SSSR v 1965 g.: Statisticheskiy ezhegodnik* (Moscow: Central Statistical Administration, 1966), p. 368.

125. *Itogi khozyaystvennoy deyatel'nosti krest'yanskikh (fermerskikh) khozyaystv Rossiyskoy Federatsii v 1994 godu,* p. 56.

126. Markin, "Stanovleniye krest'yanskikh (fermerskikh) khozyaystv v Povolzh'e," p. 28.

127. Again, profitability is not defined as in the West. Credits were considered income, and therefore if the sum of income *(dokhod)* and credits *(kredit)* exceeds expenditures *(raskhody)* a farm is considered profitable.

128. S. Sazonov, "Sotsial'no-ekonomicheskiye aspekty razvitiya fermerskogo dvizheniya v Rossii," *APK: ekonomika, upravleniye,* no. 5 (May 1995): 57.

129. "Sel'skoye khozyaystvo Rossii v 1993 godu," *APK: ekonomika, upravleniye,* no. 4 (April 1994): 18.

130. "O polozhenii v agropromyshlennom komplekse Rossii," *APK: ekonomika, upravleniye,* no. 1 (January 1995): 37.

131. *Rossiyskiy fermer,* no. 3, January 24–30, 1995, p. 1.

132. *Krest'yanskiye (fermerskiye) khozyaystva Rossiyskoy Federatsii na 1 Yanvarya 1996 goda* (Moscow: Goskomstat, 1996), p. 1.

133. See Stephen K. Wegren, "Dilemmas of Agrarian Reform in the Soviet Union," *Soviet Studies* 44, no. 1 (1992): 19–20. See also Whitefield, "Social Responses to Reform in Russia"; and Jerry F. Hough, "The Russian Election of 1993: Public Attitudes Toward Economic Reform and Democratization," *Post-Soviet Affairs* 10, no. 1 (January–March 1994): 1–37. The one exception to rural opposition to market reforms comes from Raymond Duch, who hypothesized that agricultural workers would be among "the most enthusiastic about free market reforms" (Duch, "Tolerating Economic Reform," p. 593).

134. On rural conservatism and voting data in the countryside during the April 1993 referendum, see Stephen K. Wegren, "Rural Reform and Political Culture in Russia," *Europe-Asia Studies* (formerly *Soviet Studies*) 46, no. 2 (1994): 230–32. See also Ralph S. Clem and Peter R. Craumer, "The Geography of the April 25 (1993) Russian Referendum," *Post-Soviet Geography* 34, no. 8 (October 1993): 481–96. For the December 1993 parliamentary election, see Darrell Slider, Vladimir Gimpel'son, and Sergei Chugrov, "Political Tendencies in Russia's Regions: Evidence from the 1993 Parliamentary Elections," *Slavic Review* 53, no. 3 (fall 1994): 711–32; and Ralph S. Clem and Peter R. Craumer, "A Rayon-Level Analysis of the Russian Election and Constitutional Plebiscite of December 1993," *Post-Soviet Geography* 36, no. 8 (October 1995): 459–75. For the December 1995 election, see Ralph S. Clem, and Peter R. Craumer, "The Geography of the Russian 1995 Parliamentary Election: Continuity, Change, and Correlates," *Post-Soviet Geography* 36, no. 10 (December 1995): 587–616. For the 1996 presidential runoffs, see Ralph S. Clem, and Peter R. Craumer, "Roadmap to Victory: Boris Yeltsin and the Russian Presidential Elections of 1996," *Post-Soviet Geography* 37, no. 6 (June 1996): 335–54; and Robert W. Orttung and Anna Paretskaya. "Presidential Election Demonstrates Urban-Rural Divide," *Transition* 2, no. 19 (September 20, 1996): 33–38. All of these analysts consistently found a distinctive urban-rural divide in voting patterns.

135. On the origins of AKKOR, see Don Van Atta, "Political Mobilization in the Russian Countryside: Creating Social Movements from Above," in Judith B. Sedaitis and Jim Butterfield, eds., *Perestroika from Below: Social Movements in the Soviet Union* (Boulder, Colo.: Westview Press, 1991), pp. 53–56.

136. See Craig L. Infanger, "An Inside View of Russian Agrarian Reform," *Soviet and Post-Soviet Review* 21, nos. 2–3 (1994): 189–209.

137. The concept is from Mancur Olson, *The Logic of Collective Action* (Cambridge, Mass.: Harvard University Press, 1965). It refers to a select "good" that is available to a specific population in return for something else, like political support. A selective incentive is not a "free good" that is available to all.

138. *Rossiyskiy fermer,* no. 34, September 28–October 4, 1993, p. 1.

139. See *Rossiyskaya gazeta*, December 8, 1992, p. 3.
140. Alexander Meyendorff, "Moscow Letter," *Newsletter for Research on Soviet and East European Agriculture* 14, no. 4 (December 1992): esp. 5–6.
141. The president of AKKOR, Bashmachnikov, stated in early 1994 that 1.5 million persons "worked" on private farms. *Zemlya i lyudi*, no. 7, February 18, 1994, p. 3. In 1994 there was a total of almost 40 million rural people, so private farmers were a very small percentage.
142. *Sel'skaya zhizn'*, March 14, 1995, pp. 1–2.
143. See *Krest'yanskaya partiya Rossii: ustav, programma, materialy* (Moscow: Peasant Party of Russia, 1993).
144. Unpublished letter from Yu. Chernichenko to President Boris Yel'tsin, October 6, 1993. I thank the deputy director of the Peasant Party, Aleksandr Khokhlov, for providing me with this letter.
145. Interview with the deputy director of the Peasant Party, Aleksandr Khokhlov, August 1994. Some of these organizations seemed to exist only on paper. The party headquarters in Kostroma Oblast, for example, was unoccupied and closed the entire summer of 1994.
146. Interview with the deputy director of the Peasant Party, Aleksandr Khokhlov, August 1994.
147. See Stephen K. Wegren, "Rural Politics and Agrarian Reform in Russia," *Problems of Post-Communism* 43, no. 1 (January–February 1996): 23–34.
148. Two peasant farmers, interviewed in Kostroma raion during the summer of 1994, responded that they had not received any assistance of any kind from AKKOR at any time. They dismissed the organization as corrupt and interested only in lining the pockets of its management. When asked why they had not joined AKKOR, both farmers responded they saw no point since nothing was to be gained (formally it was not necessary to be a member in order to receive loan guarantees or technical help). On the weakness of AKKOR, see the article by the vice president of AKKOR at the fifth congress of AKKOR in *Rossiyskiy fermer*, no. 46, December 21–26, 1993, p. 2; and the analysis after the sixth congress of AKKOR in early 1995 in *Sel'skaya zhizn'*, March 14, 1995, p. 1.
149. See *Krest'yanskiye vedomosti*, no. 15, April 18–24, 1994, p. 4, on tensions between AKKOR and the Yel'tsin regime. Farmers' demands are contained in an open letter to Yel'tsin and Prime Minister Chernomyrdin in the same issue (p. 2); and in an open letter to Yel'tsin in *Rossiyskiy fermer*, no. 3, January 25–31, 1994, pp. 1–2.
150. *Krest'yanskiye vedomosti*, no. 15, April 18–24, 1994, p. 4.
151. See, for example, ibid., no. 16, April 25–May 1, 1994, pp. 1–2.
152. On the creation of the union, see the interview with Bashmachnikov in *Krest'yanskaya Rossiya*, no. 2, January 16–22, 1995, p. 1.
153. *Rossiyskiy fermer*, no. 46, December 20–27, 1994, p. 2.
154. Ibid., no. 25, July 3–9, 1995, p. 2.
155. *Krest'yanskiye vedomosti*, no. 30, July 24–30, 1995, p. 1.
156. *Krest'yanskiye vedomosti*, no. 2, January 15–21, 1996, p. 2.
157. *Krest'yanskiye vedomosti*, no. 4, February 3–9, 1997, p. 2.
158. For further analysis, see Stephen K. Wegren, "The Politics of Private Farming in Russia," *Journal of Peasant Studies*, 23, no. 4 (July 1996): 106–40.
159. *Rossiyskiy fermer*, no. 22, July 6–12, 1993, p. 2.

Chapter 7. The State and Agrarian Reform
1. I have argued the point of cultural continuity in Stephen K. Wegren, "Agrarian Reform and Political Culture," *Europe-Asia Studies* 46, no. 2 (1994): 215–41.
2. This special program to develop the Russian non–black earth region was adopted in early 1994 for the period 1994-1995. See *Zemlya i lyudi*, no. 5, February 4, 1994, pp. 1–2.
3. *Reorganizatsiya kolkhozov i sovkhozov Rossiyskoy Federatsii po sostoyaniyu na 1.1.94g.* (Moscow: Goskomstat Rossiyskoy Federatsii, 1994), p. 1.
4. N. Popov, "Krest'ianskiye (fermerskiye) khozyaystva," *APK: ekonomika, upravleniye*, no. 5 (May 1996): 56.

5. V. Mashenkov and V. Malakhova, "Bezrabotitsa na sele i puti eyo smyagcheniya," *APK: ekonomika, upravleniye*, no. 3 (March 1996): 19.

6. *Trud i zanyatost' v Rossii* (Moscow: Goskomstat Rossii, 1995), p. 18; and *Statisticheskoye obozreniye*, no. 3 (March 1995): 59.

7. Within the Central region the oblasts with the highest rates of rural unemployment are Vladimir, Ivanovo, and Yaroslavl'.

8. Mashenkov and Malakhova, "Bezrabotitsa na sele i puti eyo smyagcheniya," p. 20.

9. *Itogi proizvodstvenno-finansovoy deyatel'nost' sel'skokhozyaystvennykh predpriyatiy Kostromskoy oblasti za 1995 god* (Kostroma: Goskomstat Kostroma Oblast, 1996), p. 3.

10. *Sel'skoye khozyaystvo Rossii* (Moscow: Goskomstat, 1995), p. 28.

11. A. Zaveryukha, "Problemy agrarnoy reformy v Rossii," *APK: ekonomika, upravleniye*, no. 7 (July 1995): 4; and G. V. Kulik, "Kratkiye itogi Rossiyskikh agrarnykh 'reform' v 1991–1994 godakh," *APK: ekonomika, upravleniye*, no. 12 (December 1995): 11.

12. *Zemlya i trud*, no. 45, November 7–13, 1995, p. 5.

13. *Demograficheskii ezhegodnik Rossii* (Moscow: Goskomstat, 1995), pp. 56, 67. The statistical measure indicating the "natural increase" between the number of births and deaths per thousand persons is expressed as a coefficient.

14. Ibid., pp. 34–35.

15. See William C. Thiesenhusen, *Broken Promises: Agrarian Reform and the Latin American Campesino* (Boulder, Colo.: Westview Press, 1995).

16. See Matthew Wyman, "Developments in Russian Voting Behaviour: 1993 and 1995 Compared," *Journal of Communist Studies and Transition Politics* 12, no. 3 (September 1996): 277–92.

17. See Michael Urban, "December 1993 as a Replication of Late-Soviet Electoral Practices," *Post-Soviet Affairs* 10, no. 2 (1994): 127–58.

18. On banking reforms, see Juliet Johnson, "Banking in Russia: Shadows of the Past," *Problems of Post-Communism* 43, no. 3 (May–June 1996): 49–59.

19. On the financial system and state support for agriculture, see E. Serova, "Reforma finansovo-kreditnogo mekhanizma v Rossiyskom sel'skom khozyaystve," *APK: ekonomika, upravleniye*, no. 2 (February 1996): 33–39; and E. Serova, "Osobennosti gosudarstvennoy podderzhki agrarnogo sektora v Rossii," *Voprosy ekonomiki*, no. 7 (July 1996): 88–100.

20. Richard Rose and Ellen Carnaghan, "Generational Effects on Attitudes to Communist Regimes: A Comparative Analysis," *Post-Soviet Affairs* 11, no. 1 (1995): 28–56.

21. David S. Mason, "Attitudes Toward the Market and Political Participation in the Post-communist States," *Slavic Review* 54, no. 2 (summer 1995): 385–406.

22. Arthur A. Miller, Vicki L. Hesli, and William M. Resinger, "Reassessing Mass Support for Political and Economic Change in the Former USSR," *American Political Science Review* 88, no. 2 (June 1994): 399–411.

23. Jerry F. Hough, "The Russian Election of 1993: Public Attitudes Toward Economic Reform and Democratization," *Post-Soviet Affairs* 10, no. 1 (1994): 1–37.

24. Stephen Whitefield and Geoffrey Evans, "The Russian Election of 1993: Public Opinion and the Transition Experience," *Post-Soviet Affairs* 10, no. 1 (1994): 38–60; and Hough, "The Russian Election of 1993."

Index

Age, 234, 239; demographics and importance of land leasing, 163; and land market, 154, 179; rural, 206, 219, 237
Agrarian lobby, 222; effectiveness of, 12, 14; weakness of, 115, 119, 145–46
Agrarian Party of Russia (APR), 114, 141–46, 164, 223, 235, 270$n140$
Agrarian reform, 3, 8, 13–14, 276$n98$; of collective sector, 59–107; coming from above, 6, 13, 15, 149–50, 218, 228–29; criticism of, 103, 228–29; effects of, 1–2, 77–81, 166–80, 234–36; goals of, 3, 87, 103, 227; Gorbachev's attempts at, 63–67; lack of rural support for, 15, 111–15; opposition to, 141, 235; Soviet legacy in, 57–58; statist approach to, 7, 12, 237–40; success of, 12, 133, 227–28; support for, 211, 223, 225, 235, 239
Agriculture, 1–2, 33–34, 84, 114; education in, 166–69, 275$nn82,83$, 276$n84$; effects of state interventions in, 11, 18; investment in, 48, 116–19, 121–23, 252$n106$, 254$n143$; out-migration of workers from, 39–40, 48, 55–57, 126, 237, 240; personnel, 20, 126–27, 235–36; and prices, 87–88, 119–21; profitability of, 32, 264$n2$, 284$n127$; protection of land for, 164, 166, 176–77, 181; and strength of state, 11–14, 237–38; subsidies to, 88–90, 116, 198–204. *See also* State interventions; Subsidies; Wages
Agroindustrial complex (APK). *See* Agriculture
Andreev, A., 250$n70$
Assistance, and training for private farming, 167, 169
Association of Peasant Farms and Cooperatives of Russia (AKKOR), 79, 90, 138–40, 223; and credit, 191–93, 194; criticism of, 280$n46$, 285$n148$; politics of, 114, 219–20; and Russian Farmer program, 184–86; services of, 167, 239; and state, 143–44, 224
Auctions, 258$n37$, 263$nn136-38$; in farm privatizations, 94–95, 99–101

Ballohectares, in privatization, 97–101
Bankruptcy: of collective sector farms, 64, 69; of private farms, 183, 217, 232
Bashmachnikov, Vladimir, 93, 114, 191, 219; on private farms, 266$n42$, 285$n141$; as rural spokesperson, 222–24
Bates, Robert, 112
Benediktov, I. A., 38
Berulin, Mr., 192
Bonus payments: for joint stock farms, 75; for weak farms, 27, 29, 31
Brezhnev, Leonid, 5–6, 8, 25–26, 30–31, 50–51, 109; equalizing wage differences, 20–23, 56; and personal plots, 57, 252$n107$
British Know-How Fund, 94, 101
Bryansk Oblast, 79–80
Budget deficits, reduction of, 198–204
Burykov, A., 146

Chelyabinsk Oblast, 75
Chernichenko, Yu., 221–23
Chernomyrdin, Viktor, 95, 116–17, 165, 167, 200; on farm subsidies, 89, 198; and political parties, 141, 223
Chernyshev, Alexei, 143, 166
Chubays, A., 95
Cities, 250$n77$; food for, 145, 264$n11$; land for urbanites, 159–60, 162, 176, 276$n93$; and restrictions on land use, 164, 176–77, 181; support for agrarian reform, 220–21, 229; urbanites in private

farming, 33, 169, 210, 212–13. *See also* Urban *vs.* rural
Collective Farm Charter (1988), 44
Collective farms, 163, 283*n117;* members leaving, 46–47, 257*n21*, 257*n22;* obstructions to private farms by, 150–51; personal plots on, 38–39, 249*n56;* personnel on, 43, 247*n26;* production of, 60–61, 251*n95;* combining to increase, 37; reorganization of, 59; social services on, 54, 255*n165;* under Soviets, 7–8; switched to self-financing and -accounting, 31–32; wages on, 19–25, 50–52, 124–25, 246*n12*. *See also* Collective sector farms; Weak farms
Collective gardens and orchards, 34, 213, 249*n60*, 249*n63*, 253*n114*
Collective markets, 34–36, 42, 252*n107*
Collective sector farms, 26–27, 140–41, 214, 257*n15*, 268*n89;* decrease in investments in, 116–17; democracy on, 65, 71; egalitarianism in, 230–31; lack of change after reorganization of, 71–73, 79–83, 86–90, 232–33; land shares from, 159, 161, 174–75, 274*n53*, 278*n133*, 283*n110;* members' ability to leave, 35, 65, 150; members leaving for private farms, 46–47, 158, 169–70, 172; members' unwillingness to lease land, 67–68; operations of, 64–65, 130–33; and personal plots, 34–35, 38, 42, 44, 252*n113*, 253*n118;* post-Soviet reforms, 59–107; prices on, 86–87; privatization of, 91–94, 96, 155–59, 258*n34;* problems under Soviets, 60–63; provision of services by, 80–81; reorganization of, 59, 77–79, 96, 102–03, 106, 151–52, 159, 176, 260*n72*, 261*nn80,81*, and rural social contract, 108, 112–14; subsidies to, 186, 269*n108;* unwillingness of members to leave for private farms, 80–81, 86–87. *See also* Collective farms; State farms
Collectivization, 1, 13, 48–49
Commercial banks, 185–86, 198, 204, 208, 238
Commission on Land Privatization and Reorganization, 96
Committee on Land Reform. *See* Committees on Land Resources and Land Surveying
Committees on Land Resources and Land Surveying, 72, 158
Commodity exchanges, after privatization, 14
Communists, 1; restrictions on land use under, 164–65; strength of state under, 5–6. *See also* Soviet Union

Competition, 33; lack of, 131, 269*n111;* need for, 105–07
Conservatism: and age, 239. *See also* Rural conservatives
Consumer demand, 115
Contract system, of personnel allotments, 26–27
Cooperatives: banks, 239; consumer, 250*n70;* formation of, 63–64, 81, 201; functions of, 76–77; of private farmers, 214
Corruption, 275*n79;* of AKKOR, 185, 285*n148;* in Russian Farmer program, 192–93
Credit, 163, 232, 279*n11*, 283*n117;* distribution of, 194–98; for private farms, 177, 183–98, 208, 279*n12;* subsidized, 198, 214–16, 269*n108*

Davydov, Aleksandr, 140
Decentralization: in agrarian reform, 8, 118–19; as expected effect of private farms, 149
Democracy: on collective sector farms, 65, 71, 97–99; on joint stock farms, 73
Democratization, importance of agrarian reform to, 2
Demographics, 256*n175;* in private farming, 163, 166–70, 206–08; and reluctance for privatization, 86, 209, 271*n3;* rural, 1, 179, 236–37; rural out-migration, 39–40, 48, 55–56, 57, 126, 235–36; rural *vs.* urban, 36, 39–40, 240, 250*n77*
De-Stalinization, privatization in, 112–13
Destatization. *See* Privatization
Duch, Raymond, 284*n133*

Economic Problems of Socialism in the USSR (Stalin), 35
Economy, 6–7, 265*n17;* agriculture prices relative to industry's, 119–21; cutting budget deficits, 198–204; effects of farm privatization, 93–94, 105, 149; lack of profitability of farms, 188–89, 193, 202, 208–09, 216–17, 232, 276*n98;* sources of investments in agriculture, 201–04. *See also* Rural economic environment
Education: in agriculture, 155, 166–69, 187; rural *vs.* urban, 54–55
Efficiency, of agriculture, 88–89; attempts to increase, 23, 116; disincentives for, 9, 133; effects of pricing policies on, 30, 32, 88; as goal of farm reforms, 59–63, 106, 112–13; lack of, 60–62, 134
Egalitarianism, 77, 239; in agriculture policies, 50–52, 88–90; in collective sector, 230–31; in distribution of credit, 194–98;

in distribution of land, 154, 174–75; as goal, 18–33, 56, 229–30; in private sector, 33–48, 230. *See also* Wages
Equipment. *See* Machinery/equipment

Farm debts, 28, 69–70, 147
Farm disbandment, 73, 76–77
Farmer associations, functions of, 76–77
Farmers' banks, 279n12
Farm markets, collective, 34–36, 42
Farm personnel. *See* Personnel
"Federal Program for the Development of Private Farms . . . 1996–2000," 201
Federation Council, opposition to agrarian reforms, 114
Filippov, P., 91–92
Financial levers. *See* State interventions, financial levers for
Food, 33–34, 119; for collective sector farmers, 86–87; emphasis on quantities vs. efficiency, 88–90; imported, 115–16, 144–45, 271nn154,157; shortages, 1, 35, 36, 48–49, 60, 129–30, 256n5, 269n116. *See also* Production
Food consumption, 43, 49, 60, 129–30
Food funds, federal, 134–38
Food-processing plants, 23, 115, 121
Food trade system. *See* State, procurement system of
Free trade. *See* Protectionism
Fundamentals of Land Legislation (1990), 46–47

Gaydar, Yegor, 88, 198, 199–200, 221–22
Gorbachev, Mikhail, 6, 62, 264n2; attempts at agrarian reform by, 63–67; landownership under, 155, 160–61; personal plots under, 57, 148; pricing policies under, 31, 264n11; private farming under, 34, 43–48; social justice policies of, 27–28; wages under, 23–25, 51–52, 56
Gosagroprom, 62; on land leasing, 27–28, 66

Hough, Jerry, 5
Housing, 254n132; lack of investment in, 122–24, 217; land for, 161–62; in land trades, 178–79; rural, 53–54, 236
Hunter, Holland, 49
Huntington, Samuel, 11

Ideology: of personal property, 57, 219; of privatization campaigns, 92, 112–13
Incentives, 2, 219–20; attempts to create, 107, 136; egalitarianism as goal of, 18–33; lack of, 62–63, 90–91; for privatization, 80–81, 86–87, 104–05; for production, 36, 68, 75, 252n111; and strength of state, 11, 13; and success of agrarian reform, 8–9, 58, 90–91, 229–31
Industry/industrialization, 1, 35; vs. agriculture, 48–49, 119–21, 124–25
Infanger, Craig, 114
Inflation, 49, 121, 157–58, 183
Infrastructure, rural, 54, 236; effects of poor, 83; investment in, 110, 186–88; lack of investment in, 35, 49, 121–24, 200, 204
In-kind land distribution, 71–72, 74, 158
Interest rates, 193; effects on farms, 132–33, 208, 216; state subsidies for, 184, 187–88, 198, 200
International Finance Corporation (IFC), in farm privatization, 94–96, 101, 103
Investments, in agriculture, 22, 49, 146, 252n106, 254n143, 266n47; under Brezhnev, 41, 50; capital, 117–19; in collective sector, 70; decrease in, 35, 116, 236; as lever for egalitarianism, 25–29; for private farms, 201–04; in revival of countryside, 109–10, 121–23

Johnson, D. Gale, 62–63
Joint stock farms, 13; collective labor organization on, 81, 263n142; privatization of, 92–94; in reorganization of collective sector farms, 71, 73–76, 232–33

Kalinin, Nikolay, 163, 199
Kaluga Oblast, 85
Karcz, Jerzy, 48–49
Khlystun, Viktor, 89, 95, 122, 154, 165, 259n69; and agrarian reform, 13, 114, 265n21; and collective sector farms, 78–79; and private farms, 92, 173, 175, 191–92, 219
Khokhlov, Aleksandr, 193, 221–22
Khozraschet. *See* Self-accounting and -financing
Khrushchev, Nikita, 1, 7–8, 30, 49–50, 50, 109; equalizing wage differences, 20, 56; private agriculture under, 36–39, 39–43, 57
Kostroma Oblast, 77, 176, 246n4, 260n79; demographics of, 179, 206–08; farm personnel in, 55, 126–27, 235–36; land leasing in, 67–68; land market in, 178–79; personal plots in, 43, 45, 205–11; private farms in, 72, 77, 167–68, 174, 192, 199; types of farms in, 84–85; wages in, 22–24, 32; weak farms in, 27, 32
Krasner, Stephen, 9–10
Kulakov, F. D., 41
Kulik, Gennadii, 143

Labor organization, 19, 74, 250*n70*; collective, 61, 65, 81, 92–93, 112, 260*n74*, 263*n142*; of collective sector farms, 69–71; link system of, 19, 250*n70*, 256*n10*
Labor strikes, and nonpayment of wages, 126
Land, 178–79; cultivating more, 37; obtaining for private farms, 150–52, 174–75, 201, 214
Land Code (1991), 151–54, 165–66
Land funds, 155–56, 159, 176, 273*n35*
Land leasing, 44, 102, 163, 257*n27*; on collective
 sector farms, 63–64, 66–69; for private farms, 46–47, 273*n29*, 277*n112*; to supplement personal plots, 252*n113*; *vs.* landownership, 153, 160–61
Land market, 176–80, 177, 220, 265*n18*; development of, 13, 159–66; prices, 277*n127*, 278*n133*
Landownership, 103; preferences in, 205–15; for private farms, 46–48, 152–66; and privatization of collective sector farms, 71, 98; restrictions on, 176, 221, 275*n77*; right to, 13, 33, 47–48, 231; Russian *vs.* national legislation on, 46–47; by state, 160–61; titles for, 274*nn46,50*
Land quality: criteria for "good," 84–85, 156; fertility of, 66–67, 127, 129; given for private farms, 48, 87, 171–73, 194–98, 282*n99*; relation to quantity for private farms, 174, 273*n34*
Land reform. *See* Agrarian reform
Land reserve fund, for peasant farms, 254*n133*
Land share distribution, 74, 174, 263*nn125,127*, 274*nn48,53*; in reorganization of collective sector farms, 71–72, 96–99; sale of shares, 103, 161, 259*n54*, 278*n133*; uses of shares, 74, 102, 162–66, 258*nn44,45*, 274*n49*
Land tax, 157–58, 273*n29,42*; for private farms, 200, 202
Land use, 84, 258*n44*; competition among, 87, 205–15, 212–14; restrictions on, 161–66, 176–77, 180–81
Lapshin, Mikhail, 114, 130, 270*n140*
Latin America, agrarian reform in, 3
Law on Grain (1993), 135, 137
Law on Land Reform (1990), 152–53
Law on Land (1990—USSR), 161, 257*n27*, 273*n35*
Law on Leasing (1990), 47
Law on Ownership (1990), 153
Law on Peasant Farming (1990), 76, 158, 186

Law on Peasant Farms (1990), 152, 154, 280*n30*
Law on Property (1990): in USSR, 254*n132*; in USSR *vs.* Russia, 47, 161
Law on Social Development of the Countryside, 109–11, 121–22
Leasing: of agricultural machinery, 200–01. *See also* Land leasing
Lebedev, Yuriy, 166
Legislation, influence of rural interest groups on, 11–12
Lenin, Vladimir Ilyich, 1
Ligachev, Yegor, 28–29, 44, 253*n118*
Linin, Yuriy, 191
Link system, of labor organization, 19, 250*n70*, 256*n10*
Lipton, Michael, 243*n4*
Livestock, 88; on collective sector *vs.* private farms, 88–89; decrease of, 36, 48–49; on personal plots, 37–39, 42, 45–46, 251*n95*; of private farmers, 215, 283*n124*; restrictions on personal, 38–40, 43
Living standards, 255*n165*; of private farmers, 171–72, 204; and risks of privatization, 80–81, 113; rural, 121–27, 141, 233–36; and rural social contract, 109, 145–47; rural *vs.* urban, 48, 51, 53–55, 57
Local initiative, lack of faith in, 7

Machinery/equipment: cost of, 190–91; for private farms, 77, 186–89, 200–01, 214
Machine tractor stations (MTS), 7–8, 20
Market reform, 110–11, 284*n133*
Markets, 67, 78–79, 120; collective, 34–36, 42; farmers use of against state, 133–38; lack of competition in, 105–07, 131, 269*n111*; problems of private farms in, 77, 83–85; for produce from personal plots, 35–36, 38, 42, 252*n107*. *See also* Land market
Matthews, Mervyn, 246*n3*
Media coverage, 70, 101; of private farms, 86–87, 150, 158
Medical care, rural, 54, 121, 255*n165*
Medvedev, Zhores, 93
Merl, Stephen, 49
Migdal, Joel, 9–10
Mikhailov, A., 142
Military personnel, land for, 154
Millar, James, 48
Model Collective Farm Charter (1988), 253*n122*
Model Collective Farm Charter of 1930, 34
Model Statutes, personal plot sizes in, 41
Moore, Barrington, 2, 149

Index   291

Mortgages, 162–63, 185
Moscow Oblast, 72, 177–78, 278$n129$; agricultural preferences in, 211–14
Murakhovskiy, V. S., 27–28, 44

Nadbavki (bonuses), 31–32
Nazarchuk, Aleksandr, 93, 101, 142
Nemtsov, Boris, 94, 104–05
New Economic Policy (NEP), 1
Nikonov, Viktor, 32
Nizhniy Novgorod, 13–14, 94–106
Nizhniy Oblast, 102
Nove, Alec, 48–49

"On Additional Measures to Increase Production . . . on Personal Plots . . ." (1981), 42
"On Measures for the Increasing of Production . . . and Strengthening Collective Farms" (1971), 26
"On Measures of State Support of the Agroindustrial Complex in 1992–1994," 200
"On Personal Subsidiary Plots . . ." (1977), 42
"On State Support of Small Entrepreneurs in the Russian Federation" (1995), 200–01
"On Supplementary Measures for the Stabilization of the Agroindustrial Complex in 1993," 187
"On the Procedure for Reorganizing State and Collective Farms" (1991), 69, 175
"On Urgent Measures for the Support of Peasant Farming" (1993), 188
Orel Oblast, 102

Peasant associations, 275$n79$
Peasant cooperative banks, 239
Peasant farms. *See* Private farms
Peasant Party of Russia, 220, 221–22
Pensions/pensioners, 74–76; and land leasing, 67, 163; and land shares, 175, 179–80
Personal plots, 102, 233, 249$n54$, 249$n56$, 249$n63$, 252$n107$, 253$n118$; land for, 178, 252$n113$; numbers of, *vs.* private farms, 205–12; production of, 41–43, 45, 250$n66$, 250$n72$, 251$n95$, 266$n42$, 268$n89$; restrictions on, 148, 252$n100$, 275$n64$; under Soviets, 33–48
Personnel: on collective sector farms, 26–27, 247$n26$; effects of agrarian reform on, 235–36; on personal plots, 43; on private farms, 166–70, 210, 260$n76$, 280$n30$, 285$n141$; production brigades, 19, 246$n1$, 256$n7$

Politics: of agrarian reforms, 3, 9, 15, 106, 227, 229; impact of financial levers on, 138–43; impact of reform institutions, 77–79; over restrictions on land use, 164–66; of privatization, 112–15, 149; of rural interest groups, 109, 111–12, 233, 238; subsidies in, 28–29, 218–25; urban bias in, 3–4, 235, 243$n4$
Popov, Vladimir, 113
Prices, 251$n81$, 254$n144$, 270$n125$; cost of inputs, 30–32, 189–92; deterioration of agricultural, 119–21; effects of liberalization of, 87–88, 115, 183, 209–10, 261$n86$; food, 53, 90, 129, 264$n11$; as incentive, 36–37, 62–63, 136–37; increases in, 50, 110, 216, 255$n155$, 264$n11$, 267$n55$; land, 161–62, 277$n127$, 278$n133$; paid by state, 136–37, 144, 269$n115$; for produce of private farming, 38, 67; state intervention in, 29–33; supplements for weak farms, 29, 31–32, 248$n28$; and urban bias, 4, 49, 110; zones for, 30–33, 88, 248$n42$, 248$n45$
Principles of Land Legislation, The (1969), 41
Private farms, 149, 230, 263$n142$, 275$n64$; credit for, 183–98, 214–16, 279$n12$; development of, 148–81, 282$n99$; effects of state interventions on, 166–80, 182–226, 231–32; formation from collective sector farms, 76–77, 83–85, 100, 258$n44$, 258$n45$, 261$n81$; formation of, 91–93, 282$n103$; under Gorbachev, 44, 46–47; land for, 72, 153–66, 171–73, 232–33, 254$n133$, 273$n29$, 277$n112$; numbers of, 183, 194, 205–14, 254$n134$; personnel of, 166–70, 210, 260$n76$, 276$n93$, 280$n30$, 283$n104$, 285$n141$; politics of, 138–40, 220, 221–22; problems of, 180–81, 190–91, 276$n98$; production of, 214, 276$n97$, 281$n58$, 283$n124$; profitability of, 216–17, 283$n117$, 283$n121$; reluctance to start, 86–87, 151, 211–12; size of, 174–75, 177, 180–81, 214–15, 230, 277$n112$; subsidies to, 117, 198–204, 214–16
Private sector, 230; production of, 266$n42$, 268$n89$; restrictions on, 56–57, 231; and rural economic environment, 33–48, 188–89
Privatization, 120, 231, 271$n1$; of collective sector farms, 63–64, 69–73, 231, 258$n34$; of food procurement system, 135–37; goals of, 103–04, 112–14, 229–30; of land, 13–14, 151–66; Nizhniy Novgorod as model of, 94–106; opposition to, 114–15, 218–20, 271$n3$; preferences in

agricultural types, 205–11; and state influence, 16–17, 150, 227, 238; support for, 91–94, 113, 265n20
Procurement system. *See* State, procurement system of
Production, agricultural, 138, 249n58, 256n10, 283n117; of collective sector, 78, 253n114; of collective *vs.* private sector, 85, 170, 268n89; decrease in, 117, 234–35; factors influencing, 48–49, 105–06, 214–15; impact of financial levers on, 127–29, 133; increases in, 50, 60, 102; need to increase, 36–37, 39, 43–44; of personal plots, 41–43, 45, 250n66, 250n72, 251n95; of private farms, 174–75, 180, 194–98, 215–16, 266n42, 276n97, 281n58
Production brigades, 19, 246n1, 256n7
Protectionism, 90, 144–45
Pskov Oblast, 75

Radugin, N., 122
Recreation, rural, 55
Reformed farms, state money to, 117
Reforms: and strength of state, 5–7, 10–11. *See* Agrarian reform
Regions, 234, 251n93; difference in numbers of private farms, 194–98, 208–11; difference in profitability of farms in, 30–32, 217, 281n58; difference in types of farms formed, 81–85
"Regulation of Land Relations . . ." (1993), 158–59
Rent, on land, 257n28, 273n29, 277n112
Roads, rural: lack of, 83–84, 122–24, 236; poor quality of, 171–72
Rose, Richard, 239
Roskhleboprodukt (state purchasing agency), 124, 135–37, 198, 270nn122
Rossel'khozbank (Russian Agricultural Bank), 184, 193, 198, 239, 269n108
Rostov Oblast, 72, 80, 102
Rural conservatives, 52, 164–65; influence of, 12, 14, 238; opposition to agrarian reforms, 14, 77, 114–15, 219; support for collective sector farms, 63–64, 70; trade unionists as, 138–40
Rural economic environment, 33–48, 146, 238; changes in, 18–19, 58, 108, 231–33; for collective sector farms, 61–63; effects of ending credit subsidies on, 215–17; effects of financial levers on, 130–33, 182; effects of rural social policies on, 2, 8–9, 15–16; for private farms, 180–81
Rural institutions, 7–8, 77–79, 245n19; effects of agrarian reform on, 2, 13, 14; and strength of state, 6–7; and strength of state, 11, 15
Rural interest groups, 16, 284n133; and political impact of financial levers, 138–46; and strength of state, 11, 238
Rural leaders, response to agrarian reform, 79, 111–12, 114–15, 183, 218
Rural liberals, 164–65, 222–23; and AKKOR, 138–40, 219–20; on collective sector farms, 93–94; influence of, 11–12, 238; opposition to agrarian reforms, 114–15; support for agrarian reforms, 220–22, 224–25
Rural living standards. *See* Living standards
Rural social conditions, 2, 58, 233–37; creation of new rural class, 232–33, 240; effects of state interventions on, 48–57, 166; and failure of agrarian reforms, 80–81, 91–92; influence of state on, 6–7, 9; and risks of privatization, 67, 113, 171, 179–81; weakness of, 3–4, 200, 255n165
Rural social contract, end of, 108–15, 145–47
Rural social policies, 2, 8–9, 19, 58, 145, 245n19; lack of change in, 106, 228–29
Russia, strength of state, 6–7, 10–14, 149–52
Russia Is Our Home (political party), 223
Russian Agrarian Union, 140–41
Russian Farmer program, 184–86, 191–95, 198, 219
Russia's Choice (political party), 221–22
Rutskoy, Aleksandr, 77–78, 114
Rybkin, Ivan, 143
Ryzhkov, N., 140

Sakwa, Richard, 11
Savchenko, Evengii, 271n147
Schools, rural, 54–55, 121–22, 236, 255n165
Self-accounting and -financing, 29; on collective sector farms, 31–32, 65–66, 249n53; of farmer associations and cooperatives, 76
Serova, Evgeniya, 164
Share distribution. *See* Land share distribution
Social justice policies, of Gorbachev, 27
Social Origins of Dictatorship and Democracy (Moore), 2
Society *vs.* state. *See* State, strength of
Soviet of People's Deputies, distribution of land by, 155–56
Soviet Union: food procurement system of, 134–37; legacy in agrarian reform, 57–58; rural interventions of, 18–58,

227; rural social contract of, 108–09; strength of state, 5–7, 10; urban bias in, 4
Specialists, farm, 20, 26, 169
Specialization, after farm privatization, 101–02
Stalin, Joseph, 1, 10, 109; private farming under, 34–36; urban bias under, 4, 7, 48–49
Standard of living. *See* Living standards
State: and collective sector farms, 65, 70, 104–05; in conflict between rural and urban interests, 15–16; failure to pay farmers, 70, 126, 137; landownership by, 160–61; as market, 65, 111, 133–38, 187, 200, 220, 269$n115$; and private farmers, 187, 219–20, 223; procurement system of, 14, 36–37, 65, 67, 133–38; and rural interests, 111–12, 145–46, 224–25; strength of, 9–14, 16–17, 245$n25$; strength of, *vs.* leader, 10, 12; strength of, *vs.* society, 5–7; support for agriculture, 182, 198–204; support for revival of countryside, 109–11, 121; as unit of analysis, 4–7; weakness of, 149–52, 228
State farms, 54; financial levers on, 130–31; personal plots on, 38–39, 249$n56$; reorganization of, 59, 269$n104$; wages on, 19–25, 50–52, 124–25, 246$n6$, 247$n13$, 247$n15$; workers leaving, 46–47. *See also* Collective sector farms; Weak farms
State interventions, in agriculture, 8–9, 62, 108, 111–12; effects of, 18, 127–43, 182–226, 231–32; financial levers for, 104–05, 112, 115–17, 150; objectives of, 2–4, 19–33, 227; of post-Soviet collective sector, 59–107; of private sector, 33–48, 166, 182–226; regulations as, 33–48, 68, 135–37, 166, 176; responses to, 143–46; and rural social conditions, 48–57, 121–27; setting prices as, 120–21; Soviet era, 18–58
Stolypin reforms, 1
Subsidiary farming, 33–34, 162, 249$n58$
Subsidies, farm, 74, 110, 115, 182, 186, 268$n99$; for collective sector farms, 59, 231; for credit, 132, 188; effects of ending, 70, 214–16, 222–24; and food prices, 129–30, 264$n11$; for private farms, 200, 204–05, 214–16; uses of, 88–90, 137

Taxes, 110–11, 277$n127$; advantages for private farms, 104–05, 186, 200; on farm revenues, 121, 267$n64$; on personal plots, 37–38

Trade: disadvantages to agriculture in, 119, 131, 215–16; between domestic sectors, 133–38, 148, 234, 255$n158$, 261$n86$; foreign, 65, 76, 90, 115–16, 199
Trade unions, agricultural, 138–40
Transportation, for farm products, 120–21, 171–72; in criteria for "good" land, 83–84, 156

Unemployment, 233–34
Union of Landowners, 222–23
Urban bias, 3–4, 15, 48–49, 124, 222–24, 235, 243$n4$
Urbanization, 55–56; food supplies for, 36, 39–42. *See also* Cities; Demographics
Urban *vs.* rural, 138, 166, 255$n158$; differences in living standards, 48, 233; wage differences, 50–53, 56, 124–27, 147, 234–35, 278$n2$. *See also* Cities
"Urgent Measures for the Realization of Land Reform" (1991), 158, 161
USAID, and privatization, 96, 113, 265$n20$
Uzun, V., 180

Varenov, Boris, 277$n127$
Veprev, A. F., 140
Vershinin, Vasilii, 61
Virgin Land Program, 37
Volga region, 215–16
Vorotnikov, V., 140

Wadekin, Karl-Eugen, 60
Wages, 277$n117$; of agricultural and industrial workers, 23, 124–25; in agriculture, 124–26, 234, 246$n3$; bonuses and supplements, 23, 247$n13$; on collective sector farms, 65, 86–87, 102, 109, 246$n12$, 247$n15$; differences in, 27, 230, 246$n3$, 246$n6$; and dividends from joint stock farms, 74–76; egalitarianism in, 19–25, 50–52, 56, 230–31; on private farms, 189–90, 280$n30$; urban-rural differences, 50–53, 56, 124–27, 147, 234–35, 278$n2$
Weak farms, 257$n15$; effects of pricing policies on, 30–32; number of, 131–33, 249$n53$; personnel on, 26–27, 247$n26$; privatization of, 69, 104, 157; reorganization of, 84, 269$n107$; subsidies for, 90, 231; support for, 25–29, 52, 248$n28$
Western agencies, 199, 229, 265$n17$, 265$n18$; role in farm privatization, 91, 94–96, 103, 105, 113
Wholesale trade organizations, for food purchasing, 14
World War II, effects on collective farms, 35

Yel'tsin, Boris, 265*n17;* agrarian reforms by, 13, 77, 109–10, 218–19; collective sector farms under, 69, 88; and food procurement system, 134–35; on landownership, 153, 161, 162–63, 175–77, 221; opposition to reforms of, 77–78, 141, 222–23; private farms under, 46, 183–84, 188; privatization under, 70, 91–92, 158; and restrictions on land use, 164, 166

Zaitsev, Mr., 184
Zaslavskaya, Tatyana, 19
Zaveryukha, Aleksandr, 92, 95, 114, 142, 204; on trade, 119–20, 144
Zvolinsky, Vyacheslav, 143

HD 1333 .R9 W44 1998
Wegren, Stephen K. 1956-
Agriculture and the state i
 Soviet and post-Soviet